Design Education Today

Dirk Schaefer · Graham Coates ·
Claudia Eckert
Editors

Design Education Today

Technical Contexts, Programs and Best Practices

 Springer

Editors
Dirk Schaefer
Division of Industrial Design
University of Liverpool
Liverpool, UK

Graham Coates
School of Engineering
Newcastle University
Newcastle-upon-Tyne, UK

Claudia Eckert
School of Engineering and Innovation
The Open University
Milton Keynes, UK

ISBN 978-3-030-17133-9   ISBN 978-3-030-17134-6   (eBook)
https://doi.org/10.1007/978-3-030-17134-6

This Springer imprint is published by the registered company Springer Nature Switzerland AG
The registered company address is: Gewerbestrasse 11, 6330 Cham, Switzerland

Preface

Good design is human-centered, commercially viable and technologically sound. It follows a design process and serves a well-defined purpose, while delivering a desirable experience—viable and inspirational. The experience of good design and the ownership of functional, usable and desirable products is valuable to people across their professional and domestic lives. Design has a huge impact on, and is impacted by, individuals, groups, societies, and cultures. It is at the same time a technical discipline, an art that provides opportunities for people to realize their goals and aspirations and a social activity that shapes the world around us.

The objective of this book is to provide an overview of select design disciplines, opportunities, and challenges in preparing the next generation of design workforce, and to highlight best practices of design education programs from around the world. This includes ancillary topics, such as design in the context of public policy, and standards for professional registration and design program accreditation. It serves as a handbook for design practitioners and managers, educators, curriculum designers, program leaders, and all those interested in pursuing an education or developing a career in design.

In Chapter "Teaching Without a Net: Mindful Design Education", Strong, Lande, and Adams seek to inform design instruction for engineering students that can be generalized to many design-infused disciplines. Their work draws on the experiences of reflective design educators and researchers for a multiple perspective approach to capture insights into how teachers teach design. Using collaborative inquiry and reflective practice, they look for patterns as a way of making visible the instructional moves of design educators. They examine design instructional experiences through three theoretical and empirically grounded lenses: (1) Cognitive apprenticeship: mastering requisite skills through mindful and practiced expertise, (2) Teaching as improvisation: understanding teachers as adaptive and skilled improvisers responding to unpredictable demands in the moment, and (3) Pedagogical content knowledge: means for capturing and reflecting on what we know about design in the classroom. The resulting framework offers an integrative perspective on understanding: design students as learners, effective approaches for their needs, and principles of design thinking.

In Chapter "Second-Year Engineering Design: A Use-Inspired Approach", Carberry and Brunhaver describe a use-inspired design course and its underlying project-based pedagogy offered to engineering students during the beginning of their second year at a large public university in the US.

In Chapter "Teaching Design Innovation Skills: *Design Heuristics* Support Creating, Developing, and Combining Ideas", Daly, McKilligan, Leahy, and Seifert summarize their evidence for identifying Design Heuristics in product design, and describe how they can be employed within a variety of lesson structures to foster idea generation. These Design Heuristics have been shown to support learning with innovative design outcomes across educational levels and design disciplines.

In Chapter "Enabling Meaningful Reflection Within Project-Based-Learning in Engineering Design Education", Morgan proposes that the teaching of philosophy of design combined with the use of reflective learning journals, structured using a constructivist inquiry framework, may allow students to access a deeper level of understanding of their own individual approaches to design, by enabling reflection at an ontological level. Philosophy of design serves the purpose of emancipating students from a restrictive worldview by making them aware of multiple paradigms of learning. This chapter presents the argument for such an approach to supporting group project-based learning and describes a study within a second-year engineering design course, in which this approach has been trialed. The results indicate that a positive impact on reflection and learning was achieved.

In Chapter "Contributions of Academic Makerspaces to Design Education", Wilczynski provides an overview of academic makerspaces—locations on college campuses for teams of students to design, fabricate, assemble, and test components and complete systems—and explores the impact of these spaces on design education. The spaces are unique in that they are generally open-access, much like libraries and other university-wide service-providing facilities, and available for any use. As facilities that promote design activities, academic maker spaces are having a positive effect enhancing design skills in engineering and other disciplines.

In Chapter "Engineering Capstone Design Education: Current Practices, Emerging Trends, and Successful Strategies", Howe and Goldberg present the current state of engineering capstone design education and highlight changes to capstone design practices in the US over the past 25 years. The chapter also provides recommendations for supporting engineering capstone design experiences based on the authors' vast experience teaching and managing capstone design courses and engaging with the capstone design community. These recommendations include scaffolding the design curriculum, fostering industry involvement in capstone design courses, keeping curses and faculty up-to-date with respect to current design practices, obtaining organizational support for capstone design courses, sourcing capstone design projects, and preparing students for professional practice.

The next chapter introduces the reader to bio-inspired design. In this chapter, titled "Bio-inspired Design Pedagogy in Engineering", Nagel, Rose, Beverly, and Pidaparti review current teaching practices and courses in engineering curricula for training students in multidisciplinary design innovation through bio-inspired design. Emphasis is placed on theory-based and evidenced-based approaches that have demonstrated learning impact. The significance and implications of teaching bio-inspired design in an engineering curriculum are discussed, and connections to how the essential competencies of future engineers are fostered is addressed. Teaching bio-inspired design in an engineering curriculum using cross-disciplinary approaches will not only develop essential competencies of tomorrow's engineer, but also enable students to become change agents and promote a sustainable future.

In Chapter "Designing Knowledge Sharing Interfaces with Improved Interaction: Haptics and Web3D", Hamza-Lup explores new interaction paradigms and systems enabling the shift from a 2D visual space to 3D user interfaces in Virtual Design Education. This includes the presentation of a new interaction modality, the tactile or touch modality, through haptic user interfaces and the discussion of required hardware. The chapter also includes a presentation of several successful haptics and Web3D implementations for engineering education, and a discussion of future trends in this area.

In Chapter "Development of Complex System Design Oriented Curricula: The Example of the Grande Ecole CentraleSupélec", Jankovic et al. report on the rising demand for complex systems and systems engineering curricula. This trend is particularly supported by current technology and economic developments such as the Internet of Things, Artificial Intelligence, Servitization, the need to develop Product-Service-Systems (PSS) as well as Systems of Systems (SoS). In this context, one of France's top engineering schools has been developing a complex system-focused curriculum. The authors give details of this curriculum, discuss the specifics of the French education system and its engineering schools (Grandes Ecoles), and elaborate on the process proposed to develop this new type of engineering education program.

In Chapter "A Strategic Design Approach for Anticipating the Future World", Liem presents a strategic design methodology suitable for collaborative studio settings within the context of design education. It can be used as a blueprint for planning strategic activities, collaborative constructs, and directions when strategizing for and with different types of organizations. The educational value lies in subjecting master students to situations, where they learn to balance the interests of the client-organization, end-users, and other stakeholders through deliberate and emergent strategic moves while working on design projects.

The chapter by Namasivayam and colleagues report on their work to use "Design Engineering as a Means to Enhance Student Learning in Addressing Complex Engineering Challenges".

The chapter by Woelfel and Krzywinski focus on "Industrial Design Engineering: Teaching Industrial Design as a Specialisation Within a Mechanical Engineering Curriculum". Their novel education approach involves a combination of elements from humanities, arts, and engineering sciences.

Finally, the book concludes with two chapters by Dowlen. The purpose of Chapter "Accreditation of Design Education Programmes", is to outline the meaning of accreditation as used by professional bodies, why it is important to consider, and how it is applied to design courses. Chapter "Design and Design

Education as a Profession: Professional Registration and Membership of Societies for Designers and Design Educators; Continuous Professional Development (CPD)" focuses on professional registration and membership of societies for designers and design educators, as well as Continuous Professional Development (CPD) in the design profession.

Liverpool, UK   Prof. Dirk Schaefer
Newcastle-upon Tyne, UK   Prof. Graham Coates
Milton Keynes, UK   Prof. Claudia Eckert
February 2019

Contents

Contributors

Robin Adams School of Engineering Education, College of Engineering, Purdue University, Neil Armstrong Hall of Engineering, West Lafayette, IN, USA

Cheri Beverly Learning, Technology & Leadership Education Department, James Madison University, Harrisonburg, VA, USA

Samantha R. Brunhaver The Polytechnic School, Arizona State University, Mesa, AZ, USA

John Cagnol Laboratoire Mathématiques et Informatique pour la Complexité et les Systèmes, CentraleSupélec, Paris, France

Adam R. Carberry The Polytechnic School, Arizona State University, Mesa, AZ, USA

Alexandra Coso Strong School of Universal Computing, Construction and Engineering Education, College of Engineering and Computing, Florida International University, Miami, FL, USA

Shanna R. Daly University of Michigan, Ann Arbor, MI, USA

Chris Dowlen London, UK

Didier Dumur Laboratoire des Signaux et Systèmes, CentraleSupélec, Paris, France

Valérie Ferreboeuf Laboratoire de Génie Industriel, CentraleSupélec, Paris, France

Mohammad Hosseini Fouladi School of Engineering, Taylor's University, Subang Jaya, Malaysia

Jay Goldberg Department of Biomedical Engineering, Marquette University and the Medical College of Wisconsin, Milwaukee, WI, USA

Felix G. Hamza-Lup Computer Science, Georgia Southern University, Savannah, GA, USA

Susannah Howe Smith College, Picker Engineering Program, Northampton, MA, USA

Marija Jankovic Laboratoire de Génie Industriel, CentraleSupélec, Paris, France

Jens Krzywinski Chair of Industrial Design Engineering, Technische Universität Dresden, Dresden, Germany

Micah Lande Department of Mechanical Engineering, South Dakota School of Mines and Technology, Rapid City, SD, USA

Keelin Leahy University of Limerick, Limerick, Ireland

André Liem Department of Design, Norwegian University of Science and Technology, Trondheim, Norway

Seda McKilligan Iowa State University, Ames, USA

Jayasubamani Arvi S. Moganakrishnan School of Engineering, Taylor's University, Subang Jaya, Malaysia

Thea Morgan CAME School of Engineering, University of Bristol, Clifton, UK

Jacquelyn K. S. Nagel Department of Engineering, James Madison University, Harrisonburg, VA, USA

Satesh Namasivayam School of Engineering, Taylor's University, Subang Jaya, Malaysia

Ramana Pidaparti College of Engineering, University of Georgia, Athens, GA, USA

Christopher Rose Department of Biology, James Madision University, Harrisonburg, VA, USA

Colleen M. Seifert University of Michigan, Ann Arbor, MI, USA

Douglas Tong Kum Tien School of Engineering, Taylor's University, Subang Jaya, Malaysia

Vincent Wilczynski Yale School of Engineering & Applied Science, New Haven, USA

Christian Wölfel Chair of Industrial Design Engineering, Technische Universität Dresden, Dresden, Germany

Teaching Without a Net: Mindful Design Education

Alexandra Coso Strong, Micah Lande and Robin Adams

Abstract Teaching engineering design can be an inexact art. Textbooks, tools, and methods do exist, but translation of this knowledge to design practice in the classroom setting may flounder for many reasons. There is often no script for the instructor to follow because design itself is a process that is amorphous, iterative, and coevolutionary. In this chapter, we seek to inform design instruction for engineering students that also can be generalized to many design-infused disciplines. A significant feature of teaching design involves practiced anticipation and listening to students' doing and learning to design. Sometimes, students confront unfamiliar concepts and need to be guided toward a new approach that may conflict with their prior educational experiences. For instance, students experience "threshold concepts" requiring a shift in thinking from one of engineering problem-solving to design problem-finding. Additionally, ambiguity in design can create discomfort and uncertainty, slowing students from moving forward. This chapter draws on the experiences of three reflective design educators (also design researchers) resulting in a multiple perspective approach capturing insights into how teachers teach design. Using collaborative inquiry and reflective practice, we look for patterns as a way of making visible the instructional moves of design educators. We examine design instructional experiences through three theoretical and empirically grounded lenses: (1) Cognitive apprenticeship: mastering requisite skills through mindful and practiced expertise, (2) Teaching as improvisation: understanding teachers as adaptive and skilled improvisers responding to unpredictable demands in the moment ("prepared not planned"),

A. Coso Strong (✉)
School of Universal Computing, Construction and Engineering Education, College of Engineering and Computing, Florida International University, 10555 West Flagler Street, Miami, FL 33174, USA
e-mail: alexandra.strong@fiu.edu

M. Lande
Department of Mechanical Engineering, South Dakota School of Mines and Technology, 501 E. Saint Joseph Street, Rapid City, SD 57701, USA
e-mail: micah.lande@sdsmt.edu

R. Adams
School of Engineering Education, College of Engineering, Purdue University, Neil Armstrong Hall of Engineering, 701 W. Stadium Avenue, West Lafayette, IN 47907, USA
e-mail: rsadams@purdue.edu

© Springer Nature Switzerland AG 2019
D. Schaefer et al. (eds.), *Design Education Today*,
https://doi.org/10.1007/978-3-030-17134-6_1

and (3) Pedagogical content knowledge: means for capturing and reflecting on what we know about design in the classroom. The resulting framework offers an integrative perspective on understanding: design students as learners, effective approaches for their needs, and principles of design thinking. We present suggestions for developing instructional moves to engage with your own learners at various points throughout a design project through a "playbook" for design educators.

Keywords Education as design · Teaching as improvisation · Pedagogical content knowledge

1   Introduction: A Common Scenario

A scenario: You walk into the classroom and see one of your design teams meeting before class starts. Sara, Jorge and the other members of their team have been working on a cross-disciplinary project with a client from a non-profit organization. When you check in with them and ask how things are going, Sara lays her head on the desk, while Jorge explains, "We are totally stuck." It seems that the client has been changing their mind for what they want from the student design team. The most recent iteration of the design requirements is in conflict with the user research work that Sara, Jorge and the rest of their team completed earlier in the project. They are now feeling overwhelmed. They are lacking a strategy for how to move forward and are looking to you for guidance.

What do you do, as a design educator, when you observe your students paralyzed in their design process, standing still with their wheels spinning, like in this situation? For some, this scenario may very well resemble one you have experienced as an instructor in a design course. Or this can be one that you worry that you will face in the future.

As we think about the possible next steps for the instructor, let's consider Sara and Jorge's situation more closely and ask what might happen with the team if the instructor did not engage with them on that day in class. The student team would have needed to make a decision at some point. They would have moved forward regardless of the effect of their decision on the final outcome of the project. As design educators, our goal is to support their learning and development as designers. We are there to intervene. But how might that be accomplished in a useful way? We can reflect on strategies that could be used to enable Sara, Jorge, and the team to tackle the ambiguity of their project. As responsible instructors, we could help them take a step back and consider their design challenge at a higher level perspective, or we may employ a combination of modeling and guidance to suggest what we might do if in their shoes. And we may have additional engineering professional expertise to help solve the technical problem at hand. As educators, we have many potential options for our next move. The question then becomes how do we decide which next move to make.

Teaching engineering design can be an inexact art. Textbooks, tools, and methods do exist, but translation of this knowledge to design practice in the classroom setting may flounder for many reasons (e.g., Dannels 2003; Downey and Lucena 2003;

Koen 1994; Paretti 2008). In an undergraduate context, for example, students work on design projects at a pace and frequency that are distinctly different from designers in industry (Goncher and Johri 2015; Turns et al. 2006a). In addition, we are teaching designers who are developing and growing. As learners, these students will engage with design practice in ways that are in some tension with what may be suggested in class or in ways that we might not have imagined (Ahmed et al. 2003; Crismond and Adams 2012; Newstetter 1998). In this chapter, we suggest insights from the experiences of design educators that can help educators make informed decisions as to how we might intervene to support students.

This chapter begins with an overview of how we, the authors, took a mindful approach to examining our own experiences. Our reflections and discussions led us to explore different types of **scenarios** we may experience as design educators. From those scenarios, we engage with existing work on the diversity of **roles** we could play as we guide our students. Subsequent sections will introduce a potential artifact, a **Design Coaching Playbook**, for supporting design educators that is based on taking a mindful approach to design education and reflecting on our own experiences in the classroom and studio. This Playbook represents the different **moves** that are available to us as educators given the scenario and the role(s) we might play. Lastly, we will connect the Playbook to a particular methodological and theoretical lens that grounds this instructional decision-making strategy in existing research. The chapter ends with a reflection on what this Playbook enables in terms of mindful teaching for educators and design researchers and how it may be implemented in your own context.

2 Shared Interests: Taking a Mindful Approach to Design Education

We, the authors of this chapter, drew on our experiences as three reflective design educators and researchers, leveraging a multiple perspective approach to capture insights into how teachers teach design. Each of us teaches design in an undergraduate engineering context in the United States. However, our perspectives are across differentiated institutional contexts (i.e., public/private; student populations; levels of design curriculum integration), different frames of design teaching (i.e., making/prototyping, human-centered design/large-scale systems, interdisciplinary), and are grounded in different disciplines (i.e., mechanical and aerospace engineering, human-centered design, interdisciplinary design).

The backstory behind this project was a series of informal discussions about our experiences coaching students and student teams in a variety of courses: senior design, interdisciplinary design, and introductory design experiences. Through our ongoing discussions and shared interests, we began to unpack the various roles we play within a design classroom (e.g., "coach", "model", "advisor", "grader"). We also started to critically examine the messages that are communicated to our students within the span of a learning experience in the classroom. Even though we were approaching three different design contexts within an engineering curriculum, we, ultimately, sought to develop a deeper understanding of our experiences. Our aim has been to support one another and other design educators in the day-to-day moments of teaching design. These interactions then evolved into a more formal exploration of our experiences using reflective practice (Schon 1984; Valkenburg 2001; Valkenburg and Dorst 1998) and collaborative inquiry (Bray et al. 2000; Kasl and Yorks 2002) to connect reflection, action, and sensemaking about the decisions we make as instructors in a design educational experience.

Reflective practice in design was characterized by Schon (1984) to make sense of the activities of practitioners, specifically design practitioners. Schon's analysis found practitioners who move fluidly between problem-framing and problem-solving activities, naming the aspects of the problem, generating experimental "moves" to explore and reflecting on the outcomes of those moves (Schon 1984). Reflective practice has been examined within engineering and design education (Adams et al. 2003, Turns et al. 2015) and industrial settings. Additionally, the components of reflective practice have been implemented as a research framework for exploring the activities of design teams (Valkenburg 2001; Valkenburg and Dorst 1998). Toward the development of a playbook, we used reflective practice to iteratively and dynamically explore the situations we have experienced in the classroom, naming the components of those situations, and reflecting on the moves we have tried.

Collaborative inquiry is described as a process for systematically and collaboratively learning from personal experience to inform future action (Bray et al. 2000; Kasl and Yorks 2002). With this approach, researchers create a system where participants are co-inquirers (making sense of the research question and phenomenon) and co-subjects (sharing their own experiences) (Bray et al. 2000). This experience-based approach facilitates the examination of questions focused on professional development of the individuals involved (Kasl and Yorks 2002). For instance, collaborative inquiry has been used to systematically explore and learn from the practices of qualitative researchers (Solis et al. 2016; Walther et al. 2015, 2017), educators (Bray 2002; Donohoo 2013), women scholars of color (Pritchard and Sanders 2002), a team of designers (Adams et al. 2017a, b) and others. This aligns closely with our goal to address how we might improve our ability (and that of other educators) to provide impactful learning experiences for our students in the context of a design course.

Our overall approach was comprised of regular meetings over an eight-month time frame while we were each (co-)teaching at least one design course, either at the introductory level or at the senior design level. Our shared explorations involved reflections on our experiences in our current courses and in the past. Through discussion and collaborative note-taking, we attempted to make sense of the shared and differing experiences. The result of this reflection and sensemaking was a series of common situations, which we came to call **scenarios**. These situations became the frames through which we explored different **roles** and instructional **moves**. In some cases, the discussion of moves allowed us to independently and collectively examine our previous actions in the classroom. In other cases, these discussions laid the foundation for practicing a move in a future faculty-design team interaction, either in our current course or in a future course.

3 Scenarios: Exploring Shared Situations

Collectively, we generated a set of preliminary design scenarios (see Table 1). This was done to consider specific and generalized experiences with students. It also allowed us to capture our viewpoints as designers and educators in the context of particular types of design coaching experiences. During our process, we recognized many of these scenarios were affective, associated with particular emotions like frustration, joy, rebellion, and apathy. For example, the "rebel without a cause" scenario reflects the individual student or team that was super confident in their knowledge, continuously rejecting our guidance, feedback from the client, and even sometimes the feedback given by members of their team. Jorge and Sara in our opening scenario were overwhelmed and frustrated, experiencing one of those "stuck in the mud with wheels spinning" moments; they lacked a strategy to move forward. These particular emotional associations highlight and remind us of the power and responsibility of our position and the need to recognize that these critical scenarios are more than just common misconceptions of the design process. Our moves as educators have the potential to impact not only their project, but also their self-efficacy, their sense of belongingness to the field and profession, and their ability to learn from the experience.

As a result, we saw a need to further examine these scenarios and, in particular, to capture the experiences of other design educators. Were they experiencing similar scenarios in their design classrooms and studios? As part of an engineering education conference special session (Coso Strong et al. 2018) we asked engineering design educators to share their own challenging moments when teaching design. While some of the scenarios mentioned echoed the ones we brought up in our own analysis, the participants also noted other patterns of design team behavior. For instance, many described how their students regularly jumped into "doing" without trying to understand the problem, and others discussed situations where students were working without a plan.

Generating these scenarios represented a critical first step for our explorations of our own design instructional experiences, and those of other design educators. Yet, even if a scenario had exactly the same characteristics, it is possible that we may not all approach it in the same manner. As such, we chose to more closely examine our own reactions to three predefined scenarios:

- *A Go–No Go scenario*—Consider a time where a team needed to make a critical design decision. What was the context of the situation? What did you do? Why?
- *A truly improvised move*—Consider a time when you felt like you were making things up on the fly as students struggled with a task that was unexpected or one

Table 1 Preliminary design scenario list

Scenarios	Description
Stuck in the mud, wheels spinning	The student team has hit a wall. They are overwhelmed, facing too much ambiguity and lacking a strategy for moving forward
White knuckling on a decision/idea	The student team is holding tight and unwilling to let go of a decision (whether it is implicit or explicit). They have put their feet firmly in some soil and are staying put regardless of other information. There is an overall feeling of fixation with some quality of fear of change, and, in some cases, a fear that there is not enough time to pursue a change so it's more efficient to stick with the current plan
Go/No Go moment	The student team is at a critical decision point. This moment carries a feeling of risk or a point of no return for the team and sometimes the instructor as well
"Yeah, this will work"—No justification or rationale	This student team is conducting "seat of the pants" design, pretending that design is intuition and not being able to articulate why. Many times there is a hidden agenda of reducing cognitive load (i.e., simplifying things to move forward). As a result, there is a high potential that the team is moving forward based on assumptions and bias, keeping reasoning invisible so they can change the argument as needed
Rebel without a cause	The student team is fighting feedback and guidance for reasons that are hidden and unspoken where the challenge becomes not just offering feedback but getting at what is contributing to a deadlock: "why should I document?", "I'm not motivated by this project can I do my own thing?", "this is what we want to do but it's not what the client wants", etc
Passing the buck—just tell me what to do!	The student team demonstrated that they want you, the instructor, to make the decision for them. Many times this is associated with feeling overwhelmed due to the large amount of ambiguity on the project. Yet, this scenario could also be brought on by concern that the team won't make the "right" decision, which could impact their grade and/or client satisfaction
This feels right, but I don't know why	The student team and/or an individual notices something. An idea/solution is coming together in their mind, but they are unable to step outside of the situation to fully articulate what they are experiencing and why what they are doing is working. This scenario is linked with a feeling of good work but unable to articulate why
Uh-oh, danger ahead	The student team, and in some cases, the instructor, feel a sense that a disaster is looming, anticipating a failure or situation that may be difficult to bounce back from in the time period of the project

where you lacked disciplinary expertise. What was the context? What did you do? Why?

- *A no justification or rationale scenario*—Consider a time when a team was not able to articulate the why of their design decisions. What was the context? What did you do? Why?

Using these scenarios, we individually reflected on our own experiences and collaboratively made sense of our reflections, highlighting similarities and differences, connecting to literature, identifying particular moves and roles, and examining the role of our own perspective on design on our choices of moves. Our notes and individual reflections became places of synergy, allowing situations, roles, and moves to emerge from our discussions and allowing iteration to occur around the classification of situations, moves, and roles. To provide one example of the diversity of possible actions during a given scenario, the subsequent section describes the many possible instructional decisions one might need to make to support a team in a Go–No Go moment.

Team in a Go–No Go moment

> The student team is at a critical decision point. This moment carries a feeling of risk or a point of no return for the team and sometimes the instructor as well.

There are many points of decision the student team needs to make. With limited time, budget, and cycles of attention, the closer one gets to a deliverable milestone, the more precious any action becomes. In mentoring design teams, it is useful to frequently counsel and remind oneself that this is a learning experience for the student team and that it needs to be their work and judgment that gets practiced and acted upon to make it their learning experience. Like parenting, one may foresee pitfalls and common mistakes but students may need to experience them to truly develop the knowledge to avoid such missteps in the future, in work situations where the cost may be more than their grade. Sometimes it is certainly incongruous to allow for ambiguity and uncertainty in the design process but it is part of a *mindful set of coaching strategies*. There is also an aspect that is purposefully curating the expectations and experiences such that the voice of feedback may become internalized for students as part of their understanding of what it means to be an engineer.

Consideration from the start of a project can be about what the course learning goals or individual outcomes may need to be. With a senior level, capstone engineering design course, for example, the design and engineering work provides a goal, a mission-based approach, and a project-based learning experience for the students. Their learning experience through a less structured design challenge, working together as a team and their teaming behaviors, and how to engage with stakeholder, company sponsors, and other users also may focus their attention and connect to specific prescribed student outcomes (for the undergraduate engineering program, for accreditation, etc.).

In supporting a design team then, each measure of a decision point is a trade-off. Generally, one needs to balance an approach of feasibility–viability–desirability (Feland et al. 2004). What approach is of value for the learning experience and process, and what is of value for the end product, depends on some number of factors. It is a continual consideration of not just what one can do but what one should do. Resolving this trade-off can be described not as searching out the right answer but rather, how many different answers can be generated and, from that, what criteria can help focus the team's next steps.

Any decision then, about project progress, both small and large, can then be put through this reflective engagement. Some aspect of that decision then is to give foundation to how to just start. In a human-centered design approach, for instance, that might be to go out and talk to users; similarly with an entrepreneurial-minded lean startup approach, to do customer discovery. Minor structural elements like a point of view statement (user + need + insights) focus attention to what information should be collected. There are also ways to externalize thinking to build engineering "minimum viable prototypes" or "low fidelity prototypes" to start. There can be a range of types of prototypes to employ as well: experience prototypes, critical function, functional part, or full system prototypes. It is helpful to break down the function or physicality into smaller, more actionable chunks. It also serves to focus the purpose of a prototype as some effort to answer a question that could not otherwise be addressed—a frequent move when considering the complexity of any system that has a person as part of it.

For the limitations of time, money, and effort in a classroom setting, a decision to move forward on an idea may be the most precious part of the process. A Go-No Go decision may provide an important pivot to get from divergent exploration to convergent building and action. The financial repercussions of implementing an idea from inexpensive (a word written on a post-it) to actual commitment to materials and fabrication may vary greatly. The structure of activities up until then is necessary to have thought, evidence, and perspective under consideration. It is also a transfer of expertise (or credence in expertise) for the students and how they support decisions with evidence or still rely on external feedback (from instructors, teaching team, peers) to inform their decisions.

Often, there are barriers to understanding the problem, resetting the problem space, and/or rescoping the problem space. Or, at the least, who may be appropriate to determine the same. As design instructors, our role can be to be aware of the process; in the sporting coach metaphor, we need to be aware of the score, where we are in the game, what players may need to be substituted. The instructor knows the rules of the game (learning, process vs. product) better than the student team and can advise them accordingly.

The counsel can be to focus on what is important and pertinent, and guide the student team to identify one idea out of many possible and how to operate on that. For the student team, it is a matter of how they will demonstrate their solution within the constraints. Consideration of the feedback they have received is useful but it can be the accumulated experience of the team, with knowledge brokering (Hargadon 2002) from other experiences coming into play (both advice grounded in what has been seen in previous projects and the students' own experiences). As instructors, we may also choose to help constrain what can possibly be done and offer a perspective that balances the learning process and end product. Yet, ultimately, our next move in this Go-No Go moment may depend on how we personally view our role within the design course at this particular moment.

4 Roles: Recognizing Our Own Experiences

Design educators represent a broad range of practitioners and educators, which is not surprising given the range of disciplines within which design is taught (e.g., architecture, choreography, the many subdisciplines of engineering, instructional design, product design, software design, etc.), the level of the learners, and the diversity of learning objectives. Some enter a design learning environment with years of teaching experience, but little experience in industry, while others bring years of professional experience, but are unfamiliar with the pedagogical model of a design course or learning experience. In addition, the perspectives of these educators will also vary, for example, in terms of how they make sense of design for their students (Strickfaden and Heylighen 2007, 2010). Still, similarities exist among design educators across disciplinary contexts (Adams and Siddiqui 2015). It is those similarities that we propose leveraging to explore the experiences of design educators and inform instructional decision-making in design learning environments.

In the classroom or the design studio, there can be many formal and informal interactions that support learning. Through design challenges, students and educators may take advantage of the larger learning ecology (Barron 2004) surrounding the classroom, from peers, to available space and resources. Leifer and Steinert (2011) further describe a triple learning loop where the student engages with the course content, formal learning structures, and informal learning support structures, including identified coaches. Particularly in the ambiguity of expected outcomes through design-based and making-based pedagogy (Lande et al. 2017; Sheppard et al. 2018), the roles of instructors in the classroom can be considered a broad spectrum of engagement, from responsible for delivering mostly content knowledge to coaching student teams via the intended and explicit student learning outcomes and the implicit expectations of students and instructors alike. As such, it can be instructive to move beyond a single definition for what an instructor can be and instead speak to the range of all of the varying roles in considering the different ways in which design educators could engage with their students.

For example, in the studio, design teachers fulfill a number of advice-giving roles along consultative, educational, and motivational dimensions (Reich et al. 2009). These roles are likely to change over the course of a project depending on a sense of what students need at a particular time. Reich et al. (2009) define consultative functions as problem-focused interventions where a teacher takes on the role of a source of knowledge offering feedback and recommendations drawing from their prior experience including what others have done (Cardella et al. 2014; Dannels and Martin 2008; Huet et al. 2007) or of an authority figure enforcing guidelines or rules (Reich et al. 2009). Educational functions are learning-focused interventions like imparting knowledge, expertise and how to navigate the nontrivial aspects of learning to use disciplinary knowledge in context (Wolmarans 2016), demonstrating how to perform design acts (Uluoğlu 2000), modeling how they might approach a similar situation (Cennamo et al. 2011), and providing opportunities for students to fail, succeed, and take ownership in design decisions (Yilmaz and Daly 2016). Some of these roles emphasize how a teacher moves away from a knowledge transmission role toward empowering students' self-expression and creativity (Goldschmidt 2006). Motivational functions including mentoring-focused moral support such as offering encouragement (Dickens et al. 2016), being a buddy (Goldschmidt et al. 2010), empowering students to act independently (Goldschmidt et al. 2014) and constructing their own voice as a designer (Brandt et al. 2013; McDonnell 2016; Oak and Lloyd 2014). Enacting some of these roles may involve managing competing values such as maintaining control versus enabling flexibility (Oplinger et al. 2016).

Within the learning experience of a design course, the students' expectations for the role of their professor may not align with their prior experiences. The notion of the expert of the "fountain of knowledge" may be paramount. Then, when the role of the instructor is put in the context of performance around a design challenge or project, that role of the expert may be subsumed by one of question asker or coach. The person who previously had all the answers is now raising many questions but not completing them. Similarly, for the instructor, ambiguity arises from the switch from known problems and solutions to projects framed without such certainty. The experiential nature then of such learning environments raises issues of control dynamics including the basis for grading, i.e., how do I assess learning? Similarly, the diversity of role options in this learning environment and ambiguity regarding a particular project or problem may be overwhelming to those teaching in an unfamiliar pedagogical manner (Isaacs-Sodeye and Lande 2013). Faculty do not want to look unknowledgeable in front of their students. Nor should we expect them to without support.

Yet, in a design course, that belief that we should know all of the answers doesn't align with the reality of design-based pedagogy and project courses. What might be readily apparent to the instructor as the next step in the design process or the next iteration of a design may very well be invisible to the student design team. It can be difficult to transfer one's professional vision from one's experience and judgment to the student team. In the role as an educator in this setting, one is both the instructor and the mentor and the purpose evolves over time, sometimes on the technical content and sometimes on the guidance through an engineering design process. Additionally, we may not know what the next step should be, as there may be a number of possible moves, nor can we generate the best solution for the project. In our role as design educators we can curate the learning ecology and try to set the stage for learning but that is just the start. By developing some abilities to be fluid with our feedback and to be strategic about how to deploy our counsel, we can become better educators in both large and small ways.

While efforts exist to support design educators, either through the existing community of design educators already engaged in scholarship of teaching (Turns et al. 2006b) and researchers who seek to directly support educators, (e.g., Adams and Siddiqui 2015; Coso 2014; Crismond and Adams 2012; Dym et al. 2005; Goncher and Johri 2015) a significant feature of teaching design involves practiced anticipation and listening to students' doing and learning to design (Adams et al. 2015; Turns et al. 2006a). As we have discussed, there doesn't exist a clear guide as to "what to do when", and as a result, it can feel, at times that we are teaching without a net. Yet, each day we are making instructional decisions about how to help learners like Sara and Jorge overcome and learn from obstacles in a design project. Thus, to continue the conversation of how to support instructional decision-making for design educators, we propose critically exploring the practices of design educators and integrating that exploration with our understanding of the students' design experiences to inform the development of what we call a Design Coaching Playbook.

5 Moves: Envisioning a Design Coaching Playbook

A simple search on "playbook" reveals extensive use of the idea in a variety of domains. Many associate playbooks as a book containing sports teams' strategies or a set of plays to perform (Limbert 2012), and the idea of a playbook is prevalent in other domains such as parenting, business (Carlson et al. 2013; Fallon and Verwayen 2017; Olsen 2015), software engineering (Sypolt 2017), professional development (King and West 2018), philanthropy (Alliance for Justice 2015), and design thinking (Lewrick et al. 2018) or making (Hlubinka et al. 2013). Regardless of domain, **a playbook represents strategic knowledge—a set of moves to respond to particular situations with a particular goal in mind**.

This concept of a Design Coaching Playbook provides an approach for examining not what the right move is in the scenario but rather, for the roles involved, to provide a number of possible moves that could help guide the situation at hand forward. Instead of leaning on technical content, we hope to reveal any number of process-oriented moves to engender question asking (Eris 2004), reframing, or just getting unstuck by alleviating design fixation (Jansson and Smith 1991). We have described earlier the context of the utility and importance of a flexible nature for being an instructor. These can be useful approaches for the unstructured and just-in-time nature of teaching design. We frame our playbook as a means to present multiple suggestions for instructional moves to engage with students at various points throughout a design project: a "playbook" for helping students learn to design.

There is broad utility to a playbook. Both to support instructors, and also to make evident to students that there are any number of moves to be made from scenarios. A set of possibilities makes for an interesting priming and scaffolding for their learning, providing a more expert model of attacking a design problem. In this way a playbook can make visible and shareable what instructors know and do, enabling collaborative knowledge. It can also be a way of bringing students into the thinking behind what they're experiencing—connecting the dots between what they're doing and what it means to design.

We consider the structure of a playbook akin more to the examples from sports or business. It recognizes a scenario or situation and provides multiple sets of moves depending on the context. We start to imagine the parts of a playbook to be organized and cross-referenced from a common problem that can be diagnosed but also through a number of goal orientations. Pages of a playbook, or sets of cards that can be referenced quickly is our current, imagined structure for capture, documentation, and sharing. The categories of moves–scenarios–roles can be identified. Over the course of our development of this notion of a playbook, we used a number of scenarios to discuss possible moves and outcomes. Through our reflections, we recognized that the same scenario can have many possible moves, and chains of moves thereafter. Ultimately, this chapter is the result of our process, playing through examples to imagine such a resource with more fidelity.

In the subsequent sections, we step back to connect to the broader knowledge to reveal different types of **moves** we have available to us as design educators, focusing on those that resonate with the ambiguous, opportunistic, and iterative nature of design. In particular, how do design teachers make sense of what they are seeing (or hearing) their students do and what kinds of strategies or "teacher moves" do they make to provide the kinds of help students might need? How do they do "design thinking in the moment" that translates into "design teaching in the moment"?

5.1 Through a Cognitive Apprenticeship Lens

Cognitive apprenticeship is a constructivist learning theory that emphasizes how humans learn in a social manner by observing senior members in one's community of practice (Collins et al. 1991). Coming back to our sports analogy, one might imagine how junior players on a team observe more senior players to learn from them. Within the classroom, what teachers know is not directly visible and is typically difficult to communicate or transfer to others. Instead, teachers work at making their thinking visible to students by "externalizing the processes that are usually carried out internally" (Collins et al. 1991, p. 6). By bringing their deep and internalized tacit knowledge out into the open, teachers (as master practitioners in a domain) make their underlying thinking behind their decision-making and judgment processes visible to students (as junior practitioners in a domain). This helps students examine and develop their own decision-making and judgment processes. From a Design Coaching Playbook perspective, the cognitive apprenticeship lens fits with a tradition of teaching design through experience with the aid of a teacher in the role of a master design practitioner (Adams et al. 2016). In this role, instructors may draw on a variety of general teaching moves to make their design knowledge visible to students and help students become more explicit about their evolving design knowledge. In this way design teachers continually perform their design thinking for students, making visible their accumulated experience, knowledge, and belief systems, apprenticing students into their way of understanding design (Oxman 1999; Schön 1993; Uluoğlu 2000).

As such, a playbook from this perspective may focus on the different moves teachers make as they naturally draw on and combine a variety of cognitive apprenticeship techniques (Adams et al. 2016). Teachers may (1) *model* or demonstrate a target skill or concept while thinking aloud about their heuristics or processes, (2) *coach* students by offering feedback on their performance and helping students bridge the gap between what they did as compared to what an expert might do, (3) provide instructional *scaffolding* to help students move progressively toward a more sophisticated understanding such as breaking complex tasks into smaller chunks or showing how a task builds on prior knowledge, (4) help students *articulate* their thinking so teachers can check their reasoning and help students refine their knowledge, (5) help students *reflect* on their learning by having them compare their reasoning processes to an expert or to a past experience, and (6) *give room* to students to explore and problem solve on their own. Each of these is a situation-general teaching move that seeks to enable a cognitive apprenticeship learning dynamic, and many are likely to be familiar to design teachers, whether they are aware of this framework or not. For example, all these teaching moves have been observed in formal and informal design reviews across a variety of disciplinary contexts as part of the natural ebb and flow of a master–junior practitioner dynamic (Adams et al. 2016).

5.2   *Through a Teaching as Improvisation Lens*

Teaching as improvisation builds on the idea of adaptive teachers as skilled improvisers (Sawyer 2011). Here, teaching is analogous to improvisational acting: a spontaneous and unpredictable performance that is contingent upon interactions with others and built from moment to moment but draws upon an existing repertoire of pedagogical practice (Sawyer 2004). A repertoire of practice is defined as "the sum of available tools, techniques, strategies, tactics, ways of working, expertise and know-how from which a practitioner may draw, choose from, and/or combine to suit both known and novel situations or address a particular problem" (New Learning). In other words, a repertoire of practice could represent a playbook of situations and relevant moves a teacher can make for a particular objective, such as helping a student overcome a conceptual, project, or team hurdle. Here, a repertoire is not a set script but rather fits within the model for jazz improvisation: being prepared, but not planning too much in advance, and listening in a deep way to attend to the opportunities afforded through interaction with others.

Teaching as improvisation aligns with the inherent ambiguities of design and designerly ways of knowing (Adams et al. 2016; Cross 1982). Because design projects have no single right answer, student designers iteratively and opportunistically pursue many possible pathways toward understanding a problem situation and pursuing possible solutions in ways that make it impossible for design teachers to have answers to every possible question or concern. As a result, in the design classroom, there is no consistent "one size fits all" script. To translate and convey their knowledge to students, design teachers have to be highly adaptive to the student and the situation (Goldschmidt 2006), drawing from and continually expanding their repertoires of practice.

Within a teaching as improvisation framework, a teacher's role aligns with learner-centered and constructivist approaches to facilitate students' development (Brennan 2013). The rationale is that without a script to follow students are empowered to be self-organizing and take agency in leading their own learning experiences with teachers (Crossan 1998; Sawyer 2008). Therefore, teaching moves emphasize constructivist student–teacher interactions to enable a collaborative dynamic to emerge and flow. In design, teachers might experience this as a version of codesigning with students, taking on dual roles of both master practitioner and team member. For example, a common improvisation move is "yes, and", which plays out as someone introducing an idea into the dramatic frame and another accepting that idea ("yes") and then building on that idea ("and"). These kinds of moves are easily observed in design interactions between team members (Sonalkar et al. 2016) and students and teachers (Adams et al. 2016). Another teaching as improvisation move is "breaking the fourth wall", that refers to the experience when an actor in a scene breaks the dynamic of the scene to communicate directly with the audience. Breaking the fourth wall can create a metacognitive teaching moment that has intrinsic value for the learner or field of inquiry (Sawyer 2004) such as a teacher explaining the consequences of time management to a group of students after they shared why their projects did not adhere to their planned schedules (Adams et al. 2016; Oplinger and Lande 2014). Other moves that enable a teaching as improvisation dynamic include "listen and remember", "endowing", and "playwriting" (Sawyer 2004).

As such, from a Design Coaching Playbook perspective, the teaching as improvisation lens fits with the experience of teaching design without a net (e.g., no set script for every possible contingency) and is likely to resonate with design educators that prefer constructivist teaching approaches, tolerate the uncertainty of not having all the answers for every question, and enjoy empowering students to find their own answers through collaborative sensemaking. To do so, design educators, and a Design Coaching Playbook, need a broad repertoire of practice that allows them to notice and zoom in on the help students might need and respond to these situations with a set of teaching moves that can enable learning by doing when working side-by-side with students.

5.3 Through a Pedagogical Content Knowledge Lens

Where the previous sections illustrate general instructional moves that are broadly relevant to teachers, a Design Coaching Playbook should include specialized moves that speak to what teachers know about design students as learners, what teachers want students to know or be able to do as designers, and what they've learned through experience and perhaps conversations with other teachers about ways to help students learn to design. This kind of specialized teaching knowledge aligns with the concept of pedagogical content knowledge (PCK), a learning sciences framework that represents the accumulated wisdom a teacher has developed within a specific subject or content area (Ball et al. 2008; Gess-Newsome 1999; Shulman 1987). For the case of design, design PCK provides a system view of instructional moves that integrates three kinds of knowledge: (1) knowledge about students as design learners, (2) knowledge about learning experiences that help teachers convey design knowledge and help students develop as designers, and (3) knowledge about design (Adams et al. 2016; Crismond and Adams 2012). For example, a teacher may notice that students are struggling and then draw upon their knowledge of students as learners to imagine the kinds of struggles students may be experiencing and the kinds of instructional moves that might be most effective for that situation. By turning this noticing into instructional actions, the teacher is continuing to build knowledge about what it means to teach design, which echoes our call for mindful design teaching.

In the pedagogical content knowledge framework, knowledge about students includes knowledge of student preconceptions and misconceptions, difficulties students may encounter, what students find interesting and motivating, and being able to interpret students' emerging and incomplete reasoning (Ball et al. 2008). This speaks to what teachers notice about their students—what they are doing, not doing, resisting doing, or doing ineffectively. For a Design Coaching Playbook, these may become patterns of situations teachers are sensitized to recognize, and they may have developed a set of responses to these situations, such as ways to help students productively move forward or step back to gain meta-level insight about a key design concept.

Knowledge about learning experiences can be subject matter specific. There are teaching strategies such as helping students to develop depth of understanding and make connections, and knowing when to pose new questions or tasks to students to help them deepen their understanding (Ball et al. 2008; Gess-Newsome 1999; Magnusson et al. 1999). From the perspective of a design coaching playbook, these signify various moves a teacher may use to remedy challenges students may be experiencing and likely involves invoking various design tasks for students to undertake.

Knowledge about design might include: knowledge about the inherent ambiguities and complexities of design and common pitfalls, domain knowledge such as an understanding of materials and how they behave, design process knowledge such as strategies to find and define problems or manage time and risk, and design judgment knowledge that provides the foundation for how a designer approaches a task and judges work quality in terms of novelty, technical feasibility, user desirability, economic viability, and environmental sustainability (Adams et al. 2016). This knowledge signifies the kinds of design tasks a teacher could encourage students to take that may help them start, move forward, monitor and judge their progress, iterate, and/or know if they have arrived at a desirable solution.

From the perspective of our Design Coaching Playbook, a design pedagogical content knowledge lens offers an integrated narrative that makes explicit what teachers notice about a situation and what they do in that situation in ways that collectively draws on their knowledge about students, teaching, and design (see Fincher et al. 2012). Revisiting the scenario at the beginning of this chapter, Sara and Jorge need help with getting "unstuck". For this situation, a design teacher might notice a sense of frustration, being overwhelmed, and a lack of motivation or a desire to simply give up. A teacher might interpret these signs in a variety of ways that connect to what they perceive as a root problem and potential remedies that target that root problem. For instance, a teacher may connect the sense of frustration to the conflict students are experiencing between what they envision as a desirable project direction and what the client envisions, and they might encourage students to create visual comparisons such as graphs or simple prototypes to help students and the client come to a shared understanding regarding the explicit (versus imagined) strengths and weaknesses of different ideas. It may be that the students have uncovered an important issue or design goal but that the client is unwilling to move in that direction or sees that direction as economically infeasible (e.g., nonprofit organizations have limited resources). In this situation, a teacher may need to help the students pivot quickly and turn their attention to a project that is both feasible and desirable. A teacher may connect a sense of being overwhelmed with the constant flux of client information and offer strategies for managing ambiguity. A teacher may connect the head drop on the desk as a desire to give up, and may encourage strategies that build motivation and a sense that the project is feasible and the students can succeed.

6 The End or the Beginning: Supporting Mindful Design Teaching

We began this journey as three design educators and design researchers seeking a mindful approach to design teaching. An aim was to develop a means that could capture, communicate, and share experiences for us, and also scale to include the practices of many others. Through our process, we revealed some common design teaching approaches that became a starting point for imagining an idea for a Design Coaching Playbook. In support of developing such a resource, we drew on literature from cognitive apprenticeship, teaching as improvisation, and pedagogical content knowledge to identify ways to be explicit and mindful about teaching design. Specifically, this literature and the other lenses presented provide insights into roles a design teacher may fulfill, offer situation-general strategies teachers could use that are relevant for design, and propose approaches to develop situation-specific strategies that link knowledge about students, teaching, and design.

As we adapted a playbook through our approach, the notion of a **scenario** was a means to drive a narrative of what we noticed about the situation and what we did: **roles** undertaken and the rationale/reflection therein supported discussions of possible next **moves** or those to encourage our student(s) to do. Through the interplay between examining the literature and identifying the many possible scenarios, roles, and moves around design teaching, we presented a need and generated possible pathways toward the creation of a Design Coaching Playbook that can impact educators, students, and researcher.

For educators, our conceptualization of a tool like a Design Coaching Playbook can make design teaching more explicit and shareable. Now, we can present how the often seemingly invisible decisions get made and how those moves, situations, and actions emerge in a more transparent fashion. There can be utility for practitioners who step into roles where they are asked to advise and counsel student teams through a design process. Leveraging our experience as a model, we shared our process and the importance of linking what we notice in situations to what we do, as these reflections are situated knowledge that accounts for our design pedagogical knowledge and the flexibility of how to apply our content knowledge. The metaphor of a playbook also aligns with coaching. It can provide ways to frame conversations with colleagues or other design educators on how one might overcome or approach a scenario. It is an interesting collective desire to make expertise visible in this fashion. While it may be risky to be transparent about our process and decision-making, lest we lower the need for our own expertise, it will be a step forward for the fields of engineering, art, and humanities that use design as part of their pedagogy to be more explicit about the how of it all.

For students as well, this playbook can become a useful resource to develop and practice their strategic knowledge about design. It offers an opportunity for students to be reflective about their practice, or think about how they might manage a design team themselves. With shared understanding, this playbook might disrupt design education and have students and instructors codesign the design education enterprise.

For design researchers, a Design Coaching Playbook calls for supporting design education by bridging a gap between what we know about design "doing" to enable design learning (Lande and Leifer 2010; Lande 2012; Lande 2016). A playbook can be a mechanism for allowing and acknowledging expertise of design teachers and the utility not just of the design practice but also how to manage the design learning enterprise. This suggests future work needed to support this area for design coaching itself. Whether it be the newly hired design teacher, the practitioner, or the master design teacher, the sense of a continual professional development through reflective practice is necessary. We add the idea of a suite of moves to help make that procedural knowledge become more useful in its application and deployment. There are related recommendations for a research agenda focused on design education that moves the focus from design methods or techniques to what teachers do when they help students develop as designers and learners.

Overall, this chapter offers three perspectives, and connections across those perspectives, on the act of design education, "educating the design workforce of near tomorrow." We seek to make visible the practices of educators in an attempt to support learning, where learners are broadly conceived to include students, employers, educators, and supervisors. The resulting instructional moves and recommendations are grounded in design education research. The Design Coaching Playbook described here can be a useful means to complete the research-to-practice loop and help all design teachers become more mindful and supported in their teachings.

There is not a singular Design Coaching Playbook. Rather we imagine that for each instructor there exists one for their unique context. This may make consideration for a way to be mindful of one's own practices and provide opportunities to deconstruct and discuss what potential moves that are in a local version for such a playbook. One can constructively consider not just what the right move is, in helping student teams along a design challenge. Rather, one can be generative in thinking about what are all the possible moves/prompts/probes to get students to have a better learning experience through design. Now as coaches, one can rethink their role in the learning process, and even feel more comfortable in sharing with students the ambiguity not just of design but also in navigating through the process of design, working toward a designed solution.

We are excited at the prospects for a Design Coaching Playbook. By sharing our process building up to it through scenarios, roles, and moves, we see this as the beginning of developing such a tool. Scenarios are presented here, and have been shared and piloted with the community at a recent conference (Coso Strong et al. 2018) as well. The resulting suite of roles and moves may be useful in starting a language to share and discuss design coaching itself. A Design Coaching Playbook provides a set of tools for the proverbial toolbox for design educators and instructors toward mindful design education.

References

Adams RS, Siddiqui JA (2015) Analyzing design review conversations. Purdue University Press. https://books.google.com/books?id=_ClrDQAAQBAJ

Adams RS, Turns J, Atman CJ (2003) Educating effective engineering designers: the role of reflective practice. Des Stud 24(3):275–294

Adams R, Forin T, Chua M, Radcliffe D (2015) Making visible the "how" and "what" of design teaching. In: Adams R, Siddiqui JA (eds) Analyzing design review conversations. Purdue University Press, pp 431–456

Adams RS, Forin T, Chua, M, Radcliffe D (2016) Characterizing the work of coaching during design reviews. Des Stud 45:30–67

Adams RS, Aleong RJ, Goldstein M, Solis F (2017a) Problem structuring as co-inquiry. In: Christensen BT, Ball LJ, Halskov K (eds) Analysing design thinking: studies of cross-cultural co-creation. CRC Press

Adams RS, Forin T, Chua M, Radcliffe D (2017b) Approaches for coaching students in design reviews. ASEE Conference, OH.

Ahmed S, Wallace KM, Blessing LTM (2003) Understanding the differences between how novice and experienced designers approach design tasks. Res Eng Design 14:1–11

Alliance for Justice (2015) Philanthropy advocacy playbook: leveraging your dollars. https://www.bolderadvocacy.org/wp-content/uploads/2015/09/AFJ-Advocacy-Playbook-web.pdf

Ball DL, Thames MH, Phelps G (2008) Content knowledge for teaching. J Teach Educ 59(5):389–407

Barron B (2004) Learning ecologies for technological fluency: gender and experience differences. J Educ Comput Res 31(1):1–36

Brandt C, Cennamo K, Douglas S, Vernon M, McGrath M, Reimer Y (2013). A theoretical framework for the studio as a learning environment. Int J Tech Des Educ 23(2):329–348

Bray JN (2002) Uniting teacher learning: collaborative inquiry for professional development. New Dir Adult Contin Educ 94

Bray JN, Lee J, Smith LL, Yorks, L (2000) Collaborative inquiry in practice: action, reflection, and making meaning. Sage Publications, Inc, Thousand Oaks, CA

Brennan KA (2013). Best of both worlds: issues of structure and agency in computational creation, in and out of school. PhD thesis, Massachusetts Institute of Technology

Cardella ME, Buzzanell PM, Cummings A, Tolbert D, Zoltowski CB (2014) A tale of two design contexts: quantitative and qualitative explorations of student-instructor interactions amidst ambiguity. In: Design Thinking Research Symposium. Purdue University, West Lafayette, IN

Carlson T, Cockayne W, Tahvanainen A (2013) Playbook for strategic foresight and innovation. https://www.thegeniusworks.com/wp-content/uploads/2016/01/Playbook-for-Strategic-Foresight-and-Innovation-A4.pdf

Cennamo K, Brandt C, Scott B, Douglas S, McGrath M, Reimer Y, Vernon M (2011) Managing the complexity of design problems through studio-based Learning. Interdisc J Prob-Based Learn 5(2)

Collins A, Brown JS, Holum A (1991) Cognitive apprenticeship: making thinking visible. Am Educ 6:38–46

Coso AE (2014) Preparing students to incorporate stakeholder requirements in aerospace vehicle design. PhD thesis, Georgia Institute of Technology, Atlanta, GA

Coso Strong A, Lande M, Adams RS (2018) Special session: put me in coach! Developing a design playbook for instructors to help engineering students do design. Frontiers in education conference, San Jose, CA. https://doi.org/10.1109/FIE.2018.8658648

Crismond DP, Adams RS (2012) The informed design teaching and learning matrix. J Eng Educ 101(4):738–797

Cross N (1982) Designerly ways of knowing. Des Stud 3(4):221–227

Crossan MM (1998) Improvisation in action. Organ Sci 9(5):593–599

Dannels DP (2003) Teaching and learning design presentations in engineering: contradictions between academic and workplace activity systems. J Bus Tech Commun 17(2):139–169. https://doi.org/10.1177/1050651902250946

Dannels DP, Martin KN (2008) Critiquing critiques. J Bus Tech Commun 22(2):135–159

Dickens M, Jordan SS, Lande M (2016) Parents and roles in informal making education: informing and implications for making in museums. American Society for Engineering Education annual conference, New Orleans, LA. https://doi.org/10.18260/p.25854

Donohoo J (2013) Collaborative inquiry for educators: a facilitator's guide to school improvement. Corwin A SAGE Company, Thousand Oaks, CA

Downey G, Lucena J (2003) When students resist: ethnography of a senior design experience in engineering education. Int J Eng Educ 19(1):168–176

Dym CL, Agogino AM, Eris O, Frey DD, Leifer LJ (2005) Engineering design thinking, teaching, and learning. J Eng Educ 103–120

Eris O (2004) Effective inquiry for innovative engineering design, vol. 10. Springer Science & Business Media

Fallon J, Verwayen H (2017) Introducing: impact playbook for museums, libraries, and archives. Europeana. https://pro.europeana.eu/post/introducing-the-impact-playbook-the-cultural-heritage-professionals-guide-to-assessing-your-impact

Feland J, Cockayne W, Leifer LJ (2004) Comprehensive design engineering: designers taking responsibility. Int J Eng Educ 20(3):416–423

Fincher S, Finlay J, Sharp H, Falconer I, Richards Board (2012) Change stories: a white paper from the share project. http://www.sharingpractice.ac.uk

Gess-Newsome J (1999) Pedagogical content knowledge: an introduction and orientation. In: J. Gess-Newsome (ed) Examining pedagogical content knowledge: the construct and its implications for science education. Kluwer Academic, Dordrecht

Goldschmidt G (2006) Expert knowledge or creative spark? predicaments in design education. In: Proceedings of the 6th DTRS symposium, Sydney, AU

Goldschmidt G, Hochman H, Dafni I (2010) The design studio crit: teacher student communication. Artif Intell Eng Des Anal Manuf 24(3):285–302

Goldschmidt G, Casakin H, Avidan Y, Ronen O (2014) Three studio critiquing cultures: fun follows function or function follows fun? In: Design thinking research symposium. Purdue University, West Lafayette, IN

Goncher A, Johri A (2015) Contextual constraining of student design practices. J Eng Educ 104(3):252–278. https://doi.org/10.1002/jee.20079

Hargadon AB (2002) Brokering knowledge: linking learning and innovation. Res Organ Behav 24:41–85

Hlubinka M, Dougherty D, Thomas P, Chang S, Hoefer S, Alexander I, McGuire D (2013) Makerspace playbook: school edition. https://makered.org/wp-content/uploads/2014/09/Makerspace-Playbook-Feb-2013.pdf

Huet G, Culley SJ, McMahon CA, Fortin C (2007) Making sense of engineering design review activities. AI EDAM 21(03):243

Isaacs-Sodeye O, Lande M (2013) Teaching with unfamiliar pedagogy for engineering design instructors. Frontiers in Education conference, Oklahoma City, OK, pp 1447–1449. https://doi.org/10.1109/FIE.2013.6685071

Jansson DG, Smith SM (1991) Design fixation. Des Stud 12(1):3–11

Kasl E, Yorks L (2002) Collaborative inquiry as a strategy for adult learning. New Dir Adult Contin Educ 94(94):112

King KB, West JR (2018) Futures thinking playbook: what might the future be like and what can we do to shape it? CreateSpace Independent Publishing Platform

Koen BV (1994) Toward a strategy for teaching engineering design. J Eng Educ 83(3):193–201. https://doi.org/10.1002/j.2168-9830.1994.tb01104.x

Lande M (2012) Ambidextrous mindsets for innovation: designing and engineering. PhD thesis, Stanford University

Lande M (2016) Catalysts for design thinking & engineering thinking: fostering ambidextrous mindsets for innovation. Int J Eng Educ 32:1356–1363

Lande M, Leifer L (2010) Difficulties student engineers face designing the future. Int J Eng Educ 26:271–277

Lande M, Jordan S, Weiner S (2017) Making people and projects: implications for making-based learning. In: American Society for Engineering Education Pacific Southwest conference. Tempe, AZ

Leifer LJ, Steinert M (2011) Dancing with ambiguity: causality behavior, design thinking, and triple-loop-learning. Inf Knowl Syst Manag 10(1–4):151–173

Lewrick M, Link P, Leifer L (2018) Design thinking playbook. Wiley

Limbert T (2012) Dad's playbook: wisdom for fathers from the greatest coaches of all time. Chronicle Books LLC

Magnusson S, Krajcik, Borko H (1999). Nature, sources, and development of pedagogical content knowledge for science teaching. In: Gess-Newsome J & Lederman N (eds) Examining pedagogical content knowledge. Kluwer Publishing, Boston, p 95–128

McDonnell J (2016) Scaffolding practices: a study of design practitioner engagement in design education. Des Stud 45:9–29

New Learning (n.d.) Repertoire of practice. http://newlearningonline.com/learning-by-design/glossary/repertoire-of-practice

Newstetter WC (1998) Of green monkeys and failed affordances: a case study of a mechanical engineering design course. Res Eng Design 10(2):118–128

Oak A, Lloyd P (2014) Throw one out that's problematic: performing authority and affiliation in design education. CoDesign (1–2):55–72

Olsen D (2015) The lean product playbook: how to innovate with minimum viable products and rapid customer feedback. Wiley Publishers

Oplinger J, Lande M (2014). Disciplinary discourse in design reviews: industrial design and mechanical engineering courses. In: Proceedings of DTRS 10, Purdue University, IN

Oplinger J, Lande M, Jordan S, Camarena L (2016) Making leaders: leadership characteristics of makers and engineers in the maker community. Am J Eng Educ 7(2):65–82

Oxman R (1999) Educating the designerly thinker. Des Stud 20(2):105–122

Paretti MC (2008) Teaching communication in capstone design: the role of the instructor in situated learning. Journal of Engineering Education 97(4):491–503. https://doi.org/10.1002/j.2168-9830.2008.tb00995.x

Pritchard C, Sanders P (2002) Weaving our stories as they weave us. New Dir Adult Contin Educ 94

Reich Y, Ullmann G, Van der Loos M, Leifer L (2009) Coaching product development teams: a conceptual foundation for empirical studies. Res Eng Des 19(4):205–222

Sawyer RK (2004) Improvised lessons: collaborative discussion in the constructivist classroom. Teach Educ 15 (2):189–201

Sawyer RK (2008) Learning music from collaboration. International Journal of Educational Research 47 (1):50-59

Sawyer K (ed) (2011). Structure and improvisation in creative teaching. Cambridge University Press

Schon DA (1984) The reflective practitioner: how professionals think in action. Basic Books. https://books.google.com/books?id=ceJIWay4-jgC

Schön DA (1993) The reflective practitioner: how professionals think in action basic books, New York

Sheppard M, Jordan S, Lande M, McKenna A (2018) Exploring making-based pedagogy in undergraduate mezzanine-level engineering courses. In: American Society for Engineering Education. Education and research methods division. Salt Lake City, UT

Shulman L (1987) Knowledge and teaching: foundations of the new reform. Harv Educ Rev 57(1):1–23

Solis F, Coso Strong A, Adams R, Turns J, Crismond D (2016) Towards a scholarship of integration: lessons from four cases. In: 2016 American Society for Engineering Education annual conference and exposition

Sonalkar N, Mabogunje A, Pai G, Krishnan A, Roth B (2016) Diagnostics for design thinking teams. In: Plattner H, Meinel C, Leifer L (eds) Design thinking research. Understanding innovation. Springer, Cham

Strickfaden M, Heylighen A (2007) Exploring the 'cultural capital' of design. In: International conference on engineering design, pp. 1–10

Strickfaden M, Heylighen A (2010) Scrutinizing design educators' perceptions of the design process. Artif Intell Eng Des, Anal Manuf: AIEDAM 24(3):357–366. https://doi.org/10.1017/S0890060410000247

Sypolt G (2017) Building an agile process playbook for software testing. Sauce Labs. https://saucelabs.com/blog/building-an-agile-process-playbook-for-software-testing

Turns J, Cardella M, Atman CJ, Martin J, Adams RS (2006a) Tackling the research-to-teaching challenge in engineering design education: making the invisible visible 22(3):598–608

Turns J, Adams RS, Martin J, Cardella M, Mosborg S, Atman CJ (2006b) Tackling the research-to-practice challenge in engineering design education: insights from a user-centered design perspective. Int J of Eng Educ

Turns JA, Sattler B, Thomas LD, Atman CJ, Bankhead RB, Carberry AR, Csavina KR, Cunningham P, Faust DK, Harding TS, Yasuhara K (2015) Reflecting on reflection: how educators experience the opportunity to talk about supporting student reflection. In: 2016 American Society for Engineering Education annual conference and exposition

Uluoğlu B (2000) Design knowledge communicated in studio critiques. Des Stud 21(1):33–58

Valkenburg R (2001) Schön revised: describing team designing with reflection-in-action. In: Proceedings of design thinking research symposium, vol 5, pp 1–15

Valkenburg R, Dorst K (1998) The reflective practice of design teams. Des Stud 19(3):249–271. https://doi.org/10.1016/S0142-694X(98)00011-8

Walther J, Pawley A, Sochacka N (2015) Exploring ethical validation as a key consideration in interpretive research quality. In: American Society for Engineering Education annual conference and exposition. Seattle, WA. https://doi.org/10.18260/p.24063

Walther J, Sochacka NW, Benson LC, Bumbaco AE, Kellam N, Pawley AL, Phillips C ML (2017) Qualitative research quality: a collaborative inquiry across multiple methodological perspectives. J Eng Educ 106(3):398–430. https://doi.org/10.1002/jee.20170

Wolmarans N (2016) Inferential reasoning in design: relations between material product and specialised disciplinary knowledge. Des Stud 45:92–115

Yilmaz S, Daly SR (2016) Feedback in concept development: comparing design disciplines. Des Stud 45:137–158

Second-Year Engineering Design: A Use-Inspired Approach

Adam R. Carberry and Samantha R. Brunhaver

Abstract Design is an important concept that all engineering students should experience throughout their formal education. Providing students with a design experience each semester within a design "spine" affords an explicit opportunity unmatched by a single first-year or capstone experience. The purpose of this chapter is to describe a use-inspired design course offered to engineering students during the beginning of their second year at a large public university in the United States. A description of the place, history, context, project-based pedagogy, and course details are provided as a model example for those interested in offering such a course. These details are framed within the context of limited available credits within an engineering curriculum.

Keywords Design spine · Project-based learning · Second-year · Use-inspired

1 Introduction

Integration of design within engineering curricula has become a staple of most engineering programs. Such inclusion is seen most often embedded within first-year introductory courses and final-year capstone experiences regardless of the discipline or nature of the program. These explicit opportunities to practice design speak to its importance as a topic that engineering students should learn and master before graduating (Dym 1999; Dym et al. 2005). Many have questioned if these opportunities are enough (Kotys-Schwartz et al. 2010), particularly in the development of design skills (Froyd and Ohland 2005; Savage et al. 2007).

Explicit and implicit approaches to teaching design have been included within engineering curricula because design can be applied in a variety of contexts. For example, many programs often supplement their first-year and final-year design

A. R. Carberry (✉) · S. R. Brunhaver
The Polytechnic School, Arizona State University, 7171 East Sonoran Arroyo Mall,
Peralta Hall 330G, Mesa, AZ 85015, USA
e-mail: adam.carberry@asu.edu

S. R. Brunhaver
e-mail: samantha.brunhaver@asu.edu

© Springer Nature Switzerland AG 2019
D. Schaefer et al. (eds.), *Design Education Today*,
https://doi.org/10.1007/978-3-030-17134-6_2

offerings by implicitly including opportunities woven within the mezzanine (second and third) years of the engineering curriculum. Common forms of this are the use of a project-based pedagogical approach or leveraging design thinking in the solution of an open-ended problem. It is rare that programs offer explicit design experiences regularly throughout a four-year curriculum.

Recent years have seen some early adopters going beyond the first- and final-year design offerings to institutionalize and optimize design-specific courses in the mezzanine years (de Los et al. 2010; Heitmann 1996; Pavelich et al. 1995). This approach is meant to provide students with opportunities to practice and apply their technical skills, while also developing their design and professional skills throughout the entirety of their undergraduate experience. Adding something to the curriculum usually means that something must be taken out. A major barrier to implementing regular design-specific courses throughout a curriculum has been real limits on the number of credits that engineering students can reasonably be expected to take. Some programs have begun to identify which advanced topics to remove to accommodate the extra design credits. Another proposed alternative has been to expand the engineering curricula to a fifth year (Howard and Verma 2011), but this idea has failed to gain traction like it has in other disciplines (e.g., medicine and law).

The following chapter discusses one approach taken by an engineering program to offer mezzanine level design courses as part of a four-year curriculum. The combination of these courses with first-year and capstone design courses make up the program's design spine. Background and context are provided to support the decision to structure the program using a design spine approach. A single course required of all enrolled students during the first semester of their second-year is then dissected and presented as a model example for a mezzanine design course offering. The example course is framed by project-based learning and the intention to provide a use-inspired design experience.

2 Background and Context

The engineering design course examined in this chapter is a mezzanine level offering of The Polytechnic School (TPS) undergraduate engineering program within the Ira A. Fulton Schools of Engineering (FSE) at Arizona State University (ASU). This course has become a model example of how to implement a project-based course within the second-year of the curriculum. The particular approach has been observed to be successful within the unique context of the TPS engineering program. Potential adopters will need to consider how unit characteristics may or may not impact the success of such an approach in their own program. It is with this disclaimer that we provide background for the formation of the TPS engineering program and the larger university context in which it operates.

TPS exists solely on the ASU Polytechnic Campus, separated by approximately 20 miles from the Tempe campus. Students required to take the course described in this chapter are enrolled in either the Bachelor of Science in Engineering (BSE) or the Bachelor of Science (BS) in Manufacturing Engineering. Those in the BSE program must choose one of four concentrations—mechanical engineering systems, electrical systems, automotive systems, or robotics—prior to the start of their third year. The program as it exists today has undergone many changes since its inception back in 2007. The overall goal of the program remains the same as established by the faculty who founded the program; to prioritize the use of innovative pedagogical approaches. Much of this innovation centers on the use of a project-based pedagogy, particularly within the project spine design course offerings.

TPS was established in 2014 when it became the sixth school within the Ira A. Fulton Schools of Engineering (FSE). FSE has undergone a variety of massive changes over the past decade including the expansion of programs across multiple campuses, college mergers, and rapid growth. This has led ASU to currently enroll the largest number of engineering students in the United States. The six FSE schools consist of 24 undergraduate degree programs and no departments. The sheer size of FSE means that the college structure is essentially a small university in and of itself with a hierarchy consisting of a Dean, six school Directors, and 24 program chairs.

The entirety of FSE is housed at Arizona State University (ASU), a large, public research university currently enrolling the largest number of students in the United States. The ASU mission explicitly states that it is "measured not by whom it excludes, but by whom it includes and how they succeed." The mission, which assumes "fundamental responsibility for the economic, social, cultural and overall health of the communities it serves," is guided by eight design aspirations: (1) leveraging its place, (2) transforming society, (3) valuing entrepreneurship, (4) conducting use-inspired research, (5) enabling student success, (6) fusing intellectual disciplines, (7) being socially embedded, and (8) engaging globally. FSE embraces the mission and design aspirations of the institution, which has fueled and driven endeavors such as the creation of TPS with its engineering program project-based design spine.

3 Project-Based Learning

The focus on design project courses within the TPS engineering program is geared toward providing students with as many opportunities as possible to practice what they learn in their non-project classes. A brief discussion of project-based learning is necessary to understand how such a pedagogical approach is used within the program.

The origins of project-based learning connect back to Dewey's (1959) work suggesting that students develop personal investment in their learning when provided with opportunities to engage in real, meaningful tasks and with problems that emulate what experts do in the real world. Krajcik and Blumenfeld (2006) and Greeno (2006) suggest that the benefits and use of project-based learning can be further theoretically supported by constructivist theories, active construction, situated learning, social interactions, and cognitive tools. This suggests that the overall approach leverages active, cooperative, and/or experiential learning as a means of educating students (Johnson et al. 1998).

Fig. 1 Project-based
learning model

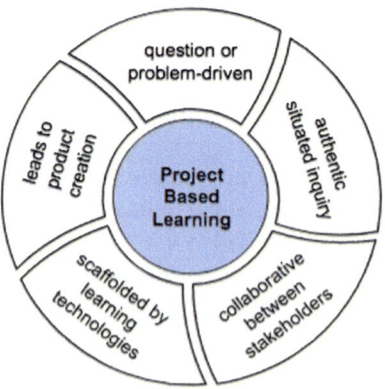

According to multiple scholars (Blumenfeld et al. 1991; Krajcik et al. 1994, 2002), project-based learning is defined by five key features as illustrated in Fig. 1: (1) driven by a question or problem to be solved, (2) explored through authentic, situated inquiry, (3) collaborative between students, teachers, and community members, (4) scaffolded by learning technologies, and (5) leads to the creation of a publicly accessible, tangible products.

The use of project-based learning in higher engineering education around the world is not a new concept and, in some instances, like the TPS engineering program, has become the primary focus of many or most course offerings (see Mills and Treagust 2003; Prince and Felder 2006). de Graaff and Kolmos (2003) add an important note that not all project courses need to be structured exactly the same way. Instructors have a choice in the degree of student autonomy embedded in the project. The choice made by the instructor can fall into three categories: (1) task project—project and approaches defined by the instructor, (2) discipline project—project subject area defined by the instructor, but students identify the specific project and approach, and (3) problem project—students choose their project and approach. The following section describes the use of project-based learning within the TPS engineering program according to de Graaf and Kolmos' framework.

4 Design Spine

TPS engineering students engage with project-based learning primarily within a series of design courses that they are required to take in order to graduate from the BSE or BS in Manufacturing Engineering programs. These courses are referred to as the "design spine" of the curriculum because students traditionally take one course in the sequence each semester, forming a backbone for the overall curriculum (see Fig. 2).

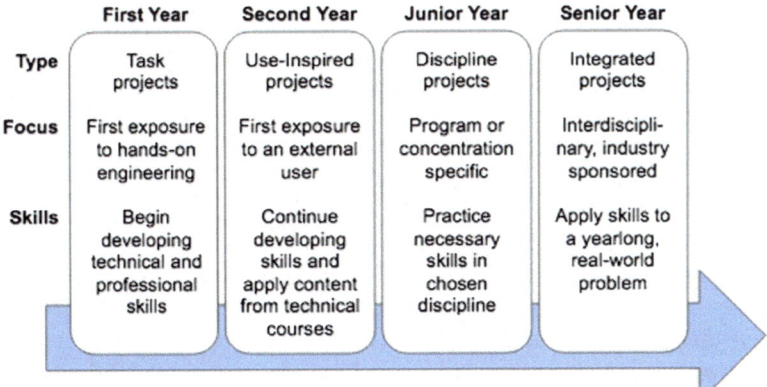

	First Year	**Second Year**	**Junior Year**	**Senior Year**
Type	Task projects	Use-Inspired projects	Discipline projects	Integrated projects
Focus	First exposure to hands-on engineering	First exposure to an external user	Program or concentration specific	Interdiscipli-nary, industry sponsored
Skills	Begin developing technical and professional skills	Continue developing skills and apply content from technical courses	Practice necessary skills in chosen discipline	Apply skills to a yearlong, real-world problem

Fig. 2 TPS engineering program design spine

The first-year and senior offerings align with most institutions' use of introductory engineering courses and a final-year capstone project. The first-year courses are designed to introduce students to multiple design challenges that help them better understand the field of engineering, gain familiarity with the different disciplines, and begin to develop their technical and professional skills. These projects fall within the task project category.

The capstone project is a yearlong interdisciplinary effort where students work in teams on a project sponsored by industry or a faculty member. As suggested by the title of "capstone", this project is meant to cap off a student's four-year education by culminating in an all-encompassing project desired by a real client. This project overlaps with the discipline and problem project categories.

The mezzanine years are where the TPS engineering program strays from traditional programs. The second-year offerings have evolved over time to focus on use-inspired design. These two courses are students' first opportunities to consider a user other than themselves or their instructor. Students work on projects that continue to build their design and professional skills, while providing a setting for them to apply and practice what they have learned in their technical courses. These projects are hands-on opportunities to learn from failure. The embedded user focus will be discussed further in the next section. Second-year projects fall in the problem project category.

It was noted before that students are not required to choose manufacturing engineering or a concentration within the engineering program prior to their third year, meaning that all engineering and manufacturing engineering students within TPS take a two-year common core of courses where they coexist in the same project courses. The third-year courses mark the first-time students are separated into their disciplines. These project courses are program specific (or concentration specific) and provide more homogenous projects that ensure students will practice the necessary skills needed in their chosen discipline. Third-year projects fall within the discipline project category. These third-year project courses are often taken in conjunction with discipline-specific technical courses. The available credits for students to take discipline-specific courses are somewhat limited by the emphasis on design courses. The program's solution to this challenge is to frame degrees around concentrations that allow for a discipline-specific focus at a reduced number of credits compared to a traditional major in the discipline.

The overall project spine is designed to introduce students to engineering, help them develop design and professional skills, provide opportunities to practice technical skills, learn to work in disciplinary, multidisciplinary, and interdisciplinary teams, and prepare for the engineering profession. It is important to also note that there have been many efforts throughout the existence of this program to ensure content is scaffolded appropriately and not repeated without purpose within the curriculum. These efforts are primarily driven by a desire among faculty to update the curriculum based on emerging engineering education research findings.

The approach taken by TPS to offer a design spine and to continue to evolve the scaffolding of content has worked in the context that exists within the unit. The structure has survived and thrived even as enrollment has increased, section offerings have expanded, class sizes have grown, and faculty have cycled in and out as instructors. This has led to a stable course structure offered in both the fall (primary) and spring (secondary) semesters at ASU.

5 Course Details

A hallmark of any design spine is the atypical requirement of second and third-year design experiences. The course to be discussed in detail for this chapter is the first semester, second-year use-inspired design course. This course marks the third project course students take as part of the TPS engineering program. Students enrolled in the course include both traditional students and nontraditional students (transfer students, veterans, full-time employed students, returners, and career changers). All students entering this course are required to have taken a full year of introductory engineering courses within or outside of the university. Students typically take four other courses in conjunction with the second-year project course. These courses vary by semester but can include electrical or mechanical engineering fundamentals, materials science and manufacturing, computer programming, engineering statistics, physics, calculus, or differential equations.

The course meets in-person twice a week for a total of two and a half hours. Growth of enrollment within the course over the past decade has forced the course to shift from one to two co-taught sections offered only in the Fall semester to six to eight individually taught sections taught in both the Fall and Spring semesters. All sections enroll no more than 45 students per section. The course uses an active, student-centered learning approach (e.g., projects, flipped classroom, small group discussions, peer teaching, etc.) and very rarely uses traditional, teacher-centered lecture. Classes are situated in studio classrooms located in near proximity to machine shops and maker-type spaces. Students are afforded ample opportunities to use a variety of open spaces to complete their assignments. The course has evolved over time to focus primarily on use-inspired design framed by the opportunity to design solutions for real users rather than instructor designed scenarios. The following subsections describe in more detail the course project structure, topics, embedded activities, and learning outcomes.

5.1 Projects and Activities

A hands-on approach is taken throughout all projects to provide students with the necessary skills and knowledge required of students who pursue any of the engineering degree options offered within TPS. The course uses a combination of multiple short-term mini-projects and a single long-term final project. All projects, regardless of the assigned time on task, are team-based, situated within a real-world context, and presented to students as either partially guided or completely open-ended. The projects are designed to provide students with an opportunity to learn, develop, and practice new skills within a design scenario. These new tools can be added to a student's "engineering toolbox" and are intended to help them build on what was learned in the first-year project courses. Students are told that the activities they undertake in the course are intended to help them develop their own personal design approach or process.

The timing and placement of each activity within a given mini-project works with a traditional engineering design process flow: identify a need/problem, research the need/problem, brainstorm possible ideas, select an idea to pursue, model the solution, construct a prototype, test the prototype, and redesign. Students are not presented with a singular design process model and instead are encouraged to discover what works best for their team and them individually. Activities included in the course to address the design process steps are product archaeology, functional decomposition, problem definition, constraint & criteria identification, morphological charts, selection matrices, engineering drawings, prototype construction, mathematical modeling, flowcharting, statistical analysis, design for assembly, life cycle analysis, reflection, written or verbal communication, and entrepreneurial mindset. A sample course schedule showing the timing and topics covered for each project is shown in Fig. 3.

5.1.1 Mini-Projects

Each mini-project lasts for a duration between two and six weeks, requiring a total of 10–30 hours of in and out-of-class work time. The course has experimented with the use of two or three mini-projects each semester. Changes in the number of mini-projects have been influenced by a number of factors, including the scheduled amount of time for the course, presence or absence of certain course modules, number of instructors teaching the course, new project ideas, and a desire by faculty to avoid

Fig. 3 Project activities in
second-year, first-semester
design course

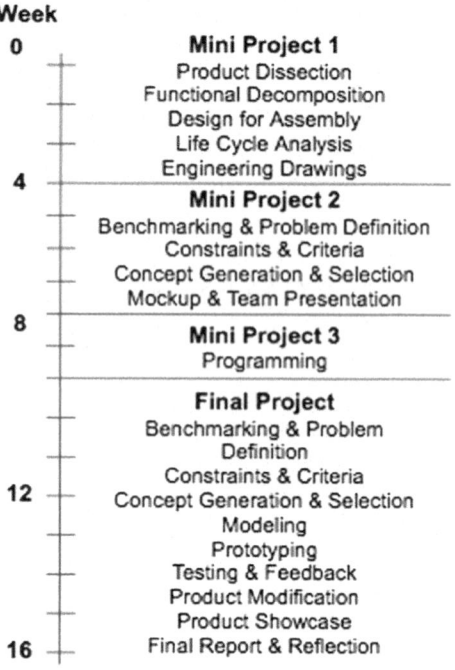

monotony. These factors have led the ever-changing teaching team to present projects embedding programming (e.g., LEGO robotics and Arduino-based challenges), product dissection, and prototype construction. Some iterations of the mini-projects were designed with a project flow around the engineering design process, while other project combinations have not required a specific order of implementation. The benefit to the latter approach was that each project could be modified and tested multiple times within a given semester as the projects rotated between sections. Recent evolutions have kept the order of projects consistent across sections and have become more thoughtful about the context of the projects. The mini-projects are now designed to align with the ASU FSE Research Themes of security, health, energy, and/or sustainability; a fifth theme, education, is the focus of the final project, discussed in more detail in the next section.

Two focus areas have persisted over the past ten or so iterations of the course's mini-projects. The first is the use of product archaeology (Carberry et al. 2015). Product archaeology is defined as "the process of reconstructing the lifecycle of a product—the customer requirements, design specifications, and manufacturing processes used to produce it—to understand the decisions that led to its development" (Simpson et al. 2011). Lewis et al. (2013) adapted this approach coined by Ulrich and Pearson (1998) for use within an engineering education setting. The project began by investigating a variety of products obtained from a donation center. Challenges faced by the variety of products analyzed led the teaching team to choose a single product for students to investigate in a given semester. The first of such products was a disposable camera. A new focus on health for this project has led to the selection of an electric toothbrush as the product with pedometers as a future potential product. The product archaeology approach embeds an opportunity for students to enter the design process at the stage of a finished product to better understand why the product was designed the way it was rather than constantly starting from an instructor provided challenge. Activities included in the project involve sketching hypothesized functionality, benchmarking related products, physically dissecting the product, conducting a functional decomposition, performing a life cycle analysis, investigating design for assembly, creating engineering drawings, and undergoing a mathematical analysis of an internal system (e.g., battery-motor functionality). Early iterations of the project required students to redesign their product for people in a different part of the world. For this part of the project, students underwent a needs analysis, defined a related problem they wished to tackle, generated a concept, and selected a concept. The results of these activities were a video presentation shared with the class and a mock-up of the new design. The videos were reviewed by classmates, who subsequently provided feedback, in written or audio form, that the receiving team then responded to (Carberry et al. 2016a, b). These activities still take place, but have been separated into an independent project using what was learned about product design in the dissection process to develop a new product for energy or sustainability.

The second focus area that has persisted throughout the mini-projects is a programming component. Early incorporation of programming centered on LabVIEW to align project activities with a concurrent, single credit module required of students enrolled in the BSE program. The need to incorporate LabVIEW led the instructors toward the use of LEGO robotic challenges. These challenges provided students with an introduction to using the programming language, while also learning how to use various sensors and interpret collected data. A shift away from LabVIEW in the curriculum and the use of Arduino in the first-year project courses resulted in adoption of Arduino-based activities in the second-year project courses. Within the first semester, second-year course, this activity has built upon the product archaeology project or been presented as a separate project. The most recent implementation of the Arduino activity involves the creation of a door alarm system to expose students to security applications in engineering.

The combination of mini-projects in each iteration is always designed to help the students prepare for the final project. Selected activities are designed to be learned and practiced during the mini-projects and later revisited so they can be improved upon. The shift in focus for each project toward an ASU FSE Research Theme helps foster the use-inspired context of the course.

5.1.2 Final Project

Students complete a final project during the remainder of the semester. The entirety of the project is framed by the activities completed during the mini-projects, providing students another opportunity to practice and demonstrate each of the associated skills. Each team is assigned a real-world partner to design a solution to their need. These partners have consisted primarily of local museums or science centers and K-12 schools, relating back to the added fifth ASU FSE Research Theme of education. The liaisons, teachers, and administrators at these collaborating institutions play a critical role in helping each team progress through the project. Teams are assigned the broad task of creating a tool or exhibit for their client. The client's participation includes an initial interview, review of student proposed solutions, feedback on the initial prototype, and attendance at the final showcase. The partners have the option at the end of the semester to keep the created prototype for use at the museum or in their classroom.

Students submit four progress reports throughout the project focusing on proposed ideas, mathematical modeling, prototyping, and testing and user feedback. The students use these reports to write-up a final report, which includes an executive summary, instruction manual, expense report, and troubleshooting guide for the user. The final demonstration is an advertised event open to the public. These events have been conducted on campus and at the partner sites. Feedback from the variety of patrons who attend is required to be collected, so that students can reflect on future iterations of their prototype.

5.2   Student Learning Outcomes

Assignments associated with project activities are assessed using a learning outcomes-based grading approach (Carberry et al. 2016a, b; Heywood 2016) structured around six learning outcomes loosely corresponding to the former ABET Criterion 3 student outcomes (ABET 2018): (1) design and build, (2) modeling, (3) systems thinking, (4) research, testing and evaluation, (5) communication and organization, and (6) teamwork. Each project assignment is designed to assess one or more of these learning outcomes. The learning outcomes are all assessed a minimum of one time per project.

The use of this grading approach has evolved since its initial use. Changes have included weighting each learning outcome based on overall importance within the course, weighting scores on outcomes within a project, and experimenting with the use of different scales (see Fig. 4 for an illustration of the current system). Less weight is typically assigned to teamwork and communication & organization than to the other four outcomes within the course. This weighting of learning outcomes based on importance is not intended to downplay the value of any given learning outcome, but rather reflects the difficulty in assessing some outcomes (e.g., understanding how well a team functions is a difficult outcome to assess because the instructor cannot observe all team interactions) and the regularity of assessment of others (e.g., communication and organization can be assessed on almost any assignment).

The weighting of scores on outcomes within a project emerged as a solution to help students better grasp the concept of learning from failure. As discussed previously,

(a)

Score	Progress Level
4	Expert
3	Advanced
2	Intermediate
1	Novice
0	Unacceptable

Grade	Score
A+	3.67 - 4.00
A	3.34 - 3.66
A-	3.00 - 3.33
B+	2.67 - 2.99
B	2.34 - 2.66
B-	2.00 - 2.33
C+	1.67 - 1.99
C	1.24 - 1.66
D	1.00 - 1.23
E	< 1.00

Project	Weight
Mini Project 1	20%
Mini Project 2	20%
Mini Project 3	10%
Final Project	50%

(b)

Mini-Project 2 Assignments	Learning Objectives & Weights					
	DB (20%)	M (20%)	S (20%)	RTE (15%)	CO (15%)	T (10%)
1. Problem definition	-	-	3	-	3	-
2. Constraint & criteria id	-	-	-	2	-	-
3. Morph chart	3	-	-	-	-	-
4. Pugh matrix	-	-	-	4	-	-
5. Team presentation & mockup	-	4	-	-	2	3
Learning Objective Scores	3.0	4.0	3.0	3.0	2.5	3.0
Project Score	3.13					
Project Grade	A-					

Fig. 4 **a** (left to right) Sample SBG rubric, project weighting, and grading distribution, **b** sample SBG score assignment and calculation

the final project essentially replicates the activities done in the mini-projects, but within a new context. The impact on students' grades of an assessment made during a mini-project is much smaller than assessments made during the final project. This approach allows students to fail early, learn from their shortcomings, and hopefully succeed later. It also ensures that students stay engaged throughout the entirety of the course and focus on their learning rather than earning points toward their final course grade.

The scale used for the course has also changed over time in an effort to simplify the assessment. A five-point scale of zero to four has been used most often in the course to denote unacceptable, novice, intermediate, advanced, and expert levels of achievement. The expectation for students in the second-year is that they reach the level of advanced or a 3.0 score on all course outcomes. This earns students an "A-" in the course overall. The biggest associated challenge with using such a scale is students' familiarity with a zero to 100-point scale. Regular reminders are needed to reassure students that receiving a score of 2 on an assignment is not equivalent to receiving a score of 50 on a traditional 100-point scale. Such a score is actually equivalent to receiving a "B-". The major benefit to this approach is that students see a breakdown of their achievement by learning outcome. Two students receiving a "B" in a course using a traditional summation of points have not necessarily learned the same content. This approach allows students to see how they did, on a scale from zero to four, for each learning outcome. Students are provided with a much more transparent understanding of how to interpret their overall grade, while providing faculty with a better understanding of where students succeeded and struggled (Guskey 1997; Wiggins and McTighe 1998; Post 2014). One drawback to this system has been that each assessment of an outcome within a project is assigned the same weight. A recent iteration has experimented with the use of a multiplier to make some assessments worth more or less depending on the amount of effort the assessment is expected to require of students.

6   Conclusion

Engineering faculty are sometimes introduced to new pedagogical advancements via research articles, reports, administrators, colleagues, etc. The typical reaction is to either assume it will automatically work or won't work for their program. What we do when we make these assumptions is neglect to ask whether or not the innovation will work at our institution, and with our faculty and students. Each program must individually assess the appropriateness of a curricular innovation for the context in which they exist.

The second-year, use-inspired design course described in this chapter is a model example of how to embed design at the mezzanine levels of an engineering curriculum at a large public institution in the United States. This course is just one of many offered at ASU as part of the TPS design spine. How this course is designed and implemented is highly impacted by the norms of the unit and institution. Those interested in offering such a course at their own institution should first identify the appropriateness of this model for their given unit and institution.

Overall, the approach taken has been well received by students. Anecdotal evidence from end of semester evaluations suggests that students enjoy the course and continue to develop their design approach or process, although the realization from students often appears to come much later in the program. The hands-on activities conducted in teams and across a range of topics and applications are one aspect that is most liked. Student recommendations have highlighted a desire to see examples of high and low-quality work and more detailed rubrics to guide their work. The instructors have also identified a need to help students navigate the ambiguity, timelines, processes, and team dynamics inherent to project courses. Efforts to address these specific concerns have been made over the last several course offerings. Insights from new instructors and other design spine course faculty have helped the continued evolution of this course as well.

It is inevitable that this course will continue to change as we learn more from engineering education research and attempt to try new pedagogical strategies. What will likely remain is the focus on use-inspired design and a continued effort to offer students an early opportunity to experience a real-world design problem.

Acknowledgements This work is supported by The Polytechnic School engineering program within the Ira A. Fulton Schools of Engineering at Arizona State University. The authors gratefully acknowledge all of the instructors who have taught the course and students who have taken the course. The ideas and feedback brought by these two groups of people are the reason the course has evolved into the worthwhile experience it is today.

References

ABET (2018) Criteria for accrediting engineering programs, 2016–2017. ABET, Baltimore, MD

Blumenfeld P, Soloway E, Marx RW, Krajcik JS, Guzdial M, Palincsar A (1991) Motivating project-based learning: Sustaining the doing, supporting the learning. Educ Psychol 26:369–398. https://doi.org/10.1080/00461520.1991.9653139

Carberry A, Brunhaver S, Csavina K, McKenna A (2016a) Comparison of written versus verbal peer feedback for design projects. Int J Eng Educ 32(2):1458–1471

Carberry A, Kellam N, Brunhaver S, Sugar T, McKenna A (2015) Excavating the impact of product archaeology. In: Research in engineering education symposium

Carberry A, Siniawski M, Atwood S, Diefes-Dux H (2016) Best practices for using standards-based grading in engineering courses. In: American society for engineering education annual conference and exposition

de Graaff E, Kolmos A (2003) Characteristics of problem-based learning. Int J Eng Educ 19(5):657–662

de Los Rios I, Cazorla A, Díaz-Puente JM, Yagüe JL (2010) Project–based learning in engineering higher education: two decades of teaching competences in real environments. Proc-Soc Behav Sci 2(2):1368–1378

Dewey J (1959) Dewey on education. Teachers College Press, New York, NY

Dym CL (1999) Learning engineering: design, languages, and experiences. J Eng Educ 88(2):145–148. https://doi.org/10.1002/j.2168-9830.1999.tb00425.x

Dym CL, Agogino AM, Eris O, Frey DD, Leifer LJ (2005) Engineering design thinking, teaching, and learning. J Eng Educ 94(1):103–120. https://doi.org/10.1002/j.2168-9830.2005.tb00832.x

Froyd JE, Ohland MW (2005) Integrated engineering curricula. J Eng Educ 94:147–164. https://doi.org/10.1002/j.2168-9830.2005.tb00835.x

Greeno JG (2006) Learning in activity. In: Sawyer RK (ed) The Cambridge handbook of the learning sciences. Cambridge University Press, New York, NY

Guskey TR (1997) Research needs to link professional development and student learning. J Staff Dev 18(2):36–41

Heitmann G (1996) Project-oriented study and project-organized curricula: a brief review of intentions and solutions. Eur J Eng Educ 21(2):121–131. https://doi.org/10.1080/03043799608923395

Heywood J (2016) The assessment of learning in engineering education: practice and policy. Wiley, Hoboken, NJ

Howard D, Verma R (2011) Should the standard undergraduate engineering education be extended to five years? IEEE Potentials 30(4):7–9

Johnson DW, Johnson RT, Smith KA (1998) Active learning: cooperation in the college classroom. Interaction Book Company, Edina, MN

Kotys-Schwartz D, Knight D, Pawlas G (2010) First-year and capstone design projects: is the bookend curriculum approach effective for skill gain? In: American society for engineering education annual conference and exposition

Krajcik JS, Blumenfeld PC (2006) Project-based learning. In: Sawyer RK (ed) The Cambridge handbook of the learning sciences. Cambridge University Press, New York, NY

Krajcik JS, Blumenfeld PC, Marx RW, Soloway E (1994) A collaborative model for helping middle grade teachers learn project-based instruction. Elem Sch J 94(5):483–497. https://doi.org/10.1086/461779

Krajcik JS, Czerniak CM, Berger CF (2002) Teaching science in elementary and middle school classrooms: a project-based approach, 2nd edn. McGraw Hill, New York, NY

Lewis K, Moore-Russo DA, Kremer GEO, Tucker C, Simpson TW, Zappe SE, McKenna AF, Carberry A, Chen W, Gatchell D, Shooter S, Paretti M, McNair L, Williams C (2013) The development of product archaeology as a platform for contextualizing engineering design. In: American society for engineering education annual conference and exposition

Mills JE, Treagust DF (2003) Engineering education—Is problem-based or project-based learning the answer. Australas J Eng Educ 3(2):2–16

Pavelich M, Olds B, Miller R (1995) Real-world problem solving in freshman and sophomore engineering. New Dir Teach Learn 61:45–54. https://doi.org/10.1002/tl.37219956108

Post SL (2014) Standards-based grading in a fluid mechanics course. In: American society for engineering education annual conference and exposition

Prince MJ, Felder RM (2006) Inductive teaching and learning methods: definitions, comparisons, and research bases. J Eng Educ 95(2):123–138. https://doi.org/10.1002/j.2168-9830.2006.tb00884.x

Savage R, Chen K, Vanasupa L (2007) Integrating project-based learning throughout the undergraduate engineering curriculum. J STEM Educ 8(3):15–27

Simpson TW, Okudan GE, Ashour O, Lewis K (2011) From product dissection to product archaeology: exposing students to global, economic, environmental, and societal impact through competitive and collaborative 'digs'. In: ASME international design technical conferences-design education conference

Ulrich KT, Pearson S (1998) Assessing the importance of design through product archaeology. Manage Sci 44(3):352–369

Wiggins G, McTighe J (1998) Understanding by design. ASCD, Alexandria, VA

Teaching Design Innovation Skills: *Design Heuristics* Support Creating, Developing, and Combining Ideas

Shanna R. Daly, Seda McKilligan, Keelin Leahy and Colleen M. Seifert

Abstract An innovative idea generation process explores a variety of diverse design ideas. While challenging to achieve even for expert designers, support tools can assist designers as they learn to generate more ideas, and more varied ideas, throughout their idea generation process. In this chapter, we summarize evidence identifying *Design Heuristics* in product design, and describe how they can be employed within a variety of lesson structures to foster idea generation. We describe alternative lesson plans using the *77 Design Heuristics*: specifically, in initial idea generation, idea development, generation of subcomponent ideas, and team design. The *Design Heuristics* have been shown to support learning and innovative design outcomes across educational levels and design disciplines.

Keywords Design pedagogy · Idea generation · Innovation tools

1 Introduction

Generating new ideas during the initial phases of design has significant impact on the success of a product and the potential for innovation (Römer et al. 2001). Developing a larger, more diverse pool of options for evaluation and concept selection increases the potential for innovation. To visualize the search for innovative ideas, design

S. R. Daly (✉)
University of Michigan, 2350 Hayward Avenue, Ann Arbor, MI 48109, USA
e-mail: srdaly@umich.edu

S. McKilligan
Iowa State University, Ames, USA
e-mail: seda@iastate.edu

K. Leahy
University of Limerick, Limerick, Ireland
e-mail: Keelin.leahy@ul.ie

C. M. Seifert
University of Michigan, 530 Church Street, Ann Arbor, MI 48109, USA
e-mail: seifert@umich.edu

© Springer Nature Switzerland AG 2019
D. Schaefer et al. (eds.), *Design Education Today*,
https://doi.org/10.1007/978-3-030-17134-6_3

researchers often talk about a "design space" (following Newell and Simon's (1972) "problem space"). Some ideas in this space are easy to find because they are obvious, or they have been seen before in existing products. Other, less obvious ideas require more effort to identify. Ideally, this search for less obvious, more innovative ideas would entail visiting all feasible ideas in the design space. The resulting set of potential candidate designs is better informed by understanding more possibilities.

However, both novice and experienced designers often struggle with identifying alternative designs (Ball et al. 2004). For novices in particular, limitations in technology or technical expertise make it difficult to generate multiple different ideas. Attempts at diverging from existing solutions may result in only minor tweaks to known designs, limiting the chances of innovation. Design fixation, or an attachment to the early ideas generated, has often been observed (Crilly 2015; Smith 1995; Ball et al. 1994), and since early ideas are only rarely successful, this leaves novice designers more likely to fail in creating innovative solutions (Fig. 1).

A variety of strategies have been proposed to help designers explore design spaces; in fact, Smith (1998) identifies over 150 different approaches to creative strategies. For example, brainstorming (Osborn 1957) and brainwriting (Geschka et al. 1973) are intended to facilitate the flow of ideas during idea generation. Analogical design (Casakin 2004), morphological analysis (Allen 1962), and Synectics (Gordon 1961) propose to stimulate the formation of an initial idea. Other methods such as lateral thinking (de Bono 1999), Scamper (Eberle 1995), and TRIZ (Altshuller 1984, 2005) provide ways to transform and improve upon existing ideas. However, TRIZ and other methods require extensive training; ideally, a training tool would be easy to learn in a short session. Most importantly, these methods are not based on studies of what designers naturally do when they attempt to generate new designs.

By studying how designers create a variety of concepts, we set out to learn about the strategies they find useful. Our goal was to examine *how designers think* when generating ideas, and to identify the strategies evident in their creation of new ideas (Fig. 2). By systematically comparing their steps in idea generation, we uncovered underlying cognitive strategies; in turn, these strategies provide guidelines for

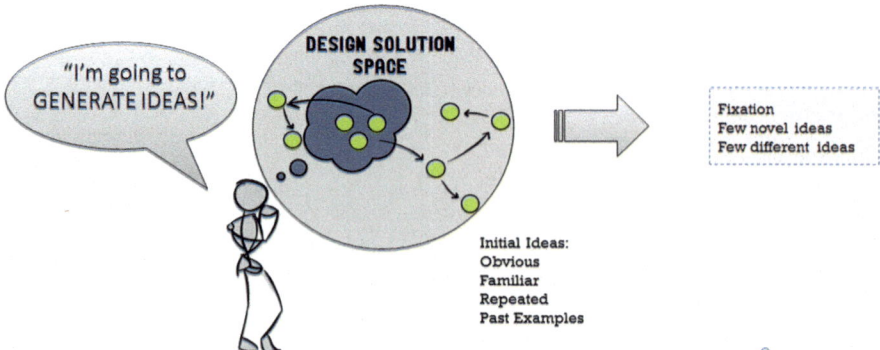

Fig. 1 The problem: A designer may generate only a few ideas from the space of possible solutions

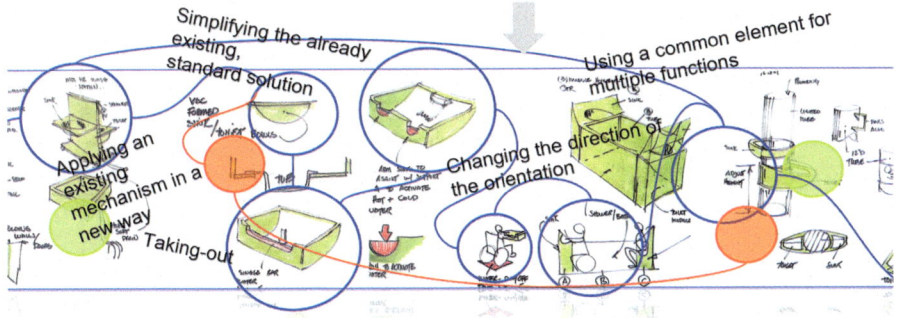

Fig. 2 A designer's series of concepts with annotations of design strategies identified in the work

other designers to use when generating new ideas. By consolidating results across four empirical studies of concept generation with varied contexts and more concepts sampled (Yilmaz et al. 2016), we detected a broad array of strategies for innovation in design.

To help design students and practitioners alike broadly explore solution spaces and generate novel ideas, we introduced *Design Heuristics* (Design Heuristics Inc. 2009). They advance engineering education by grounding innovation strategies in research evidence about successful design. Studies of designers across engineering and industrial design investigated how ideas were generated while working on many different product design problems. The results identified successful strategies designers used to help them create new and original ideas. *Design Heuristics* capture these "lessons learned" from many different designers about how to create more, and more varied ideas. Drawing from the "best practices" of product designers, *Design Heuristics* provides an effective pedagogy for giving students a "jump start" into successful idea generation.

Next, we set out to determine whether this method was effective through empirical studies, and whether it can be effectively taught in the engineering classroom. In this chapter, we describe the *Design Heuristics* tool—"77 Cards for Inspiring Ideas"—and a pedagogy for their use in design education. These lessons include (1) initiating ideas, (2) developing ideas, (3) designing subcomponents, and (4) team ideation. The pedagogical lessons are supported by empirical studies with student designers who followed these lesson plans to generate new ideas in classroom settings.

2 What Are *Design Heuristics?*

Design Heuristics are "prompts" that encourage exploration of a greater variety of ideas during design (Daly et al. 2012; Design Heuristics Inc. 2012; Seifert et al. 2015; Yilmaz and Seifert 2011; Yilmaz et al. 2016). As defined in the field of design, a heuristic is a simple "rule of thumb" used to generate a judgment or decision

(Cross 2011; Lawson 1980). Cognitive heuristics are not guaranteed to lead to a determinate solution; rather, they describe specific methods for generating "best guesses" at potential solutions (Yilmaz and Seifert 2011). Experts across domains use cognitive heuristics constantly and effectively, and their efficient use of domain-specific heuristics distinguishes them from novices (Klein 1998). *Design Heuristics* provide a specific set of 77 "rules of thumb" that have been shown in studies to help product designers and engineers generate new solutions (Daly et al. 2012; Kramer et al. 2015; Yilmaz et al. 2015).

The complete list of 77 *Design Heuristics* is available in a journal article (e.g., Yilmaz et al. 2016) and published in a pack of 5 × 7 paper cards (Design Heuristics Inc. 2009). A sample card front is shown in Fig. 3. On the front of each card, a descriptive title, graphic image, and action prompt provide specific instructions on how to apply the heuristic. On the back of the card, two existing products are shown to illustrate the application of each heuristic to the same product (a seating unit) on every card and to one of a variety of consumer products. The examples demonstrate how each heuristic can be applied to product design to create new ideas.

How do designers use *Design Heuristics* to create novel designs? Consider this scenario: You are tasked with generating initial ideas for a new product line of children's footwear. After coming up with some ideas involving placing animal features on the shoes (tiger stripes, zebra stripes), you consider applying *Design Heuristics* to your designs. How might the prompt, *Use Opposite Surface*, suggest new places to search for innovative ideas? The opposite surface of the shoes (the bottom) might also serve as a space to add more animal features to your ideas (see Fig. 4).

The same heuristic can be applied to more than one focal area, feature, or function of a problem because the application is nondeterministic (Yilmaz et al. 2010). That

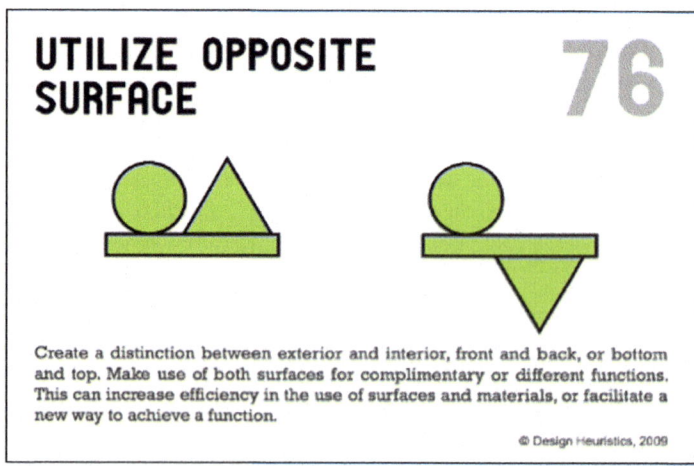

Fig. 3 Front of a *Design Heuristics* card (#76): *Utilize Opposite Surface,* presenting a written description of the heuristic and a graphical image on the front. (Courtesy of Design Heuristics Inc.)

Fig. 4 Children's shoe design incorporating the use of a *Design Heuristic, Utilize Opposite Surface,* to place animal footprints on the soles

means that there are a variety of ways to use a single Design Heuristic within a single design problem. In the shoe design example, the same heuristic (*Utilize Opposite Surface*) can be applied again to generate the idea of using the underside of the shoe to tighten and stabilize the lacing (see Fig. 4). The back side of each *Design Heuristics* card provides two existing product design examples where the heuristic is evident. For *Utilize Opposite Surface*, the second example illustrates using the area under the back of a chair (an "opposite surface") is used to provide additional storage (see Fig. 5).

Where do *Design Heuristics* come from, and how do they aid in generating designs? *Design Heuristics* were derived from empirical studies of designers generating new ideas (Yilmaz et al. 2016): (1) behavioral studies of students and experts designing new concepts for consumer products (Daly et al. 2010, 2011, 2012; Yilmaz et al. 2015); (2) analyses of idea development from existing consumer products resulting in award-winning concepts (Yilmaz et al. 2016a, b) (3) a case study of designs generated during a 2-year project by a professional designer (Yilmaz and Seifert 2011).

Accumulating evidence across these three studies, 77 unique *Design Heuristics* were identified (listed in Fig. 6). An important feature of this compilation of heuristics across studies is that each heuristic was observed multiple times (at least four) in different consumer product concepts, and all were observed in designs created by multiple designers.

Next, we created an instructional tool—the "77 cards"—to allow even beginning designers to access and apply the heuristics during their idea generation sessions,

UTILIZE OPPOSITE SURFACE

76

FARALLON CHAIR
fuseproject
The back side of this chair has a pocket
for storage.

980 TATOU
Annika Luber
The laces wrap around the bottom of
this shoe and connect with the sole.

Fig. 5 Back of a *Design Heuristics* card (#76): *Utilize Opposite Surface*, presenting examples of two existing consumer products where the heuristic is evident. (Courtesy of Design Heuristics Inc.)

1 Add levels	27 Cover or wrap	53 Reduce material
2 Add motion	28 Create service	54 Repeat
3 Add natural features	29 Create system	55 Repurpose packaging
4 Add to existing product	30 Divide continuous surface	56 Roll
5 Adjust function by movement	31 Elevate or lower	57 Rotate
6 Adjust functions for users	32 Expand or collapse	58 Scale up or down
7 Align components on center	33 Expose interior	59 Separate functions
8 Allow user to assemble	34 Extend surface	60 Simplify
9 Allow user to customize	35 Flatten	61 Slide
10 Allow user to rearrange	36 Fold	62 Stack
11 Allow user to reorient	37 Hollow out	63 Substitute mechanism
12 Animate	38 Impose hierarchy on functions	64 Synthesize functions
13 Apply mechanism in new way	39 Incorporate environment	65 Telescope
14 Attach independent functions	40 Incorporate user input	66 Twist
15 Attach product to user	41 Layer	67 Unify
16 Bend	42 Make components attach/detach	68 Use common base
17 Build user community	43 Make multifunctional	69 Use continuous material
18 Change direction of access	44 Make product recyclable	70 Use different energy source
19 Change flexibility	45 Merge surfaces	71 Use human-generated power
20 Change geometry	46 Mimic natural mechanisms	72 Use multiple components in one
21 Change product lifetime	47 Mirror or array|	73 Use packaging as component
22 Change surface properties	48 Nest	74 Use repurposed or recycled
23 Compartmentalize	49 Offer optional components	75 Utilize inner space
24 Contextualize	50 Provide sensory feedback	76 Utilize opposite surface
25 Convert 2-D to 3-D object	51 Reconfigure	77 Visually distinguish functions
26 Convert for second function	52 Redefine joints	

Fig. 6 Descriptive titles for the 77 *Design Heuristics* presented in alphabetical order

leveraging the heuristic patterns to spur idea searches in ways proven useful to other designers. To promote their use in design education, we conducted training workshops with educators at national design and engineering conferences, and established an online dissemination platform (www.designheuristics.com) with videos of the *Design Heuristics* pedagogy in use in classrooms, research papers, and examples of the use of *Design Heuristics* in idea generation. *Design Heuristics* have been documented in training by over 500 design educators, practitioners, and students in over 300 classrooms in 97 different universities, high schools, and industries, and in 163 different locations spanning countries around the world.

To examine how to train designers in the use of Design Heuristics during idea generation, we conducted further empirical studies with designers at beginning, senior undergraduate, and professional levels. Using classroom and practice settings, we conducted short training sessions on *Design Heuristics* followed by free-flowing idea generation sessions and open-ended surveys. Across these studies, we found that training on the use of *Design Heuristics* can be accomplished in a 10-min. session; that even beginning designers can use these heuristics to generate more, and more varied, ideas; and that students and practitioners alike find them easy to use and helpful to their idea generation processes.

Next, we created lesson plans for training designers on the use of *Design Heuristics* in educational settings. We developed multiple lessons that leverage *Design Heuristics* to facilitate idea generation and development based on our studies of learning through the incorporation of the tool into varied contexts. These lesson plans include: (1) Idea Initiation (the original *Design Heuristics* pedagogy), (2) Idea Development, (3) Subcomponent Design, and (4) Team Design.

These four lesson plans for training students in the application of *Design Heuristics* fit naturally into contexts typically found in design education. The lessons can be leveraged throughout a design course or projects when ideas are developed. They can be used to support the generation of initial ideas, as well as to develop and refine existing ideas. They can be used once during a course or project, or multiple times throughout the design process. *Design Heuristics* have been incorporated into instruction with preengineering, undergraduate, and graduate students, as well as in professional training for engineers and industrial designers. These lessons, and the synthesis of initial studies on their impact, are a new contribution to understanding how *Design Heuristics* can be used to support ideation in educational settings.

3   Lesson Plan 1: *Design Heuristics* in Initiating Ideas

The formation of initial ideas is a generative process (Finke et al. 1992) characterized by creating ideas "from scratch." In the lesson plan for *Design Heuristics* in idea initiation, students are asked to use *Design Heuristics* cards to prompt their generation of novel ideas. The training includes a short (10 min.) introduction to ideation, an introduction to the *Design Heuristics* and how they were developed, and practice using a few cards on a simple design task. An important contribution of these heuristics is their efficacy in communicating new concept generation principles to designers. Ideally, these heuristics distill knowledge of precedents in product design into generative constraints that are easy to learn and apply. Our goal is to help even novice students experience a "flow" of ideas success in producing many candidate ideas. This lesson plan emphasizes developing skills to continue generating new and different ideas and to follow the recommended "best practice" of considering multiple candidate ideas.

We performed extensive testing to compare the use of *Design Heuristics* for initiating ideas to other idea generation approaches. The studies showed that students and experts in both engineering and industrial design can learn to use the *Design Heuristics* cards with just a short instructional session, and then go on to successfully create their own novel and varied ideas (Daly et al. 2016a).

In one study, 102 first-year engineering students were introduced to one of three different ideation techniques—*Design Heuristics*, Morphological Analysis (Allen 1962), or Individual Brainstorming—and asked to generate solutions to a given design problem in a 25-min. session (Daly et al. 2016a). Using an adapted version of the consensual assessment technique (Amabile 1982), all concepts were rated for creativity, elaboration, and practicality, and all participants' concept sets were rated for quantity and diversity. All three techniques produced creative concepts averaging near the scale midpoint. The elaboration of the concepts, however, was significantly higher with *Design Heuristics* and Morphological Analysis techniques, and practicality was significantly higher with *Design Heuristics*, suggesting that *Design Heuristics* facilitate more detailed and feasible outcomes.

Another study tested 48 first-year engineering students in an 80-min. classroom session (Daly et al. 2012). The students were each given a different subset of 12 *Design Heuristics* and were asked to create concepts for a portable solar oven. Of the 161 designs generated, 55% showed evidence of *Design Heuristics*. The concepts resulting from the application of *Design Heuristics* were rated by trained coders as more creative (averaging 3.6 on a 7-point scale) than those without heuristics (averaging 2.7). *Design Heuristics* also facilitated novelty and elaboration. Examples of the kinds of ideas generated with and without heuristics are shown in Fig. 7. As one can see in these examples, there are distinctions among the ideas generated with heuristics and ideas generated without heuristics.

A third study of twenty second-year industrial design students resulted in 59 new design concepts with heuristics and 19 concepts without heuristics (in an 80-min. class) (Yilmaz et al. 2012). The average creativity rating of the concepts with evident heuristics was 3.7 (on a 7-point scale), and 2.3 without heuristics. Students were also observed applying multiple heuristics to find alternative concepts, leading to more complex and developed solutions, as shown in Fig. 8. The first design is a packaging box another function as a stand for the magnifying cube and metal bowl. This design uses a combination of heuristics, including Repurpose Packaging, Reverse the Direction, and Make Components Multifunctional. The second design is a set of square mirrors sewn together, rolled and used to concentrate sunlight, using the heuristics, Mirror and Roll.

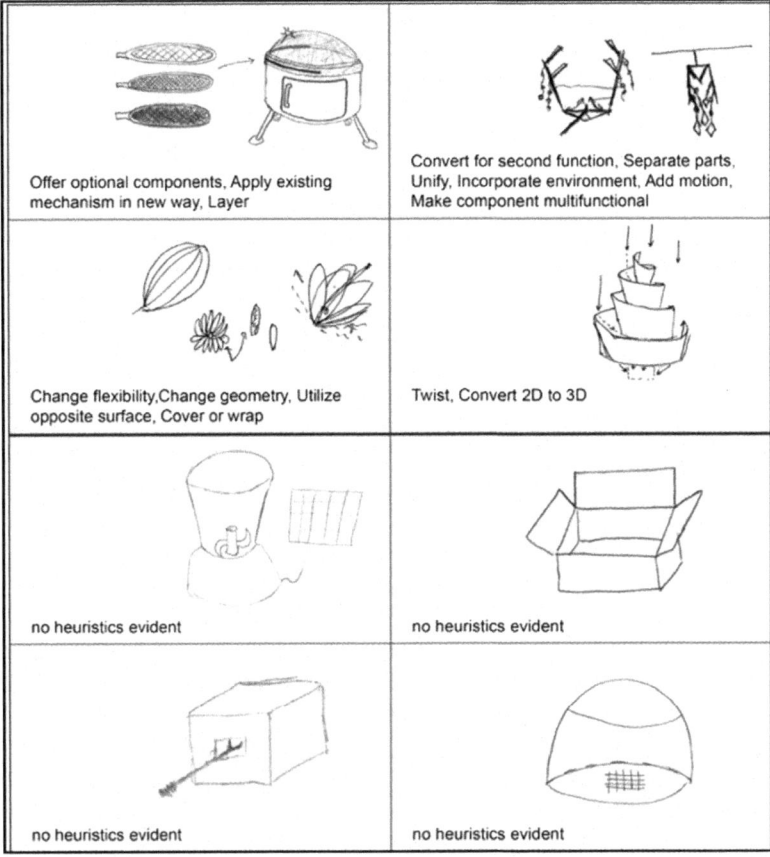

Fig. 7 Examples of ideas generated with (top 4) and without (bottom 4) *Design Heuristics*

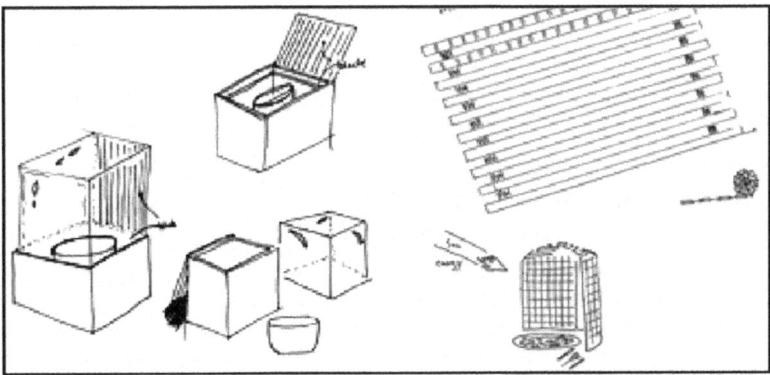

Fig. 8 An industrial design student's concept for a solar oven. The design combines the heuristic, *Mirror*, with another, *Roll*, to produce an array of mirrors rolled around a focal point to concentrate reflected sunlight

To determine whether use of *Design Heuristics* improves idea generation for more advanced students, a further study examined how engineering students use *Design Heuristics* in their senior capstone projects. This study traced the changes in initial concepts based on *Design Heuristics*, and followed their evolution throughout the course (Kramer et al. 2014, 2015). Analysis revealed that all eight teams carried their heuristic-inspired concepts to their latter stage designs, with seven teams carrying their heuristic-inspired concepts through to their final prototypes. As all the teams were working on different, team-specific open-ended design problems, these findings demonstrate the utility and practicality of *Design Heuristics* across varied design problems.

Another study of *Design Heuristics* with upper level students in a design course found that students added this additional technique to their repertoire for generating ideas to solve the complex projects proposed by their industry client; and, they used the cards early in their design processes (Koty-Schwartz et al. 2014). These studies found that *Design Heuristics* can be successfully used to initiate new ideas by engineering students in senior-year project-based courses, and can facilitate producing positive outcomes for idea generation.

Across studies, these results indicate that concepts generated using *Design Heuristics* resulted in more creative designs and facilitated greater variety in designs. Designers found the cards easy to use following a just short introduction. These studies documented the use of every card in the set of 77 in multiple design problems by multiple individual designers, showing that they capture knowledge about creating designs that is helpful to a wide range of designers working on a wide range of design problems. While these studies focused on the process of idea initiation for individual designers, *Design Heuristics* can be used in multiple ways to assist ideation. For example, they can be used to support individual designers as well as within team design settings. In the following sections, we describe multiple lesson plans for using *Design Heuristics* to support idea development, design of subcomponents, and team ideation, as well as research studies supporting each plan.

4 Lesson Plan 2: *Design Heuristics* in Developing Initial Ideas

In the Idea Development Lesson, students generate their own initial ideas, and then apply *Design Heuristics* to add more ideas. In this way, the fixation arising from the presence of prior examples can be overridden by transforming typical initial ideas into novel ones (Leahy et al. 2018a). In early studies, we saw students applying the heuristics both to initiate and transform ideas (Christian et al. 2012), so the application of heuristics for transformation makes intentional a strategy some students were doing naturally. The goal of this lesson is that a single idea can be the source of interesting novel ideas through variations suggested by the *Design Heuristics*. Students learn how to break free of stereotypical first ideas by adding their own, novel changes to create new ideas (Leahy et al. 2018a).

During idea development, students can apply the same Design Heuristic card to prompt multiple variations of an idea. For example, Fig. 9 represents a student's work as she designed ways to cook food using solar energy. She used the same heuristic, Change Geometry, twice to develop new versions of an original concept: first one from a narrow horizontal rectangle to a wider vertical one, and second one from a rectangle to a pyramid.

Alternatively, different heuristics can be used to transform an initial idea to multiple new ideas. Figure 10 represents an example of this where the designer began by attaching two existing components to each other—a magnifying glass and a griddle—to create a surface with focused sunlight. In her second concept, she transformed the magnifying glass to a square magnifying glass attached to the griddle. In the following concept, she made the lens height adjustable, and, in the fourth concept, she added sides to it to maintain the heat more effectively. She then considered portability by adding a rigid handle, which was changed to a flexible handle. The final concept also included an attachment that held utensils and a spout for draining fluids from the cooking surface.

These examples demonstrate how one concept can be transformed with serial application of single heuristics; for example, five cards can be serially and separately applied to an original idea to produce five different potential solutions not considered before. In addition, students can try applying the same card multiple times to create more ideas. While each card provides a clear prompt to guide in the generation of

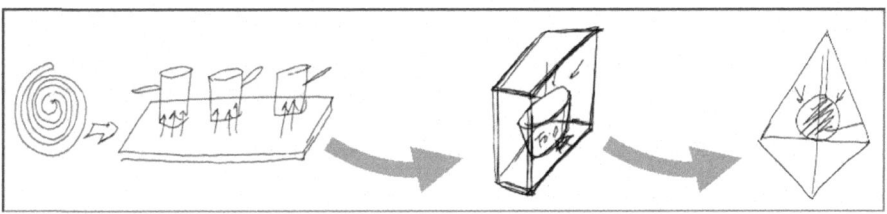

Fig. 9 Transforming an initial concept twice using the same Design Heuristic, *Change Geometry*

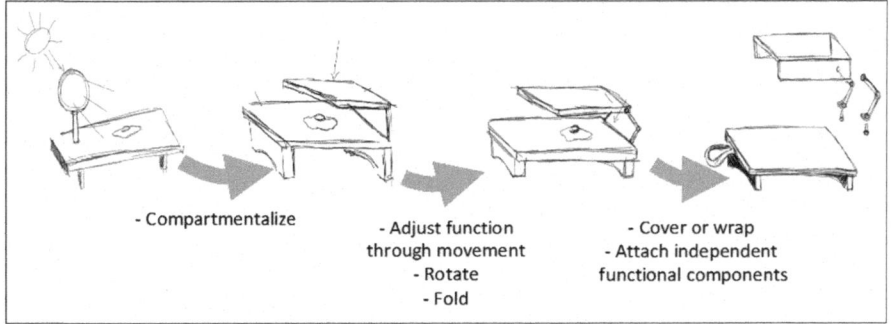

- Compartmentalize   - Adjust function through movement   - Cover or wrap
                      - Rotate   - Attach independent
                      - Fold   functional components

Fig. 10 Transforming an initial concept with serial application of multiple heuristics

ideas, the designer must also select how and how much to apply each heuristic. In an early study of designers using the heuristics, we discovered that designers often developed more designs by applying multiple heuristics within a single new idea (Daly et al. 2012a, b). Finally, students can change the heuristic they are considering and where in the existing design they choose to apply it, resulting in an idea production session maximizing the variation in designs serving as a base and in heuristics applied. This lesson plan is structured for the student as:

1 Pull a card at random.
2 Generate a solution by transforming an existing concept.
3 Shuffle the cards and randomly pull another.
4 Apply the card to the same or another of your existing solution.

An organizing framework, the concept tree, can be added to the idea development process to organize and reflect on the relationships between ideas. As shown in Fig. 11, concept trees map the relationships among proposed ideas. In an alternative lesson plan for idea development, the use of *Design Heuristics* is combined with the structure of a concept tree. The instructor begins the lesson by setting a goal for the number of branching ideas to generate (we recommend at least 10), and students select what they believe to be the best of their original ideas to further develop. Alternatively, the instructor can recommend an idea from the student's set (especially one on which a student is fixated). Students then build off the original idea, developing subsequent ideas by applying *Design Heuristics* cards (one or more) to each of the resulting ideas to create more branching ideas. They continue to use the last concept as the base idea and repeat the process of developing each idea using *Design Heuristics* in repeated iterations, culminating in varied alternative concepts.

After the ideation phase, the class can discuss the variations in developed concepts based on the original ideas. This lesson plan helps to make variations within concepts prominent and highlights how designs can be transformed during development to lead to differing outcomes. Another key point is for students to recognize there are still more characteristics that can be varied or further iterated upon if they continued developing their ideas, even if they base all of their new designs on the same original concept.

In a classroom study on pedagogy for using *Design Heuristics* to transform existing ideas, senior engineering students created initial concepts, and then used *Design Heuristics* to transform these concepts into alternative solutions, resulting in more variety to choose and develop (Leahy et al. 2018a). For some concepts, students applied a single heuristic, and in other cases, they applied multiple heuristics to the same transformed concept. The concept sets generated were analyzed, and eight types of developments were identified, including the enhancement of aesthetics, features, functions, settings, materials, sizes, organizations, and usability.

The outcomes of this pedagogy study showed that heuristics facilitated exploration of possible concepts in diverse ways, resulting in variations in designs to achieve the desired functions, as well as changes in aesthetics and usability. *Design Heuristics* did not lead students to follow the same trajectory of development, suggesting the heuristics provide direction for concept development without prescribing

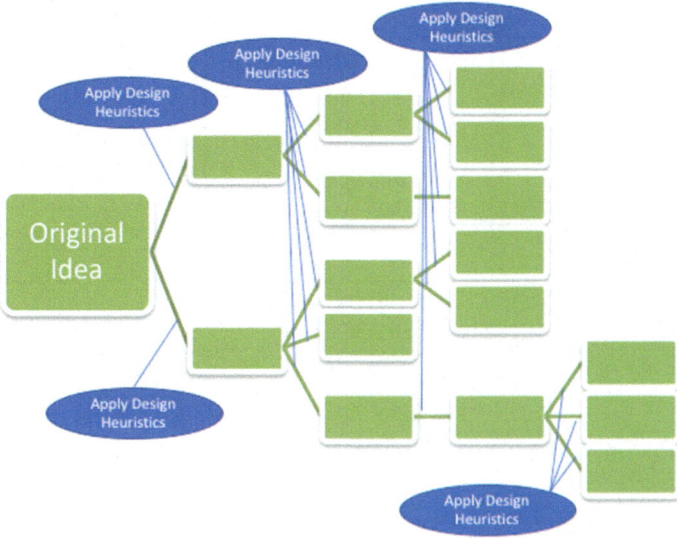

Fig. 11 Concept tree generated using repeated applications of *Design Heuristics* to develop an existing idea into transformed ideas

a particular way to implement that feature within a design. As a result, students pursued deeper explorations of alternative concept designs by pushing their initial idea through further development. *Design Heuristics* supported students' idea development by providing on-point suggestions about ways to iterate on their initial concepts to lead to variations in ideas. As a result, students explored alternative concepts by producing iterations on their early designs and were more likely to select these more-developed concepts as their most creative, unique, and favored designs (Leahy et al. 2018a).

5 Lesson Plan 3: *Design Heuristics* in Subcomponent Design

This lesson builds on a curricular goal in most engineering design courses; namely, redesigning products using incremental changes to improve product components. In industry, this allows companies to continue production while simultaneously bringing new-generation products to the market. However, when students are trained to analyze components, they are not given instructions on how and when to separate designs into components and how to tackle design issues independently. This lesson is framed around the differences in designing an entire product versus making modifications to its components. In this lesson, students decompose existing products, redesign individual components using *Design Heuristics*, and suggest new versions of the product based on combinations of the redesigned components. This lesson teaches students to generate ideas through decomposition and recombination.

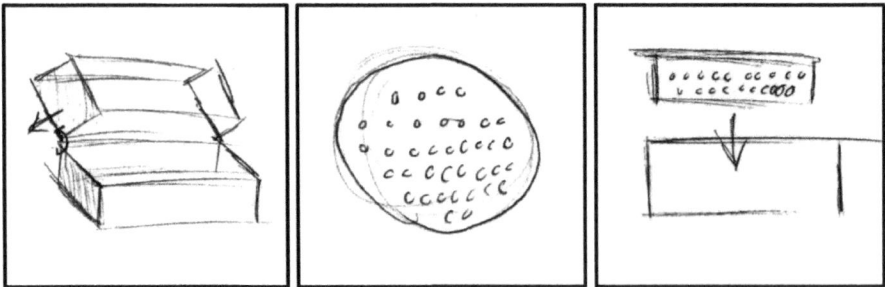

Fig. 12 Evolution of an industrial design student's concepts from subcomponent design to the recombination of the concepts into a whole concept

The lesson plan involves asking students to follow a sequence of steps implemented as pages within an idea generation workbook. The steps included:

1. Complete a functional decomposition of a design problem into its subfunctions (the time required to do this varies greatly with the complexity of the problem).
2. Generate ideas for the subfunctions using the *Design Heuristics* cards, focusing on generating as many alternatives for these subcomponents as possible.
3. Create whole concepts by recombining selected subcomponent ideas.

In a classroom study investigating pedagogy for subcomponent design using *Design Heuristics*, industrial design students were asked to apply functional decomposition to a given problem, then apply *Design Heuristics* to the individual components identified, and then to reconstruct the concepts into a "whole" solution (Gray et al. 2015). The initial functional consideration of the problem space allowed the students to productively generate diverse concepts using the heuristics, within a focused design space by using a selected function as an explicit constraint. Students also used *Design Heuristics* in distinct ways reflecting their growing understanding of the range of concepts that might exist within a solution space, suggesting the application of *Design Heuristics* for exploratory, iterative, reframing, and synthetic activities.

Figure 12 shows an example of a process for an industrial design student's ideation addressing a design problem focusing on food preservation and storage. The student created a concept for a product with a "living" pop-out flexible hinge (using *Animate*) in the ideation stage to facilitate placement in a dishwasher. In the iteration stage, he added rubber nibs to further aid in cleaning (using *Change Surface Properties*); and in the recombination stage, he added similar nibs to flexible parts from another earlier concept, improving grip and discoverability of functionality (again using *Change Surface Properties*; *Create System*; *Use Packaging as a Functional Component*).

Idea Generation for Subcomponent Design allows students to also leverage the use of another ideation method, Morphological Analysis (Allen 1962), alongside of the *Design Heuristics*. Students create a chart as they would if they were doing Morphological Analysis. They write subfunctions in the first column and then ideas generated using the *Design Heuristics* for each subfunction along the row. The ideas should be drawn and described, the table cells sized accordingly. Students themselves create combinations using their charts, and then draw the resulting idea based on the combination. The morphological chart allows students to see that many different ideas are possible.

In a study with engineering students, this lesson plan was conducted in a classroom with upper level engineering design students. The results revealed that nesting the use of *Design Heuristics* within Morphological Analysis promoted students' abilities to elaborate on features and consider additional aspects of the context as compared to their initial ideas (Leahy et al. 2018b). Morphological Analysis facilitated the decomposition of complex artifacts into separable functions, and *Design Heuristics* facilitated the generation and exploration of multiple, diverse ideas. Both tools also support elaboration on designs. Students were able to combine the two tools of Morphological Analysis and *Design Heuristics* within a relatively short period of training and use. Additionally, students reported that combining these tools was relatively easy.

A second variation of the Lesson Plan for Subcomponent Design encourages hands-on discovery of subcomponents through the physical dissection of an existing product. First, instructors ask teams to carefully dissect an existing product, sorting the components into categories or types. Then, each team member is asked to select a single component (or set of components). Students use the *Design Heuristics* to generate ideas for their component(s). Finally, the teams combine their subcomponent ideas to generate whole ideas for the product.

Lesson Plan 3 for Subcomponent Design draws students' attention to important subgoals within a design and allows them to consider ideas that work well for a given subfunction. This "local focus" for design allows the designer to optimize the features of the subcomponent designs to maximize function. Then, the recombination stage draws the students' attention to the tradeoffs in design that may be necessary to fit a subcomponent's design into a whole concept. For example, a solar oven design may include a local focus on retaining heat in the cooking surface, suggesting a solid metal cooktop; however, when recombined into the portable solar oven concept, the cooktop may be altered to address the need for a lightweight design. Lesson Plan 3 works to provide practice in designing for local and global design goals.

6   Lesson Plan 4: *Design Heuristics* in Team Ideation

Many design activities involve teamwork, especially as the complexity of the design problem increases. The team approach requires benefitting from the knowledge, expertise, and contributions of all members while avoiding conflict resulting from differences in perspectives. Research has shown both positive (Sutton and Hargadon 1996) and negative (Diehl and Stroebe 1987) impacts on ideation outcomes from working in team settings. In some cases, as a team develops more concepts, the quality of the concepts created by a team improves (Rowatt et al. 1997), and the team helps in selecting the best among multiple ideas.

In a field study, we investigated whether *Design Heuristics* pedagogy could support a team's ideation process. The team consisted of professional engineers as they worked collaboratively on concepts for their company's existing product line (Yilmaz et al. 2013). The engineers met in a 4-hour "innovation workshop" facilitated by an external consultant meeting over a 2-day period. During ideation, the team worked together to generate multiple ideas by reviewing one heuristic card at a time. The individuals "called out" their ideas for new concepts and "added on" to the ideas of others while recording their designs on easel pads, calling out new ideas and building on "called out" ideas from others. An example from the audio transcription of their interactions includes the following segment:

Engineer 1: "Consider whether they are purchased separately or included with the product, and where they will be stored when not in use." (reading aloud from card, *Offer Optional Components*)

Engineer 2: "I mean, we have talked about the existence of a couple of interfaces that could make that quite interesting. One is your skateboard interface, right? So anything that's pulled by a device, it could be fertilizers or other things that people want to do in the same process."

Engineer 3: "Or even a brush to clean your driveway when there's a lot of sand or something in the spring."

This transcription of their workshop session illustrates how discussing *Design Heuristics* facilitated the generation of novel concepts for the existing product line they had been working on for many years. The team members at times used a prompt from the cards to initiate new whole ideas, generate new ideas for subcomponent designs, and to transform their ideas for both whole and subcomponent designs by applying the heuristic. Additionally, the team used the heuristics to better define existing needs within their product line so as to better explore the "real problem" by identifying novel views of their existing design problems. These professional engineers later completed individual surveys about their experiences with the workshop. The engineers stated that the heuristic cards forced them to stay on track and helped them focus on one direction of idea generation at a time. They also reported that using the heuristic cards helped them to further explore the problem space—what are the real needs to be addressed?—as well as create innovative solutions they had never before considered despite years of work on these design problems.

Based on this study, we identified a lesson plan for using *Design Heuristics* in a team ideation session. The alternative versions of the plans are included to suggest ways of varying teamwork using the *Design Heuristics* during ideation. These alternatives allow the instructor to focus on the needs of the teams and their stage of the early design process, along with allowing variations for repeated ideation sessions. The four alternative lesson plans are the Team Workshop, Idea Rotation, Card Rotation, and Team Jigsaw lesson plans.

"Team Workshop"

Following this "workshop" model of *Design Heuristics* training, a group ideation lesson can be set up in a design group with one individual taking on the role of a facilitator. The facilitator brings each individual card to the attention of the team by reading the card aloud and explaining the example products illustrating each heuristic. Once everybody on the team agrees, they begin applying the heuristic to their existing problem while discussing ways to apply it. The group maintains a focus on that particular heuristic card with the help of the facilitator. When the team exhausts all the ideas they can generate through applications with that card, the facilitator moves to the next heuristic. This workshop model appears to work well when the team is very familiar with the design problem and is comfortable building collaborative ideas with each other.

"Idea Rotation"

Another approach to group lessons using *Design Heuristics* combines training with *Design Heuristics* and responding to team members' ideas. A common difficulty with group ideation methods is production blocking, where team members must wait while others present ideas before presenting their own (Diehl and Stroebe 1987). At the same time, the advantage of the teamwork session is the exchange of ideas leading to new concepts created "across" individual designers. To accomplish both of these goals, a variant of the Brainwriting technique (Geschka et al. 1973) can be employed to combine the prompting of ideas using *Design Heuristics* with building onto others' ideas. In the "Idea Rotation" procedure, each student on the team is given a single, different *Design Heuristic*, and they study the card to become the local "expert" on that one prompt. After generating their own design solutions working individually for a given time interval (i.e., between 5 and 10 min.), the team members pass their concept on to another group member and receives someone else's—without *any* conversation. Then, each team member creates a new concept by applying their assigned heuristic to the design created by someone else. After sharing the concepts across several team members holding different heuristics, the resulting concepts are elaborated and transformed into more distinctive and elaborated ideas.

"Card Rotation"

In a similar Brainwriting format, a team can begin with one heuristic card per member. After applying this first heuristic to initiate ideas for a set interval (e.g., 5 or 10 min.), the team member "passes on" their card and receives a new card from another member. Using the set of ideas they created, each team member then adds onto their ideas by using the received *Design Heuristic* to transform their ideas or to develop subcomponent designs. In this session, each student applies new heuristics to their ideas one after another and tries out new ways of developing their ideas with their own concept set. And, each team member is exposed to the same subset of heuristics seen by the others. At the end of the session, the team members select ideas to discuss with their team, and together they discuss how each differently applied the same heuristics to generate alternative ideas. The overlap in their use of the subset of *Design Heuristics* appears to facilitate discussion of the generation process and helps team members to identify the evolution of their ideas as well as the unique qualities they each contribute to the designs.

"Team Jigsaw"

Another example of a group lesson focuses on subcomponent design while allowing teams to work together in generating ideas. After creating a list of subfunctions as a group, the teams assign each subfunction to a single (or pair of) team members. Each team member then works to generate concepts to address their assigned subcomponent individually while applying *Design Heuristics*. After a longer interval (e.g., 20 or 30 min.), the teams reform to confer about their ideas for subfunctions. Finally, the team then selects and combines the subcomponent designs to develop whole concept ideas. This method combines individual learning about ideation using *Design Heuristics* within the interdependent design structure of "real world" design teams. The team members are exposed to many different ideas generated by their members in a brief interval, and each can see the varied subcomponent problem settings where team members successfully applied the *Design Heuristics*. Then, the team's combined ideas carry forward the distinct contribution of each individual team member while joining their ideas into a shared overall design.

There are several advantages to these lesson plans for training with *Design Heuristics* in a team setting. First, each individual experiences a time interval where they individually consider how to apply a given *Design Heuristic* to a specific problem. This effort takes them through the steps of (1) understanding one heuristic and its examples, (2) seeking ways to apply that heuristic to a problem, and (3) generating ideas using the prompted approach, for a complete (though shorter) lesson on how to use the heuristics. In addition, each individual is exposed to multiple heuristics either through their own longer work session or through reports from their teammates about their work sessions. Finally, each individual sees multiple ideas (their own or their teammate's) created using the same heuristic, allowing them to gain an appreciation of the generality of the *Design Heuristics*. To this, these team lesson plans add opportunities to appreciate ideas from other team members, exposing the individuals to many alternative ideas and providing a glimpse of how others came up with multiple ideas. The Team Ideation Lessons also accomplish collaborative work on developing the team's ideas and include sharing and selecting viable ideas by all of the team members.

7   Discussion

The 77 *Design Heuristics* examined in these studies of idea generation instruction are the only set of strategies available that were identified through systematic, empirical studies of design practices (Yilmaz et al. 2016). These heuristics were observed in studies of award-winning products and in student and practicing designers' cognitive processes as they worked to solve design problems. The heuristics have also been documented in use across highly varied product design settings (Yilmaz et al. 2016).

The TRIZ method (Altshuller 2005) is the only other approach to idea generation strategies that has also analyzed successful product designs; however, there is limited evidence validating its effectiveness for use by practicing designers (Hernandez et al. 2013). The empirical basis for the *Design Heuristics* ensures these strategies are tied to design practices and represent a varied set of heuristics captured across multiple problems and multiple designers.

Given the empirical evidence presented above for the successful use of *Design Heuristics* through short instructional lessons, the success in teaching designers (both novice and experienced) to use these heuristics in creating new concepts and diverging from their fixated solutions is extensive and varied. Use of the heuristics was associated with more original and creative concepts, resulting in more elaborate and practical candidate concepts to choose from (Daly et al. 2012, 2016a). These findings show that the *Design Heuristics* greatly enhance the design process for both novice and experienced designers, providing them with a collection of strategies to leverage across product design idea generation. Further, the findings demonstrate that these heuristics can improve the quality of their design outcomes.

One open question about *Design Heuristics* is how to decide which heuristic to apply in any given design context. The data from existing designs collected in these studies suggests the heuristics are applicable across many design problems. In studies with *Design Heuristics*, providing a subset of cards selected at random has produced improved concept outcomes (Daly et al. 2012; Yilmaz et al. 2012; Yilmaz et al. 2010) perhaps by focusing the designers on using a subset of cards within the short time interval provided. Not every heuristic is used each time it is considered, but the combination of an open-ended consideration of the heuristics with a smaller subset of heuristics in any one session (based on available time) appears to benefit designers.

The prompts provided by the 77 *Design Heuristics* as an external representation through cards may be used to pace ideation sessions through card selection. The *Design Heuristics* cards can be shuffled to consider each when selected at random. In the studies testing their use in lessons, selecting a subset at random to consider during a single session was effective. If more concepts are desired, more cards can be considered. However, it is possible that further research might identify cues to indicate when these heuristics are most relevant for application in a given problem. Using a criterion of efficacy, the studies on the Design Heuristic pedagogy show that the set of 77 *Design Heuristics* captures design variations at a level useful in concept generation.

Another question is whether more strategies may be uncovered in future research on design. A more specialized field of product design was examined in a study of patents awarded for microfluidics medical devices (Lee et al. 2018a, b). Examining 235 patents, researchers found 36 design strategies in these technical devices, 19 of which (53%) were also observed among the 77 consumer product *Design Heuristics*. Future research on other design domains such as service, software, and apps may uncover similarities and specialization in heuristics evident for each domain. Further, new design goals and solution contexts may give rise to new heuristics as the design field evolves over time. In addition, exploring design through diverse cultures may result in the identification of more heuristics less common in U.S. engineering and design schools. It is important to expand the observation and scientific study of designers within the many varied settings where design takes place, and to use this knowledge of common methods and practices as instructional resources for training designers in idea generation.

What is the best level of heuristic definition; that is, the level that provides helpful guidance while avoiding becoming overly specific and therefore less applicable across problems? Comparing approaches to heuristics in design, it appears there is a continuum from the very general to the very specific. Product design strategies may be captured through a large number of very specific heuristics; for example, the TRIZ method includes 39 design features (e.g., temperature, ease of repair) combined in a 39×39 matrix to suggest which of principles may be useful in devising a solution (Fogler and LeBlanc 2013). Or, strategies may be captured more generally, resulting in a smaller number of more general ones; for example, the SCAMPER strategies (Eberle 1995) include seven very general strategies; namely, substitute, combine, adapt, modify, put to other uses, eliminate, and rearrange/reverse. The benefits of a smaller set of more general heuristics are apparent: access to the set is easy. However, the application of these heuristics then requires more cognitive effort (e.g., Modify what? Eliminate what?).

Design Heuristics offer an intermediate level of description where the heuristic is already abstracted away from the concepts where they appeared. The relevant principle is provided at a level of description that facilitates implementing the heuristic in a new problem context. The needed information about how to create a new concept is readily available within the heuristic. Of course, many decisions must be made about how to apply the heuristic within a specific design, leading to possibility of reapplying the same heuristic to the same problem and creating a different concept, as was observed in the studies (Daly et al. 2012). Other researchers have documented the value of intermediate support structures. This includes *bridging concepts* between empirically grounded theory and practical use (Dalsgaard and Dindler 2014), and *strong concepts,* a form of intermediate knowledge describing core design ideas that are inherently generative (Höök and Löwgren 2012).

Observing successful designs and extracting heuristic principles to generate ideas has been demonstrated as a successful technique in several other research programs. For example, TRIZ (Altshuller 2005; Altshuller and Rodman 1999; Hernandez et al. 2013; Savransky 2000), transformative design heuristics (Singh et al. 2007, 2009; Skiles et al. 2006; Weaver et al. 2008, 2010), and other approaches (Cormier et al. 2011; Haldaman and Parkinson 2010; Perez et al. 2011; Saunders et al. 2011) all point to the usefulness of observational study. The methodology used in this approach emphasized the collection of observations during the idea generation process. The generation of initial concepts while attempting to create multiple, different designs for consideration appears to give rise to patterns not evident in final designs. Consequently, it is important to observe designs created within a design session or project in order to capture the ways in which designers generate multiple concepts. Through systematic observation of the creation of many concepts by multiple designers in varied design settings, we can attain a deeper understanding of the role of *Design Heuristics* in helping students learn to generate variation in design concepts.

These results provide evidence of an effective tool to aid designers in the early phases of design. Best practices in design suggest that generating as many concepts as possible will lead to better design process outcomes (Akin and Lin 1995; Atman et al. 1999; Liu et al. 2003); in addition, the generation of more and different designs will logically increase opportunities to consider novel and innovative designs. Training in the use of *Design Heuristics* stands to benefit both novice and practicing designers working on any product design task. The use of *Design Heuristics* has been shown to facilitate idea generation by opening areas of the design space not previously explored. *Design Heuristics* help to identify new designs across design problems and designers, providing a set of general strategies for idea generation that also allow individuals learn to create original and varied designs through their use.

8   Conclusions

Research has documented a lack of systematic, empirically validated instruction for ideation in design education. We have developed pedagogical approaches that leverage the empirically driven and validated the 77 *Design Heuristics* cards as a tool to support skill development in idea initiation, idea development, subcomponent design, and group ideation. Using the tool in multiple ways helps instructors incorporate instruction on idea generation over multiple sessions, allowing students to experience success and become confident in their ability to create many designs. For students, the *Design Heuristics* serve as a foundational support for their increasing exploration of more innovative ideas in both classrooms (Daly et al. 2014, 2016b) and in their future design tasks.

Acknowledgements This material is based upon work supported by the National Science Foundation under Grant Nos. 1323251 & 1322552.

References

Akin O, Lin C (1995) Design protocol data and novel design decisions. Des Stud 16:211–236
Allen M (1962) Morphological creativity. Prentice-Hall, New Jersey
Altshuller G (1984) Creativity as an exact science. Gordon and Breach, New York, NY
Altshuller G (2005) 40 principles: TRIZ keys to technical innovation, extended edn. Technical Innovation Center, Inc., Worchester, MA
Altshuller G, Rodman S (1999) The innovation algorithm: TRIZ, systematic innovation and technical creativity. Technical Innovation Center, Worchester, MA
Amabile T (1982) Social psychology of creativity: a consensual assessment technique. J Pers Soc Psychol 43(5):997–1013
Atman CJ, Chimka JR, Bursic KM, Nachtman HL (1999) A comparison of freshman and senior engineering design process. Des Stud 20(2):131–152

Ball LJ, Evans J, Dennis I (1994) Cognitive processes in engineering design: a longitudinal study. Ergonomics 37(11):1753–1786

Ball LJ, Ormerod TC, Morley NJ (2004) Spontaneous analogising in engineering design: a comparative analysis of experts and novices. Des Stud 25(5):495–508

Casakin H (2004) Visual analogy as a cognitive strategy in the design process: expert versus novice performance. J Des Res 4(2)

Christian JL, Daly SR, Yilmaz S, Seifert CM, Gonzalez R (2012) Design heuristics to support two modes of idea generation: initiating ideas and transitioning among concepts. In: Proceedings of the annual conference of the American Society of Engineering Education, San Antonio, Texas. https://peer.asee.org/21152

Cormier P, Literman B, Lewis K (2011) Empirically derived heuristics to assist designers with satisfying consumer variation in product design. Paper presented at the ASME international design engineering technical conferences and computers and information in engineering conference, Washington, DC, 28–31 Aug 2011

Crilly N (2015) Fixation and creativity in concept development: the attitudes and practices of expert designers. Des Stud 38:54–91

Cross N (2011) Design thinking: understanding how designers think and work. Berg, New York, NY

Dalsgaard P, Dindler C (2014) Between theory and practice: bridging concepts in HCI research. Paper presented at the annual ACM conference on human factors in computing systems, New York, NY

Daly SR, Christian JL, Yilmaz S, Seifert CM, Gonzalez R (2011) Teaching design ideation. In: Proceedings of the annual conference of the American Society of Engineering Education, Vancouver, BC. https://peer.asee.org/18507

Daly SR, Christian JL, Yilmaz S, Seifert CM, Gonzalez R (2012a) Assessing design heuristics for idea generation in an introductory engineering course. Int J Eng Educ 28(2):1–11

Daly SR, Mosyjowski EA, Seifert CM (2014) Teaching creativity in engineering courses. J Eng Edu 103(3):417–449

Daly SR, Seifert CM, Yilmaz S, Gonzalez R (2016a) Comparing ideation techniques for beginning designers. J Mech Des 138(10):101108

Daly SR, Mosyjowski EA, Seifert, CM (2016b) Teaching creative process across disciplines. J Creative Behav

Daly SR, Yilmaz S, Christian JL, Seifert CM, Gonzalez R (2012b) Design heuristics in engineering concept generation. J Eng Educ 101(4):601–629

Daly SR, Yilmaz S, Seifert CM, Gonzalez R (2010) Cognitive heuristic use in engineering design ideation. In: Proceedings of the annual conference of the American Society for Engineering Education, Louisville, Kentucky. https://peer.asee.org/16280

de Bono E (1999) Six thinking hats. Back Bay Books

Design Heuristics Inc. (2009). www.designheuristics.com

Diehl M, Stroebe W (1987) Productivity loss in brainstorming groups: toward the solution of a riddle. J Pers Soc Psychol 53(3):497–509

Eberle B (1995) Scamper. Prufrock, Waco, Texas

Finke RA, Ward TB, Smith SM (1992) Creative cognition: theory, research, and applications. MIT Press, Cambridge

Fogler HS, LeBlanc SE (2013) Strategies for creative problem solving, 3rd edn. Prentice Hall, Upper Saddle River, NJ

Geschka H, Schaude GR, Schlicksupp H (1973) Modern techniques for solving problems. Chem Eng 6(80):91–97

Gordon WJJ (1961) Synectics. Harper & Row, New York

Gray CM, Yilmaz S, Daly SR, Seifert CM, Gonzalez R (2015) Supporting idea generation through functional decomposition: an alternative framing for design heuristics. In: Proceedings of the international conference on engineering design, Milan

Haldaman J, Parkinson MB (2010) Reconfigurable products and their means of reconfiguration. Paper presented at the ASME international design engineering technical conferences and computers and information in engineering conference, Las Vegas, Nevada

Hernandez NV, Schmidt LC, Okudan GE (2013) Systematic ideation effectiveness study of TRIZ. J Mech Des 135(10):101009

Höök K, Löwgren J (2012) Strong concepts: intermediate-level knowledge in interaction design research. ACM Trans Comput Hum Interact (TOCHI) 19(3):23

Klein G (1998) Sources of power: how people make decisions. The MIT Press, Cambridge, MA

Kotys-Schwartz D, Daly SR, Yilmaz S, Knight D, Polmear M (2014) Evaluating the implementation of design heuristic cards in an industry-sponsored capstone design course. In: Proceedings of the annual conference of the American Society for Engineering Education (ASEE), Indianapolis, IN. https://peer.asee.org/20435

Kramer J, Daly SR, Yilmaz S, Seifert CM (2014) A case-study analysis of design heuristics in an upper-level cross-disciplinary design course. In: Proceedings of the annual conference of American Society of Engineering Education (ASEE), Indianapolis, IN. https://peer.asee.org/19915

Kramer J, Daly SR, Yilmaz S, Seifert CM, Gonzalez R (2015) Investigating the impact of design heuristics on idea initiation and development. Adv Eng Educ 4(4)

Lawson B (1980) How designers think: the design process demystified. Architectural, London, UK

Leahy K, Daly SR, Murray J, McKilligan S, Seifert CM (2018a) Transforming early concepts with design heuristics. Int J Technol Des Educ 1–21

Leahy K, Daly SR, McKilligan S, Seifert CM (2018b) Supporting idea generation with design tools: nesting design heuristics within morphological analysis. Global J of Eng Educ 20(3):159–170

Lee JW, Daly SR, Huang-Saad A, Seifert CM, Lutz J (2018) Using design strategies from microfluidic device patents to support idea generation. Microfluidics Nanofluidics 22(70)

Lee JW, Ostrowski A, Daly SR, Huang-Saad A, Seifert CM (2018) Idea generation in biomedical engineering courses using design heuristics. Eur J Eng Educ 1–19. https://doi.org/10.1080/03043797.2018.1514368

Liu YC, Bligh T, Chakrabarti A (2003) Towards an 'ideal' approach for concept generation. Des Stud 24(4):341–355

Newell A, Simon HA (1972) Human problem solving. Prentice-Hall, Englewood, NJ

Osborn A (1957) Applied imagination: principles and procedures of creative problemsolving. Scribner, NY

Perez A, Linsey JS, Tsenn J, Glier M (2011) Identifying product scaling principles: a step towards enhancing biomimetic design. Paper presented at the ASME 2011 international mechanical engineering congress and exposition, Washington, DC

Römer A, Weißhahn G, Hacker W (2001) Effort saving product representations in design—results of a questionnaire survey. Des Stud 22(6):473–490

Rowatt WC, Nesselroade KP, Beggan JK, Allison ST (1997) Perceptions of brainstorming in groups: the quality over quantity hypothesis. J Creat Behav 31(2):131–150

Saunders MN, Seepersad CC, Hölttä-Otto K (2011) The characteristics of innovative mechanical products. J Mech Des 133:021009-021001-021009

Savransky SD (2000) Engineering of creativity: introduction to TRIZ methodology of inventive problem solving. CRC Press, Boca Raton

Seifert CM, Gonzalez R, Yilmaz S, Daly SR (2015) Boosting creativity in idea generation using design heuristics. In: Product Development and Management Association (ed) Design and design thinking: essentials in the PDMA's new product development series. Wiley, pp 71–86

Singh V, Skiles SM, Krager JE, Wood K, Jensen D, Sierakowski R (2009) Innovations in design through transformation: a fundamental study of transformation principles. J Mech Des 131(8):081010-081011-081018

Singh V, Walther B, Krager JE, Putnam N, Koraishy B, Wood KL, Jensen D (2007) Design for transformation: theory, method and application. Paper presented at the ASME 2007 international design engineering technical conferences and computers and information in engineering conference, Las Vegas, Nevada, 4–7 Sept 2007

Skiles SM, Singh V, Krager JE, Seepersad CC, Wood KL, Jensen D (2006) Adapted concept generation and computational techniques for the application of a transformer design theory. Paper presented at the ASME 2006 international design engineering technical conferences and computers and information in engineering conference, Philadelphia, PA, 10–13 Sept 2006

Smith SM (1995) Fixation, incubation, and insight in memory, problem solving, and creativity. In: Smith SM, Ward TB, Finke RA (eds) The creative cognition approach. MIT Press, Cambridge, pp 135–155

Smith GF (1998) Idea-generation techniques: a formulary of active ingredients. J Creat Behav 32(2):107–134

Sutton RI, Hargadon A (1996) Brainstorming groups in context: effectiveness in a product. Adm Sci Q 41:685–718

Weaver J, Wood K, Crawford RH, Jensen D (2010) Transformation design theory: a meta-analogical framework. J Comput Inf Sci Eng 10(3):031012

Weaver JM, Wood KL, Jensen D (2008) Transformation facilitators: a quantitative analysis of reconfigurable products and their characteristics. Paper presented at the ASME international design engineering technical conferences and computers and information in engineering conference, Brooklyn, NY

Yilmaz S, Christian JL, Daly SR, Seifert CM, Gonzalez R (2012) How do design heuristics affect design outcomes in industrial design? In: Proceedings of the international design conference, Dubrovnik, Croatia

Yilmaz S, Daly SR, Christian JL, Seifert CM, Gonzalez R (2013) Can experienced designers learn from new tools? A case study of idea generation in a professional engineering team. Int J Des Creat Innov 2(2):82–96

Yilmaz S, Daly SR, Seifert CM, Gonzalez R (2015) How do designers generate new ideas? Design heuristics across two disciplines. Des Sci 1:1–29

Yilmaz S, Daly SR, Seifert CM, Gonzalez R (2016a) Evidence-based design heuristics for idea generation. Des Stud 46:95–124

Yilmaz S, Seifert CM (2011) Creativity through design heuristics: a case study of expert product design. Des Stud 32(4):384–415

Yilmaz S, Seifert CM, Daly SR, Gonzalez R (2016) Design strategies in innovative products. J Mech Des 138(7)

Yilmaz S, Seifert CM, Gonzalez R (2010) Cognitive heuristics in design: instructional strategies to increase creativity in idea generation. J Artif Intell Eng Des Anal Manuf 24(3):335–355

Enabling Meaningful Reflection Within Project-Based-Learning in Engineering Design Education

Thea Morgan

Abstract Group project-based-learning (PBL) is a form of experiential learning in which students develop tacit knowledge about creativity, critical thinking, collaboration and communication, in addition to deepening and contextualising core subject knowledge. Students construct this tacit knowledge by reflecting on their lived experience of meaningful group project work, or rather on their 'perceptions' of this lived experience, meaning their prior knowledge, worldview(s) and previous experience will have a strong influence on the outcomes of learning from group PBL. Students of engineering design are heavily influenced by the positivist learning paradigm of engineering science, and so many students struggle to learn effectively from group PBL design experiences, because the constructivist paradigm that underpins this type of learning is not in accord with their cognitive structure. Pedagogical aids to reflection are required to support learning in group PBL design courses in engineering design education. Aids that reveal the underlying learning paradigms within this subject, and their conflicting nature, allowing students to place their own learning in the appropriate epistemological context. It is proposed here that the teaching of *philosophy of design*, combined with use of reflective learning journals structured using a constructivist inquiry framework, might potentially allow students to access a deeper level of understanding of their own individual approaches to design, by enabling reflection at an ontological level. Philosophy of design serves the purpose of emancipating students from a restrictive worldview by making them aware of multiple paradigms of learning. This chapter presents the argument for such an approach to supporting group PBL and describes a study within a second-year PBL engineering design course, in which this approach has been trialled. The results indicate that a positive impact on reflection and learning has been achieved. Philosophy of design appears to give the students a language and a conceptual structure with which to reflect on personal design activity in constructivist terms, and the reflective learning journals provide an appropriate means to externalise and enhance this reflection through the representation of learning.

T. Morgan (✉)
CAME School of Engineering, University of Bristol, Clifton BS8 1TR, UK
e-mail: thea.morgan@bristol.ac.uk

© Springer Nature Switzerland AG 2019
D. Schaefer et al. (eds.), *Design Education Today*,
https://doi.org/10.1007/978-3-030-17134-6_4

Keywords Project-based-learning · Reflection · Engineering design · Philosophy of design

1 Introduction

Engineering design as a subject encompasses two distinct, and essentially opposing schools of thought on the nature of knowledge and its acquisition. This manifests as contradictory learning paradigms within engineering design education. The dominant learning paradigm within schools of engineering is the positivist paradigm of engineering science (Downey and Lucena 2003), as articulated through the teaching of core engineering science subjects. Knowledge within these science-based subjects is largely explicit, i.e. it can be codified into spoken, written, or mathematical language and readily transferred to the learner. Students commonly acquire this knowledge through direct teaching methods such as lectures, where knowledge is transferred systematically from teacher to student as 'facts' or 'truths'. Students quickly come to understand not only what they need to know, but how they can come to know it.

Creative design, however, is most helpfully understood from within the constructivist learning paradigm (Schön 1984). This is because design knowledge is largely tacit (cannot be codified into language), and, therefore, must be acquired experientially by the learner themselves. Students typically gain design knowledge by undertaking design projects, or 'project-based-learning' (PBL) design experiences. In engineering design education, however, PBL is generally not supported with pedagogical approaches that highlight the differing nature of engineering science knowledge and design knowledge. Instead PBL courses tend to involve instruction in the use of prescriptive design processes and design methods. These have their place in design education, but when taught in isolation only serve to reinforce the positivist learning paradigm of engineering science. The constructivist learning paradigm is not well articulated or pedagogically supported in schools of engineering, in fact there is little said on the constructivist nature of learning creative design (Downey and Lucena 2003).

In order to learn effectively from PBL design experiences, students need to be able to reflect meaningfully on those experiences. Research shows that engineering design students often struggle to reflect appropriately on personal or group design activity, due to a lack of understanding of the differences between design and science at a philosophical level (Morgan and Tryfonas 2011; Morgan 2017). Instead, engineering design students tend to treat design as a form of rational problem-solving (Downey and Lucena 2003). Previous research (Morgan 2017) shows that engineering design students in a PBL setting intuitively set about design activity using a constructivist inquiry approach, but are largely unaware of it. Their conscious view of their own design activity does not match their observed behaviour. This has also been observed in studies in industry (e.g. Morgan and Tryfonas 2011).

The conflicting learning paradigms of engineering design education are not revealed by current approaches to learning and assessment, and this limits the ability of students to reflect meaningfully during PBL. The students effectively experience an 'epistemological block to reflection'. We need to sufficiently widen the worldview of engineering design students, to encourage an openness to learning design experientially, and to reflect meaningfully on that learning. In other words, the aim is to emancipate engineering design students from a restrictive worldview about the nature of knowledge and its acquisition. New educational approaches, which reveal the differing paradigms of engineering science and design, and enable deeper reflection and metacognition, are required to support PBL in engineering design.

2 Background

2.1 Project-Based-Learning and Constructivism

PBL was developed in the early 1970s as a new approach to education that added to a theory of learning with cognitive, emotional, and social dimensions (Illeris 2007). PBL is now practiced all over the world, and engineering education tends to follow the Aalborg model (Savin-Baden 2003). According to Edström and Kolmos (2014) PBL projects 'are the platform for students to achieve competences, and to relate disciplines to each other in analysis and identification of problems as well as the problem-solving process. Process skills such as self-directed learning, project management, collaboration, communication, and collaborative knowledge construction are taught in an integrated way by letting students reflect upon their practice. A fundamental principle is that the students are owners of the learning process and the facilitator guides the students by presenting several ideas, methods, and tools'. As a pedagogical model PBL is built upon well-established theories of learning, such as experiential learning (Kolb 1984), cooperative learning (Johnson et al. 1998), and reflective practice (Schön 1984). It encompasses key design activities, such as 'gathering and evaluating information, framing problems, generating and testing solutions, making and communicating decisions, and iterating, that allow students to experience complete learning cycles' (Atman et al. 2015).

The number and curricular sequencing of these PBL experiences still vary widely. According to Atman et al. (2015) engineering design education over the last two decades has tended towards three curricular models, each incorporating PBL in a different way. The first model, and the most well-established, is the final year PBL design experience (or capstone course in the US), often done in collaboration with industry (Dym et al. 2005). Historically the purpose of this late sequencing was to allow students to acquire a solid foundation in engineering science in the earlier years of the programme, before applying this knowledge in a quasi-real context (Sheppard et al. 2008). The second model involves a design experience in the first year of the academic programme, as well as the final year, with the remainder of the intervening time filled with traditional engineering science courses. Dym et al. (2005) suggest that the first year PBL experience serves a variety of purposes, such as familiarising students with conceptual design and teamwork, as well as improving retention rates in engineering programmes. Both of these models, however, are increasingly seen as inadequate as our understanding of design evolves, and the range and complexity of design skills required by graduate engineers become apparent (Dym et al. 2005). A more integrated approach to design experiences within the engineering curriculum is emerging in many institutions, with programmes that include first-year and final-year PBL experiences, but also other design learning experiences woven throughout. The current trend in engineering design education is towards this more integrated model. An example of such an integrated curriculum is the Conceive, Design, Implement, and Operate (CDIO) curriculum initiative, that contains significant PBL elements, which has been adopted widely by engineering institutions around the world (Crawley et al. 2011).

In an ethnographic study by Holland et al. (2012) aimed at understanding and improving student learning in PBL engineering design courses, several areas of 'troublesome knowledge' were identified, including; engineering science, project management, tacit skills, domain knowledge, and tools and equipment. These were areas of knowledge that the students had particular difficulty in dealing with during design projects, but were not well supported or assessed within existing PBL course pedagogies. Holland et al. subsequently developed conceptual question-based tools for assessing student learning and evaluating pedagogical improvements, and an iteratively expanded pedagogical framework intended to guide the development and evaluation of learning environments in PBL courses, built upon these identified problem areas of knowledge (Holland 2014). This research draws attention to the 'fundamental obstacles to learning in [...] project-based engineering design courses [...], including challenges in providing structure for students while supporting flexible exploration of potential designs' (Holland 2014). However, Holland and colleagues do not attempt to address the fundamental paradigm conflict at the heart of engineering design education, and how this impacts on group PBL design experiences.

PBL is a constructivist approach to learning. Constructivism holds that learners are determinants of what is learnt, and that their existing ability and knowledge shapes the nature of that learning. In this context, teachers operate as facilitators of learning, not as conveyors of knowledge. Constructivist approaches utilise experiential methods such as prototyping, experiments, and real-world problem-solving, to encourage students to create new knowledge and to reflect on how this knowledge changes their understanding. From a cognitive perspective, existing knowledge is organised within the mind in what is known conceptually as a 'cognitive structure', a network of 'facts, concepts, propositions, theories, and raw perceptual data that the learner has available to him at any point in time' (Ausubel and Robinson 1969). The cognitive structure continually changes as new knowledge is constructed, during 'processes of mutual accommodation of the cognitive structure and [...] new material of learning' (Moon 2013). Because new learning is linked to existing knowledge in this way, learning is always meaningful. Some students will be naturally receptive to new materials of learning, even if this involves radically altering their cognitive structure, whilst others will tend to reject new learning unless it is largely compatible with what they already know. Or as Moon (2013) puts it 'what we already know, in its guiding and organizing role, can act as a gate-keeper to learning'. This indicates that encouraging an openness to learning is a key factor in supporting PBL. Moon goes on to suggest that 'the mission of adult education is to emancipate people from a self-imposed restrictive view of the world, to one that is open to new ideas […]'.

Students may also use either a 'deep' or 'surface' approach to learning in PBL courses. The deep approach involves an intention by the learner to understand ideas for themselves, by 'relating ideas to previous knowledge and experience; looking for patterns and underlying principles; checking evidence and relating it to conclusions; examining logic and argument cautiously and critically; becoming actively interested in course content' (Moon 2013). Conversely, the surface approach to learning involves an intention to 'cope with course requirements, by studying without reflecting on either purpose or strategy; treating the course as unrelated bits of knowledge; memorizing facts and procedures routinely; finding difficulty in making sense of new ideas presented' (Moon 2013). The deep approach to learning is associated with more significant changes to the cognitive structure, as new material of learning is accommodated. It is also more strongly linked to relevant existing knowledge, and is, therefore, more meaningful.

Representation of learning (Eisner 1993) also plays a key role in PBL. That is, how learning is externally expressed by the learner. This could be, for example, in the form of a written report, a verbal presentation, or a reflective learning journal. Representation of learning plays its own role in learning itself, in that 'learning occurs in the process of representing the learning (Eisner 1993)'. The learner is able to reflect on how different aspects of their newly constructed knowledge link together and to previous learning. By making further accommodations between the cognitive structure and the materials of learning, the process of representation of learning can upgrade surface learning to deep learning. Deep learning is associated with higher quality representation of learning.

Successful outcomes in PBL are also closely linked to the concept of 'self-regulated learning' (SRL). SRL can be considered as 'the extent to which learners are meta-cognitively, motivationally, and behaviourally active in their own learning process' (English and Kitsantas 2013). English and Kitsantas claim that 'to effectively engage in PBL, students must become responsible for their learning and actively participate in the processes of constructing knowledge and making meaning'. They suggest that 'for many students, this role conflicts with deeply ingrained habits they have developed through more familiar classroom experiences, in which they have been passive recipients of knowledge'. SRL is, therefore, a key skill for effective learning in PBL and one that must be supported and developed in engineering design students. A key aspect of developing SRL (amongst others) is enabling deep learning through reflection.

2.2 Reflection and Experience

Reflection is defined by Boyd and Fales (1983) as 'the process of internally examining and exploring an issue of concern, triggered by an experience, which creates and clarifies meaning in terms of self and which results in a changed conceptual perspective'. Schön (1984) introduced us to the idea of reflective practice in design. He described reflection-in-action, a kind of spontaneous, improvised reflection that occurs during practical activity, and guides the development of that activity. He also described a higher level reflection, on reflection-in-action itself, namely reflection-on-action. A categorisation of reflection has also been proposed by Van Manen (1991), based on four levels of increasingly higher abstraction: 1. Thinking and acting in a common-sense manner on an 'everyday' basis. 2. Reflection that is focused on events or incidents. 3. Reflection on personal experience and that of others. 4. Reflection on the manner of reflection. The first two levels are roughly equivalent to Schön's (1984) reflection-in-action and reflection-on-action, respectively, whereas the third level of reflection is more general and not linked to a specific event. The fourth level is meta-reflection—reflecting on the very nature of knowing and reflecting. In other words, reflecting at an ontological level.

Reflection is a fundamental process in experiential learning, as highlighted by Kolb's (1984) experiential learning cycle: experience, reflect, conceptualise, experiment. Experience in this sense is not limited to the immediate activity being undertaken. Understanding from previous experience, as well as prior knowledge about theories and ideas that relate to this activity, will also be considered for reflection. Representations of learning produced from this experiential learning, will involve further reflection and, therefore, further learning (effectively a second cycle of Kolb's (1984) experiential learning cycle). Therefore, 'few situations of learning from experience will be pure activity, or absolutely distinguishable from all aspects of classroom teaching. It seems reasonable to suggest that no learner will come to a situation without any prior knowledge and experience', indeed 'the nature of learning itself can be defined as the active construction of meaning by a learner building on and modifying meaning that they have arrived at in prior learning' (Moon 2013).

Reflection and learning from experience cannot be forced, they can only be encouraged and guided within a learning environment. Assessment can provide a strong incentive to learn and reflect, and also shape the form of that learning and reflection. Deep learning is defined, in part, by the occurrence of meaningful reflection, therefore, the enablement of reflection in PBL, through assessment and other means, is fundamental in encouraging a deep learning approach by students.

It is worth clarifying that students do not actually learn directly from experience, rather they learn from their 'perceptions' of that experience. The outcomes of PBL will be strongly influenced by the prior experiences, knowledge, and worldview(s) of students. Moon (2013) suggests that educators involved in experiential learning 'need to pay more attention to the prior experiences of the learner that will affect their initial perceptions of the experience'. This has significant implications for engineering design students taking PBL design courses, whose perceptions of that experience will be strongly influenced by prior and contemporaneous engineering science learning and its related positivistic worldview.

Several studies describe possible approaches to encouraging reflection during PBL design courses. Whilst it is hard to evaluate the impact and value of these approaches, especially in quantitative terms, the studies reveal an apparent positive value in the use of these approaches by the learners involved. A study by McDonnell et al. (2004) explored the construction of video-stories to encourage reflection in design education, using small teams of final year industrial design students. The students spoke of 'a genuine struggle to make sense of the discrepancies between education and practice, to resolve the objective views of designing products through rational problem-solving and structured process stages with their practical experience as creative individuals cooperating socially in design activity'. Elsewhere, Jaskiewicz and van der Helm (2017) have introduced 'progress cards' as a tool for supporting student reflection on design activity during PBL in industrial design education. The cards, completed daily by students, contain 'space for a visual and textual summary of the latest version of the designed experience and a summary of progress regarding the new knowledge generated in the design process'. Despite some initial resistance, the authors claim that the cards show value in supporting the students to 'analyse and manage their own design process from a holistic perspective'.

Reflective learning journals are the most commonly used means of encouraging reflection in formal education (Chappell 2006; Thorpe 2004). Reflection occurs in the process of representing learning in written form, and then reading this representation back. Reflective learning journals also enable the upgrading of learning, from surface to deep learning 'where unconnected areas of meaning cohere, and deeper meaning emerges' (Moon 2013). Other key purposes of reflective learning journals are; to enable the learner to understand their own learning process, to increase personal ownership of learning, to explore the self, personal constructs of meaning, and understand one's view of the world. Reflective learning journals can be structured or unstructured, with structured forms containing questions, exercises or guidance about particular issues to reflect on. In a study by Chappell (2006), reflective learning journals were used to support a PBL module, in the first year of a Geography programme. Chappell describes how the students were seen to undergo a 'grieving process' in the transition from 'teach me' to 'help me to learn' attitudes, and that 'the students learning journals demonstrated a critical engagement with their learning environment'.

2.3 Philosophy of Design

Moon (2013) suggests that students be encouraged to reflect by including teaching of philosophy in formal education… 'Following a course in philosophy legitimises and enables practice in the questioning of assumptions that underpin any other learning—the process of reflection'. Therefore, a possible pedagogical aid to reflection in engineering design education is the teaching of 'philosophy of design', before or alongside PBL design courses. Philosophy in this context is defined as 'a group of theories and ideas related to the understanding of a particular subject' (Cambridge Dictionary 2015), in this case the subject of design, where 'philosophy of design' can be considered as 'the pursuit of insights about design by philosophical means' (Galle 2002). In other words, exploring the learning paradigms within which different problem-solving approaches sit, including the differing nature of knowledge and its acquisition, and how creative design activity can be understood in terms of philosophical constructs such as paradoxes, themes and frames (Dorst 2015), which make sense of design in constructivist terms (further detailed in Sect. 3.3). This would help to make visible the dominant positivist learning paradigm of engineering science, and allow students to understand how problem-solving in creative design is different. Downey and Lucena (2003) suggest that 'reform in engineering education may have to move beyond expanding and enhancing design education to address the very distinction between science and design'. Galle (2002) suggests that '[philosophy of design] serves the end of helping, guiding, suggesting how the [designer] comes to understand what he is doing, and not simply how he comes to do what he is doing'.

2.4 Complexity and Constructivist Inquiry

Findeli (2001) has proposed an alternative epistemology of creative design practice based on complexity theory, and a theoretical model of design for education conceptualising a design project as a complex system. Within this model a new structure of the design process would be: instead of a problem, we have state A of a system; instead of a solution, we have state B of the system; the designer and the user are part of the system. Findeli suggests that 'the designer's task is to understand the dynamic morphology of the system, its "intelligence". One cannot act upon a system, only within a system'. Because the designer and user are part of the system, they change state too, i.e. they are learning. Mathematical or formalistic approaches are not appropriate, he argues, for a complexity-based model of design activity in design education, due to their 'objective' nature. Rather, 'a system, and especially a human or social system, is best understood from within, through a constructivist, phenomenological, approach'. Wang (2010) supports the idea that a design project can be conceived as a complex system and points to the fact that, within complexity theory, 'there is no distinct separation between knower and knowledge. A learner does not assimilate objective truth; instead, he or she creates subjective meanings and values from the educational experience'. Like Findeli, Wang concludes that learning is research-like and that 'under this paradigm research is mainly concerned with constructivist understanding of the relationship among all the parts of the system. In other words, such research is sociocultural instead of scientific'.

In a recent paper Adams et al. (2018) propose 'collaborative inquiry' as an appropriate sociocultural framework with which to understand design activity, as a social process of 'creating a collective pool of insight and building coherence to construct valid knowledge'. It is a framework embedded in an epistemology of experiential, subjective and situated knowing and 'the accumulation of knowledge as co-creation within an inquiry group'. Collaborative inquiry is an experiential learning process, involving cyclical episodes of reflection and action as the group learns their way towards a mutually agreed outcome. The key findings of previous research by the author (Morgan 2017; Morgan and McMahon 2017), involving participant observation of student groups during PBL design experiences, similarly conceptualise intuitive student design activity in this context as a form of collaborative constructivist inquiry. Five major themes relating to student design activity emerged from the observation studies; *gathering design information, decision-making using design methods, identifying project themes, developing and testing concepts,* and *telling the design story.* Student groups were seen to engage spontaneously and non-linearly in these core activities, which are directly analogous to those undertaken in formal constructivist inquiry; *collecting data, analysing and interpreting data, identifying themes, theory-building and testing, and telling the story* (Simons 2009). The students shifted frequently and intuitively between these activities, responding appropriately to the evolving demands of their projects. For a more in-depth analysis see Morgan (2017), Morgan and McMahon (2017). Constructivist inquiry makes sense of design as a highly situated activity. It embraces complexity, it is not prescriptive, or linear, and accounts for the uniqueness of every design situation. It includes the designer themselves as part of the design situation and acknowledges that the experience of designing actually changes the designer, i.e. they are learning.

3 Enabling Reflection in PBL Design Courses: A Study

It is proposed here that the teaching of philosophy of design, together with the use of reflective learning journals structured using a constructivist inquiry framework, might potentially allow engineering design students to access a deeper level of understanding of their individual and collaborative approaches to creative design during PBL, by enabling reflection at an ontological level. This section describes an educational intervention that has been carried out within a second-year PBL engineering design course, using this approach.

3.1 Study Context

The intervention study took place within the CAME School of Engineering at the University of Bristol (UK), focusing on the 5-year MEng Engineering Design degree programme (which includes a year in industry in the third year). The educational context of the undergraduate MEng Engineering Design programme is that of an integrated PBL approach to learning design. As an institution, the University is also a member of the international 'Conceive, Design, Implement, Operate' (CDIO) curriculum initiative.

The particular focus of the intervention was a second-year group PBL design unit, 'Design Project 2'. A mandatory core unit within the programme, spanning 24 teaching weeks across the academic year. At the time of the study, there were 31 students registered on the unit, with a near equal gender balance. The students were required to carry out 3 group design projects, of increasing complexity, with the final project involving both design and make aspects. The unit is taught by three members of academic staff (including the author), each with a minimum of 3 years' experience teaching on the course. Reflective learning journals have not been used previously in this unit or elsewhere in the degree programme.

3.2   The Design Project

The focus of the study was the third and final design project of the three undertaken during the unit, and the only one to involve both design and make elements. The students had not been asked to keep or submit reflective learning journals for the previous two projects. The 31 students were each assigned to one of 6 project groups; 5 groups of 5 students, plus 1 group of 6 students. Allocation of students to groups was done by unit staff and designed to ensure that students worked with a variety of different peers over the three projects, and to achieve a good mix of gender and intellectual ability. The groups were given 11 teaching weeks in which to generate and develop a concept, build and demonstrate a prototype, and to submit a 20-page group design report and CAD model of the final design. They were also required to submit a structured summary of their individual reflective learning journal one week later, along with a peer assessment form. The students were supported by three academic staff during the project, plus several technicians during the build phase. They also had the opportunity to talk to the previous year's student cohort about their project, which had a not dissimilar brief. Examples of good project reports submitted by the previous cohort were also made available to the current students.

3.2.1   The Project Brief

Each group was given the same project brief at the outset; to design and prototype an 'industrial plant disaster recovery robot'. The problem context was summarised as follows:

> The occurrence of both natural and man-made disasters and their impact on increasingly complex industrial facilities and modern infrastructure has focused the need for autonomous vehicles in the recovery operations. Where the situation remains unsafe for human intervention, due to structural instability or release of hazardous substances an autonomous robotic platform can perform essential remedial actions.

Fig. 1 Wooden block representing sensor

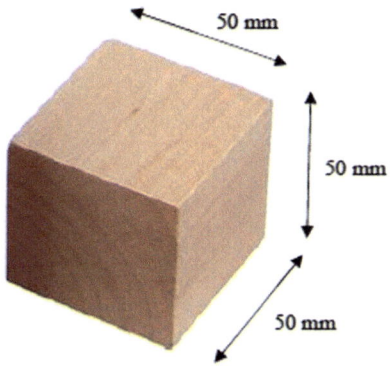

50 mm

50 mm

50 mm

The project groups were asked to consider themselves as competing design consultancies and each challenged to design an autonomous robotic concept in response to this problem situation. The brief went onto specify that the design of any such autonomous robotic platform must be capable of access to restricted spaces and, therefore, must fit within a size envelope of 0.3 m width × 0.4 m length × 0.15 m height. It must also fulfil the following key functions:

Operation over rough, debris covered surface: The robot must be able to traverse surfaces covered with debris resulting from the disaster incident. To prove this a test terrain using randomly positioned paving blocks will need to be negotiated. The blocks will be placed in their normal orientation and will not be stacked. The block dimensions will typically be 50 mm high, 100 mm wide and 200 mm in length.

Placement of remote sensor: A key task of the robot is to facilitate the ongoing monitoring of the hazardous environment. The sensor is a square cube with sides of 0.05 m and weighing 50 g (Fig. 1). The robot system must be capable of placing the sensor at a location between 150 and 400 mm from ground level and at a distance of 250 mm from the extremities of the robot chassis and drive gear.

Closure of a critical valve: The robot will be required to close a valve (Fig. 2) by rotating its handle until the closed position is reached (i.e. no further rotation is possible). This may require one or more complete turns.

3.2.2 Project Assessment

The project was assessed in three parts, each carrying a different weighting towards the overall mark. The written design report/CAD model and demonstration prototype were awarded group marks, and the reflective learning journal summary an individual mark.

- 60% Written design report and CAD model (group)
- 30% Prototype (group)
- 10% Reflective learning journal summary (individual).

Fig. 2 Valve requiring
closure

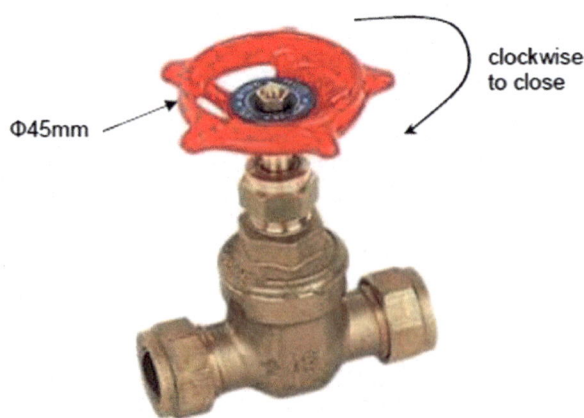

Φ45mm

clockwise
to close

These 3 marks were then combined, and adjusted according to peer assessment to give a total individual project mark.

3.3 Philosophy of Design Lectures

During the first half of the unit, the students received a series of 3 lectures on philosophy of design, alongside their studio-based design work. The purpose of the lectures was to open-up the worldview of the students, to allow them to place their own learning within the appropriate epistemological context, and to reflect on their experience of designing at a deeper and more meaningful level. These lectures were titled; 'The nature of design', 'Paradoxes, frames, and themes' and 'Learning to design'.

Philosophy of design I: The nature of design

The first lecture asked the question 'what is design'? Then, by way of answer, introduced students to a range of ideas intended to aid an understanding of the fundamental nature of design. The stated aim of the lecture was: to provide an awareness and understanding of the underlying nature of design, from a philosophical perspective, and how this nature differs from engineering science. The lecture content included the following topics:

- *Engineering science and creative design*: Key differences between engineering science and creative design. Explicit knowledge and tacit knowledge. Different worldviews.
- *The nature of design problems*: Problem contexts are always unique. Design problems only make sense in relation to their solution, so to learn about the problem you must explore possible solutions. Design problems are open, complex, dynamic, and networked.

- *Design reasoning*: Introduction to different types of reasoning; deduction, induction, abduction and design abduction. Framing as a way to reason from a desired outcome backward to a possible solution. Problem and solution 'spaces' and the concept of co-evolution.
- *Design interaction*: How design takes place in teams. Design as talk, the productive force of language, storytelling in design, negotiation and shared understanding.

Philosophy of design II: Paradoxes, themes and frames

The second lecture provided an introduction to the philosophical constructs of paradoxes, themes and frames, which describe design activity in constructivist terms (Dorst 2006, 2015). The stated aim of the lecture was: to provide an understanding of creative design in terms of resolving paradoxes by searching the broader problem context for themes with which to create new frames. The lecture content included the following topics:

- *Fixation*: Three different types of fixation; problem fixation, process fixation, solution fixation. Examples of each type. Fixation can be avoided through questioning of underlying assumptions by using a frame creation approach.
- *Paradoxes*: Defining paradoxes in a design sense—a 'design problem' can be considered a paradox, made up out of the clash of conflicting discourses within a design situation.
- *Themes*: Themes are patterns across data sets that are important to the description of a phenomenon. Themes emerge in design projects which can be used to develop a new frame.
- *Frames*: A frame is a way to think from consequences (e.g. a need to be addressed) back to causes (the designed objects, systems, services) and working principles (the way things work). A new frame can help resolve the central paradox of a design situation.
- *Frame creation*: A model of design activity within a project. Themes inform the development of a frame that articulates a response to the central paradox of the problem situation. The 9 stage frame creation process.
- *Case studies*: Frame creation used in real projects.

Philosophy of design III: Learning to design

The third and final lecture focused on design ability and expertise, and how design is learned within an engineering design context. How this learning can be supported and enhanced. The stated aim was: to provide an understanding of design ability and expertise, and how these are acquired. To highlight issues and opportunities for learning design in engineering design education. The lecture content included the following topics:

- *Design ability and expertise*: An introduction to 'designerly ways of knowing' (Cross 2007). The common traits and stages of design expertise. Expertise as a combination of natural aptitude, learned skill and personal experience.

- *Experiential learning*: Kolb's learning cycle. Constructing knowledge by reflecting on lived experience. Adapting/replacing mental models and experimenting with new knowledge. PBL design experiences as experiential learning.
- *Different paradigms of learning*: Outline of positivism and constructivism, how they are different and why they must coexist in engineering design education. Engineering science and creative design present different problems to solve.
- *Design activity as constructivist inquiry*: An introduction to constructivist inquiry and how this is a helpful way to think about design activity. Complexity as meta-paradigm.
- *Supporting PBL*: An introduction to the reflective learning journals and how these may help to support design learning, in conjunction with the philosophy of design lectures.

3.4   Reflective Learning Journals

The students were each asked to keep a reflective learning journal throughout the project, and were advised that it should include:

- Notes on the development of the client brief, project objectives and requirements specification.
- Notes on, and results of, concept generation methods, including sketches and descriptions.
- Notes on, and results of down-selection methods, including references to files or spreadsheet versions and revisions.
- Project planning information (e.g. schedule of work, work distribution).
- Details of research, including different sources of information and data: papers, textbooks, standards, manuals, patents, personal communications, etc.
- Results of calculations and experimental/modelling work.
- Details of prototyping activities.

The students were also encouraged to record any thoughts and observations about their design activity, and any aspects of the project that went well and any that did not. Each entry was to be dated and they were asked to bring the journal to every studio session and design meeting.

They were then asked to submit a 3-page summary reflection, drawing directly on their learning journals and the philosophy of design lecture material, in which they were asked to reflect on their experience of designing written from the perspective of a constructivist inquiry researcher exploring a complex system. They were asked to structure their reflective summary using the 5 headings shown in Table 1. These headings echo the themes of intuitive student design activity identified in Morgan (2017), and are directly analogous to formal constructivist inquiry methodology.

Students were provided with extensive guidance notes on how to complete the summary reflection, including prompt questions for each of the 5 sections. No example answers were given, however, to avoid students fixating on these. The guidance

Table 1 Constructivist inquiry activities and equivalent sections of the summary reflection

Core activity of constructivist inquiry (Simons 2009)	Equivalent section heading of summary reflection
Collecting data	Gathering design information
Analysing and interpreting data	Decision-making using design methods
Identifying themes	Identifying themes and a central paradox
Developing and testing theory	Developing and testing concepts
Telling the story	Writing the design report

notes also referred the students back to the philosophy of design lecture materials (available to them online). The three basic assessment criteria were also included, these were; engagement with learning, quality of written summary (representation of learning), and depth of insights into personal design activity. It was emphasised that this should be a personal reflection, with no right or wrong answers. Not a piece of formal academic writing.

The purpose of the reflective learning journals (and the summary reflections) was to deepen learning from experience, by encouraging the students to develop meaning about that experience through reflection. The reflective learning journal was also a place for each student to exercise reflexivity during design activity, in order to explore the self and personal constructs that influence how they view the world. Assessment of the structured summary reflections was intended to provide motivation for the students to engage with, and learn from, the process of meaningful reflection.

3.5 Analysing the Dataset

The submitted summary reflections formed a dataset with which to evaluate the impact and potential value of the novel combined approach. This qualitative dataset has been analysed and interpreted with respect to the assessment criteria outlined in Table 2. The five distinct sections of the summary reflection were given individual scores, which were then combined to give a total percentage score for each submission. Three key criteria were chosen to form the basis of the assessment; engagement with learning, quality of representation of learning, and depth of insights into personal design activity in relation to philosophy of design materials of learning.

Engagement with learning: This is defined here as the extent to which the student has attempted to respond meaningfully to the task. Even if ultimately, they have not produced work of good quality or insights of any depth. Engagement demonstrates that students are motivated and open to learning, which is a prerequisite to reflection. Reflection and learning from experience cannot be forced. If students are not open to learning, then little or no reflection or learning will take place. This criterion serves as an indicator of the extent to which the students were open to this new approach to supporting PBL.

Table 2 Assessment criteria and corresponding scoring bands for the summary reflection submissions

Scoring bands (%)	Criteria to be met for each band
>40 (fail)	Little or no engagement with learning, poor quality of representation and little or no reflective insights into personal design activity
40–50	Good engagement with learning, but poor quality of representation and little or no reflective insights into personal design activity
50–60	Good engagement with learning, good quality of representation, but little or no reflective insights into personal design activity
60–70	Good engagement with learning, high quality of representation, but little or no reflective insights into personal design activity
70–80	Excellent engagement with learning, good or high quality of representation and good reflective insights into personal design activity
80–90	Excellent engagement with learning, high quality of representation and deep reflective insights into personal design activity

Quality of representation of learning: This is defined here as the extent to which the representation of learning is well-written, well-structured, and provides an articulate recount of design activity undertaken during the project. The quality of representation of learning is an indication of deep or surface learning, with a higher quality of representation of learning associated with deep learning. This criterion, therefore, serves as an indicator of depth of learning associated with this new approach to supporting PBL.

Depth of insights into personal design activity: This is defined here as the extent to which the student has made meaningful connections between their personal design activity and philosophy of design materials of learning. The representation of learning must go beyond a basic recount of project activities, and stating philosophy of design constructs in isolation is insufficient. It is the meaningful connection between the two to form new insights that is key. Relating personal design activity to philosophy of design materials of learning demonstrates accommodation of the cognitive structure and the new materials of learning, and an ability to reflect at a high level of abstraction, in constructivist terms. This criterion, therefore, serves as an indicator of the extent to which students were enabled to reflect meaningfully using this new approach to supporting PBL.

4   Results and Discussion

In this section, excerpts from the summary reflections are presented and discussed. The selected quotes generally represent the more insightful (and higher scoring) submissions, and it should be noted that not all the submissions were of this level. Roughly half of the cohort have been quoted here. These excerpts serve as an illustration of what is possible, rather than a representative spread across the cohort. A discussion of scores achieved across the cohort is discussed in the final Sect. 4.6.

4.1   Gathering Design Information

The students were asked to describe and reflect upon the differing sources and uses of design information gathered through the course of the project. The majority of students were able to identify and discuss different types of design information, for example, verbal, visual (2D/3D spatial), or written information and how these were used in different ways and at different times in the project. They understood the sheer volume and variety of information available, not all useful, not all easy to interpret, and that choosing which information to use is a skill. They appeared to understand the importance of talk in shaping design, e.g. group discussion, and information gained through talking to staff and previous cohorts:

> At the beginning of the design project it was notable that the amount of verbal sources used in comparison to visual and written was a lot more. These sources provided guidance on how to go about starting the project and general points to look out for throughout. Although the sources were reliable since they included experienced lecturers in design, engineering and past students, there was a huge quantity of it. I found that trying to take in every piece of information and fully comprehending it was a difficult task. Therefore, brief notes were taken to allow me to look back later on. The actual information gathered was often useful but vague and so required further examination. For example, one piece of information from a previous year 2 student was "when building the prototype don't permanently join things together until absolutely necessary". It took a further thought to realise why this was and how exactly it might cause problems and where.

> Verbal sources that influenced this project included students from previous cohorts who have completed a similar task, university staff running the unit and members of the group. The practical advice received from students through their experience was useful in terms of learning to avoid common pitfalls, but was biased to their personal situation and different people could contradict one another.

> When using verbal sources, a certain amount of trust is placed into what is said and it is up to the individual to decide not only how valuable the information is but how reliable the source is and how the information should be used.

> Conversing with group members combined our collective knowledge of different areas and pooled any specialist information which resulted in more informed decisions throughout the design process. Speaking with cohorts that had previously undergone similar design projects was very useful as it provided guidance and direction in the initial stages of the project. However, an important consideration of verbal sources is their subjective nature.

> Group members mainly passed on information from other sources that they had encountered, this was good for provoking discussion in the group.

Many students demonstrated an understanding of the role of visual (2D/3D spatial) information sources in design, such as sketches, photographs, videos, prototypes, etc. For example, as a way to convey complex ideas, to communicate, to inspire, and to create shared understanding. Or how they can be a source of information that highlights issues that other sources do not:

> When working as a group, sketching proved a useful tool to convey complex ideas quickly and accurately to ensure a shared understanding of an idea, in particular during group discussion. When researching new solutions to the problem, images on the internet allowed the group to quickly understand possible design solutions, as well as inspiring creative thinking.

> Visual sources were also used during the project as they provided inspiration for designs and a way of presenting ideas. Photographs and videos of similar robotic devices were very useful for visualising initial concepts and concept generation. Observations of past prototypes also informed our concept generation and helped identify key features for the detailed design phase. Despite being more objective, visual sources can be interpreted in different ways by the individual which allowed more aspects of existing designs to be highlighted and compared. Sketches provided not only another source of information but an effective way of communicating complex ideas which was used throughout the project, supporting primarily verbal communication.

> The prototype was the best visual source of checking all the dimensions. It highlighted key areas where the CAD did not. A couple components which did not conflict in the CAD model, did in reality on the prototype and so it was invaluable in overcoming these problems before completing the final design.

Several students reflected that certain design information had the potential to cause fixation:

> Another main influence on the direction of the project were the previous cohorts' reports available to us. The previous years' designs definitely influenced our own design generation, and most likely created a design fixation on certain elements such as the mobility system as the reports were reviewed early in the design process.

> The process of data gathering can affect the quantity of different ideas that are generated. This is due to subconscious fixation on previous and already seen solutions, whether this is from previous years' prototypes or from videos of other products completing similar tasks. The fixation can even be caused by experiences prior to learning the brief. For example, some of the ideas generated were like robots from pop culture such as the hand design from the movie WALL-E but also from recent Boston Dynamics projects. This prevents the possibility of new innovative designs being produced and they may not actually meet the specification. We tried to prevent this from affecting the outcome and ensure that design abduction thinking was achieved.

> We also imposed some other fixations on ourselves when researching and designing, such as seeing the end effector as trying to replace what a human hand would do. Next time perhaps a focus should be on what needs to happen to the valve, instead of what action we would take to solve the problem.

> Advice from members of previous cohorts was particularly beneficial to the design process and was taken on-board with serious consideration. These students had been through the same design process, merely with an altered brief, and so had the benefit of hindsight to help assess what did and did not work for them, giving credibility […]. On reflection, it could be argued that there was an element of process fixation on the information we were given, with it being early on in the project when there were still many unanswered questions. The fact that we were being offered some kind of direction from what we viewed as a credible source made us susceptible to giving this information perhaps a greater weighting than was advisable.

Some students showed tentative understanding of the different learning paradigms and were able to reflect on their own learning experience in relation to these:

> Various reports, industry standards and websites were used to refine the design. This is because, a high level of technical guidance can be attained very quickly this way. Knowledge acquired in this way is due to the positivist paradigm, as it is seeking objective truth that can be acquired directly from teaching. These truths were then applied to improve the design. Verbal sources of information took the lead for gathering information when considering the manufacturing of the prototype. The technicians in the workshop, due to their high level of experience were invaluable when it came to assembling/altering the design. The knowledge acquired from such sources were developed via manufacturing of the prototype itself. It is for this reason that this is an example of the constructivist paradigm, where knowledge is developed directly from experience. In this context, knowledge is treated as a skill, it can only be improved via practice.

4.2   Decision-Making Using Design Methods

The students were prompted to reflect on the various design methods they used within the project, both formal and informal, and how these shaped the design decisions made. Many students reflected that seemingly objective deign methods were not objective at all, but that they served the purpose of promoting debate, structuring the solution space and facilitating the design process:

> This systematic approach to down selection [MCDA/pairwise] was performed to highlight the key solutions that seemed to fulfil the criteria, however, it is important to note that this method does not result in the identification of the perfect solution. Solutions that score similarly cannot be ranked by score alone as the results can be very sensitive to slight changes in the weighting of the criteria. If the selected criteria do not reflect the requirements of the project, then undesirable solutions will be presented in a desirable manner. Despite this, MCDA was a useful tool to narrow the potential solutions and promoted further discussion.

> During the down-selection process, constant group evaluation of conceptual design through debate and deliberation was crucial to selecting the final design. Whilst the use of Multi Criteria Decision Analysis (MCDA) charts for each sub-system was effective at providing either the best single or two selections, it is now clear there was a disadvantage of this that was not obvious at the time, in that the weightings of each criteria were vulnerable to the personal bias of the group. [...] This layer of bias is also apparent through the use of the pairwise comparison that dictated the weightings of the MCDA categories. The reason this was not identified as a possible issue may be down to the fact that because the down-selection methods being used appeared systematic and thorough, it gave us no cause to anticipate prejudice creeping into our designs. In reality, this had already occurred earlier on in the design process. [...] The very essence of group discussions means that final decisions are inherently predisposed to the use of personal experience and judgement to decide on a final design.

> Due to the subjective nature of pairwise comparisons it was essential the choices were based on research; in cases where factors were related, such as durability and reliability, it was important to merge them together to prevent bias. This led to heated debates within the team. This itself was a useful tool in exposing the subjectivity in the design process and thus led to the incorporation of a sensitivity scale (different pairwise weightings) to offset this- thus

trying to make the process more reliable. Sometimes, with an MCDA it was found people biased their scores to ensure a certain proposal came out on top.

Students frequently reflected on the importance of team discussion as an informal design method, and that discussion can be influenced by personality and disagreement:

> In some instances, team discussion seemed the only way to resolve differences of opinion about certain aspects of the design. Although discussion is usually an effective platform for communication, the possession of strong views by individuals can lead to an argument. It can be difficult to make common-sense decisions when individuals disagree on what exactly falls under common-sense.

> Emotions and attachment to initial designs led to the neglect of logical conclusions in some instances and fixation in design solutions. Fundamental differences in opinion, resulted in a negativity in the group and thus productivity and communication dropping. It is important to use dis/advantage tables to compare the relative merits of a design and ensuring the decision process remains logical, rather than engaging in emotionally driven arguments.

Several students also reflected on the interplay between informal design methods, such as group discussion and common-sense decisions, and more formal design methods such as brainstorming, pairwise comparison, and multi-criteria decision analysis (MCDA) when making design decisions:

> Throughout the process, we seem to have naturally interspersed formal methods with informal, for instance using a mind map to come up with initial ideas, then screening using common-sense decisions. This helps keep the project focused, as informal methods and debates are better connected with the specific constraints of a project, whilst formal methods provide a better structure to analyse the costs and benefits of designs that appear to perform similarly.

> In the cases were discussions did not have a design method associated with them, they tended to get wayward, and away from the key discussion points. They also tended to be fairly roundabout, with a lot less practical thought, and more assumptions with little evidence. This could have been solved by having a "chair", or designated team leader, however this was less in-keeping with the team dynamic. Using design methods to structure discussions offered a much gentler and natural was of keeping the discussions focused and on topic.

> Initially, the main methods involved were brainstorming/sketching, these were utilised to generate many concepts rapidly. It was quickly discovered that formal methods such as these were greatly enhanced if accompanied by an informal method. For example, after a period of brain sketching, coming together as a collective to discuss ideas aided in development of the concepts which greatly progressed preliminary designs. Comparing the proposed designs proved to be difficult. Therefore, structured formal methods were used to help ease this process. The MCDA coupled with a pairwise comparison was used to great effect for this. This was successful as it enabled each aspect of the design to be quantified, allowing for direct comparison. Again, this was followed with a group discussion, allowing for the results to be sanity checked (which was required). Evidentially, problems that arose in the design section were initially solved via the use of a formal design tool followed by discussion.

4.3   Identifying Themes and a Central Paradox

The students were prompted to reflect on any conflicting requirements within the project and to identify a 'central paradox' of the design situation. They were also prompted to reflect on any key project themes that emerged and how they used these themes to resolve conflicting requirements. Most of the students reflected success-fully on conflicting requirements within the project, and were able to identify an appropriate key conflict that dominated (the central paradox):

> Because the tasks the robot had to complete were so complicated, the robot itself had to be complex. Because the robot has to cope in a disaster situation it has to be simple and robust. Because the robot is complex it is likely to break in a disaster situation as it is not robust enough.

> Because of the existence of debris in disaster area it is essential that the robot must be able to fit into restricted spaces in order to reach critical areas of the plant. Because the robot must fit into a restricted space, it may not be able to contain the necessary hardware for completing functions stipulated in the brief. Because the robot may not be able to perform the specified functions, it may not fulfil the brief of closing the valve and stabilising the disaster zone.

> Because the robot should be as versatile as possible, this increases the size of the components and overall size of the robot. Because the brief has size constraints on the robot, each individual function was restricted in its capabilities due to space constraints.

> Because the robot must solve a complex task it must itself be a complex design. Because it is a complex design it will be difficult/expensive to manufacture. Because it is in an industrial disaster and the environment will be hazardous the design should be as simple as possible.

Many students were also able to reflect on the connection between a central paradox, project themes, and a new frame:

> The central paradox of this design is that these robots must be small to access areas humans cannot, however as humans cannot access the area they often can't see the conditions there and therefore the robot must be versatile enough to cope with many different situations. Therefore, two key themes in this project are then compactability and versatility. [...] The project frame became adaptive robotics, i.e. what if instead of needing one robot that had to be able to react and work in any environment, you had a basic robot that could be sold with add-on packages, such as the tracked flippers, and different sensors that could be used depending on what is known of the factory environment.

> The central paradox was identified through the analysis of the project brief supplied by the client. The group interpreted this paradox as follows: Because the robot must meet the brief, it must fit within a fixed envelope and be robust. Because the robot must be able to reach objects outside the range of the envelope, it must have moving parts that can extend and retract. Because the robot must have moving parts, the robot is less robust. Key themes were identified through further research in conjunction with the brief, these were interpreted as challenges the robot would need to overcome, including: traversing obstacles, manipulation of valves varying in diameter and shape, placement of a sensor, coping with extreme environments, remote control from a safe distance, operational for long periods of time. By creating a frame for the solution, the group decided that the same subsystem would be used to both place the sensor and manipulate the valve, thus reducing the number of moving parts and increasing the robustness of the robot. This was a successful solution, as it decreased the complexity as well as completing the proposed task.

Throughout the project, several key themes emerged that held a significance over the design. Firstly, the theme of 'Extension' presented itself as despite the various technical features of the robotic device, the primary objective is to act as an extension of the user and allow them to act remotely as they otherwise would. This means that the design must be intuitive and easy to control. If the design is versatile, it will provide the user with the most options to deal with any situation that may occur, and it will not inhibit their ability to deal with the issue in an effective manner. This theme was deduced from development of the requirement specification through written and verbal sources. Its importance was highlighted by the critical analysis of the criteria selected to down select the different ideas. The second theme to be identified was 'Space', which referenced the idea that all the components needed to be efficient and fit into a small and compact space. It emerged during discussion about the different methods of performing the required functions from the functional analysis that some elements of the design were more 'valuable' and could have a greater theoretical space set aside whereas some would physically require more space being allocated to them. The central paradox of the design problem was ultimately related to the space constraints: Because the robotic device needs to be an extension of the user, it requires certain elements to achieve this. Because all these elements are needed, there is a constraint on the space available. Because the space constraints for different elements of the design are unknown, these elements cannot be designed or sourced. Because the elements haven't been designed or sourced, the robotic device cannot be an extension of the user. The paradox was resolved by framing the design situation as 'versatility efficiency', where the more versatile and space efficient an element, the higher priority it was given, and this created an order to take the elements further into detailed design. This allowed the device to be an extension of the user whilst meeting the specification.

A few students also reflected that the brief itself was over-constrained in defining the form of the solution as an 'autonomous robotic device':

I believe the frame could have been made broader from 'how can a robot perform the necessary functions' to 'how can a disaster-affected zone be accessed and made safe'. This opens up the solution to the problem coming from sources other than a robot. For example, it may be determined that a protective suit should be developed.

The key type of fixation experienced during the design process was solution fixation. This was due to the fact that the brief was already bounded to the production of a robot of certain dimensions with specific requirements. As very similar products are already in existence for the same and similar applications, these were looked at for possible solutions that could be used to meet the requirements of the specification. Although it meant creativity was stemmed, it increased reliability of producing a robot that met the specifications.

It is clear that the boundaries defined in the design brief created a solution fixation, so impacted the approach taken towards concept generation and therefore the final product produced.

4.4 Developing and Testing Concepts

The students were asked to describe and reflect upon the approaches and tools they used to develop and test their concepts. Most students were able to give a good recount of the ways in which they developed and tested their design concepts. They described how they gathered further information through research, modelling, and by doing calculations. Many students reflected that prototyping and CAD modelling were effective for detailing and testing ideas, and that even sketching and group discussion served this purpose. They reflected that these activities were often about gaining tacit knowledge, that they revealed issues (and triggered learning) that were not initially apparent and not revealed through calculations alone:

> The next step that was taken was to prototype the final design. This involved both the creation of a CAD model and a physical prototype. These methods of development were the most useful and produced the largest improvements to the final design. The process was a completely constructivist way of solving the problem. Each component of the overall design went through many stages of development (e.g. the chassis was changed and reprinted 5 times). At each stage of the development of the prototype parts, there would be a group discussion about what worked well, and the changes needed to improve it. The part would then be redesigned with the slight changes in an iterative process until all the specification demands were met and the group agreed that it was complete. These changes could then also be added to the CAD model at a higher level of manufacturing quality and precision. The overall prototype was then tested in a realistic scenario to assess the its ability to perform the specified tasks. Even the final test highlighted many areas where the prototype could be improved and were changed in the final design. This method of development was by far the most intuitive as the design could be handled and tested in the similar manner to its intended final use which meant that issues could be easily identified.

> The majority of initial concepts were developed through team discussion and sketching. This process was very subjective, but it was able to quickly highlight a broad range of ideas which would have been prohibitive to fully develop individually. Prior to down selection, suitable ideas were researched in order to evaluate them. Upon selecting the final concept, further proving calculations and CAD models were generated, allowing the group to develop a robust solution to the problem presented. Iterative designs were modelled in order to decrease weight and increase strength as well as improving the overall performance of the robot leading to the final design. As well as using CAD to visualise the design and determine dimensions, a working prototype was manufactured and tested to highlight areas for improvement. For example, the chassis was initially too weak to support the moment applied by the wheel axles, leading to an improved mounting method for the final design.

> Designs taken forward for further investigation were developed via the following stages: detailed sketches, research, calculations, CAD/prototyping. When a concept looked promising, the first stage of development required a more detailed sketch; outlining the finer details, this weeded out some of the designs that appeared to be less plausible. [...] Creating the CAD for the final design and the manufacturing of the prototype were invaluable for the development of the design. As fine features that were missed from the previous development stages were incorporated. For example, the prototype grounded due to a design flaw, therefore a vital alteration was made to the design that could have possibly been missed. This development process lead to a strong final design.

> Initial concepts were developed through many means including further research, sketches and CAD which proved useful in visualising the concept and was an effective tool to develop the design. Where appropriate, calculations were performed to test the feasibility of certain aspects of the design and this ruled out some concepts. Some aspects of the control were tested on a small-scale articulated arm to inform the design process and to create a smoother transition to a full-scale prototype. After evidence of the concept's suitability was collected, the final decision was the product of team discussion. Despite producing a suitable concept, this process does not seem objective as individuals were more willing to do further research and calculations about concepts that they supported.

The development and testing of design concepts was a key learning point for the group. It was quickly discovered that implementing detailed design was not possible before large assumptions were tested by rough calculations. For instance, we required a calculation for the worm gear and motor for pivoting the linear actuator using estimated masses before being able to proceed with it as a viable solution. Most of the initial development was performed through group discussions and research. Once a rough outline of the design was determined, the general method of testing design concepts was through prototyping. This initially stemmed from not using CAD to determine the design with so many unknowns being present. There were both advantages and disadvantages to this way of testing our concept. It allowed us to gain a lot of advice from expert machinists leading to a design optimised for manufacture. It also allowed for a great deal of flexibility in the design as different elements of the design changed. However, certain parts of the manufacturing process could have been sped up and a more elegant and rigorous solution could have been achieved if a CAD model was being used as a reference.

Research was combined with detailed sketches that forced us to spend time thinking about how the specific mechanisms might operate the moving parts in the system. For example, how the gripping mechanism might rotate 360 degrees or how the drive-train might turn on the spot, requiring more in-depth thought about the position and number of motors required and where they might sit. [...] by going through the detailed sketching process, it provoked this thought process that molded and refined ideas into developed concepts. These sketches were supported throughout with the use of key calculations that quickly eliminated unfeasible options.

Initial ideas were developed relatively slowly, often with an early common-sense-based discussion about how likely the design was to succeed. Once this was completed, a select few designs were chosen for further development. This was done through sketches and further research, rather than CAD as the focus was on developing the idea, and CAD would have taken more time, and required focus on the details, such as manufacture rather than the overall design. Although no detailed calculations were done, as parts could be chosen to the requirements, some consideration was given to weights, torques required and general viability. Often this was very subjective, as it was based on what people thought, not on evidence. This process could have been made quicker with slightly more thought-out calculations to give more empirical bases to discussion. Ideally, rough prototyping could have been carried out on key ideas much earlier in this stage to reduce speculation, and offer more objectivity, when it was eventually carried out it proved the viability of designs and immediately changed the discussion from subjective opinions to objective facts. This made designs much easier, but also meant they were far more likely to be correct.

Although CAD was helpful when ensuring that the subsystem fitted together as a whole, it was often unhelpful when designing components. The CAD model made it very easy to lose a sense of scale meaning that in initial components were largely too weak, but appeared to be reasonably sized. This was coupled to another problem, that our calculations often lacked accuracy. Although the methods were accurate there were too many variables to give useful answers. This meant during the prototyping often calculations gave an accurate order of magnitude, but then it was down to testing to make parts functional, such as increasing gear ratios on the lift. This was solved through testing and improving designs, making the prototyping invaluable.

4.5 Writing the Design Report

Students were asked to reflect on how and why they chose to write their final design report in the way that they did, and if this 'story' of the design project was different from the chronological reality of their experience. Many students maintained that their design report was a true and accurate account of reality, albeit a summarised one. Some students, however, reflected that the narrative of the report was a construction for the benefit of the client, and that reality of the project was quite different:

> The choice of structure and content of the report was chosen to highlight the methodology of the design process and to justify the design decisions taken. To that extent, a great deal of figures were used to give an insight into the design process. As a consequence of this, much of the proving calculations, costing and research carried out has been placed in the appendices as it was thought of as an essential supportive role in the "narrative" of the report but not a necessary to understand the choices made. The report differs in its structures to the chronology of the project in that it shows a more sequential design process. Many of the stages of the design process were occurring simultaneously. This allowed for greater flexibility in the design and lead to a more developed design, although the constant revisions were time consuming. In addition, the categorisation of the report implies that many of the design decisions were made in isolation. The decisions taken in the design development stage were constantly influencing each other. Finally, the layout of the report indicates a linear approach to the project when the approach was far more iterative, with greater understanding of the design paradox leading to changes in the design that influences subsequent alterations made later.

> The structure of our design report portrays a very linear development of the robot design from start to completion, giving the impression that all stages lead naturally from one to the next with minimal deviation. However, during the actual design process, a number of iterations of the design were required, with several key proportional changes necessary to allow all parts to fit within the confines of the robot. The design of the gripper was completed quickly and required very little alteration through the rest of the project (the only physical change was to the position of the mounting hole), other sections though, such as the arm and tracks continued to develop alongside the construction of the prototype as it became clear that the initial design had not provided sufficient provision for tensioning of the tracks or drive belts. The initial design also failed to allow for the pathing of the drive belts when the arm was folded down, and so the width of the upper arm had to be modified to allow for this error. We settled on the structure of linear development for the report because, although not entirely factually accurate, it made the adjustments made to the design much easier to follow for the reader.

> This design report was written for the client, therefore it was essential to focus on showcasing the design clearly and accurately [...] The main difference in the order of the report to the actual design process was the large influence that making the prototype had on the final design, as some problems with the design only became apparent when trying to realise them physically. Additionally, some component dimensions, for example those of the gripper and the chassis, were changed upon the final assembly of the design in CAD. This was due to their design not being considered as part of the whole unit, so were optimised for individual performance. These had to be altered and in some places compromised to achieve a cohesive end product. This is not apparent in the report, which shows the parts having originally been designed as they are in the final assembly. Some iterations of the design process were not mentioned in the final review. These tended to not have any impact on the development of the overall design, therefore would not have contributed to the progress of the report.

The final design report was structured in such a way that outlined the key decisions and justifications for design decisions whilst ensuring that one section would lead into another. It was important to introduce certain topics in a select order to make sure the reader could follow the steps taken between them. In practice, the structure and content differ from the chronological reality of the project as much of the final design and prototyping was done in parallel, both feeding back into each other. Research was conducted throughout the project and some processes were revisited when the concepts were found to be unsuitable.

It could be argued that the process we took in reality was not similar to the double diamond process we laid out in the report since many aspects had to be "re-expanded" after realising that the chosen route was not suitable. For example, after researching ways to attach the motors to the chassis in the prototype and then selecting one option which was produced, it was realised that this option was not in fact suitable. Therefore, the design had to be re-looked at and developed.

Whilst a final evaluation was conducted, in reality evaluations were carried out throughout the process. In the report the prototype was used as a method of evaluating the final design, however in actual the final design was not completely completed until after manufacturing the prototype. Things learnt from manufacturing the prototype influenced many of the final concept refinements, for example the gripper design used on the prototype was not very effective and was changed for the final design.

A few students made the astute observation that the 'client' in this project was in fact the trio of academic staff who were assessing the submitted work, and that this influenced the narrative of the final design report:

The structure of the report reflects the mark scheme for a student project over the requirements of the client. The report would differ in that it would be much more focused on the actual design, manufacturing and proof of functionality of the prototype rather than the route from concept to the final design. As a client, they would expect to know more about the final product. This could include feasibility of manufacture and analysis of price in batches, detailed drawings of bespoke parts and reliability test results rather than the weighting criteria for different concepts.

The choice of structure and content of the report was based largely on previous years and assessment of what needed to be mentioned by our Professor, who has overseen many such reports over the years. The neatly structured report doesn't accurately reflect how messy and confusing much of the initial concept generation and down selection was. As mentioned previously this processed included much back-and-forth discussion and changing our minds rather than the neat function chart and description of this process in the report. In addition, several choices such as the motors for the dual intake tracks, were chosen based on the perceived strength of motors we borrowed from friends and could test ourselves, rather than calculation. This proved far more effective because there was an overwhelming choice of motors available online. We chose to leave these details out as they proved informal in written language and the calculation section demonstrates that the motors fulfil their function.

The design report was written to describe the process and result of designing a relief robot for disaster areas. Given the client for the task as an assessed student project was the assessor, the report was written to explain the final product from a designer's perspective.

4.6 Summary Discussion

All 31 students submitted a summary of their reflective learning journal and successfully completed these according to the constructivist inquiry framework set out in the guidance notes. All students demonstrated good engagement with learning, a prerequisite to meaningful reflection. This may be in part because the journals were assessed, providing a strong incentive to engage, and because the intake for this particular degree programme tend to be high achievers. It may also be partly due to the structured nature of the journals and their link back to specific lecture material, giving students a reference for their reflections and a foothold to get them started. The general quality of these representations of learning was high, indicating that for the majority of students deep learning has occurred. Key philosophy of design concepts introduced in the lectures, such as; fixation, framing, problem/solution 'spaces', design abductive reasoning, and constructivist/positivist learning paradigms were used to support reflection in 23 of the 31 submitted journals, even though these particular concepts were not explicitly mentioned in the journal guidance notes. Arguably the most challenging element of the constructivist framework was identifying themes and a central paradox, as this was a new experience for these students. 24 of the 31 students were able to identify appropriate themes and a central paradox for the design problem (with the remaining 7 students attempting to, but falling wide of the mark).

As a minimum all students were able to give a good recount of their project work, and to do so using the requested structure. So, it can be inferred that all students will have deepened their experiential learning associated with the project to some degree. For some, this may have involved upgrading their learning from surface to deep learning. The lowest score achieved was 58% and the highest 86%. The average score was 70%, with 19 of the 31 students achieving this score or higher. In order to achieve a score of 70% or more students needed to demonstrate not only engagement with learning, but also good or high quality of representation of learning, *and* meaningful insights into their own design activity in relation to philosophy of design materials of learning. These results indicate, for a majority of the students, a critical engagement with the new materials of learning relating to philosophy of design, and their use in aiding reflection on personal design activity. This implies a mutual accommodation of the cognitive structure with the new materials of learning, apparently leading to meaningful reflection at a high level of abstraction. A tentative conclusion, therefore, is that this approach to supporting group PBL design experiences in engineering design education appears to have had a positive impact in terms of widening the students' worldview on the nature of knowledge and its acquisition, and in deepening their reflection on experiential learning of creative design. Philosophy of design appears to give the students a language and a conceptual structure with which to reflect on personal and group design activity in constructivist terms, and the reflective learning journals provide an appropriate means to externalise and enhance this reflection through the representation of learning.

5 Conclusion

The value of using philosophy of design teaching, and reflective learning journals structured using a constructivist inquiry framework, as the basis for pedagogical aids to reflection in group PBL design courses is fourfold. First, constructivist inquiry makes sense of design as a highly situated activity. It embraces complexity, it is not prescriptive, or linear, and accounts for the uniqueness of every design situation. It includes the designer themselves as part of the design situation and acknowledges that the experience of designing changes the designer. Therefore, there are positive implications for ethics and reflexivity in design education. Constructivist inquiry, within a complexity paradigm, acknowledges the influence of the self in design and the need to be reflexive. The designer has a responsibility to users, the wider population, society and the environment. Second, philosophy of design, complexity and constructivist inquiry are established areas of academic scholarship. There is a wealth of published literature available, and an established core of knowledge. Thirdly, this approach to enabling reflection in PBL design courses should be straight forward and practical to implement, even in a saturated curriculum. Reflective learning journals are a well-established method of encouraging reflection, and philosophy of design may be taught effectively through conventional lecture-based methods. Finally, and perhaps most crucially, philosophy of design has the potential to resolve the paradox at the very heart of engineering design education, the clash of constructivist and positivist paradigms of learning. Complexity as a meta-paradigm is compatible with both constructivism and positivism. Students can become aware of, and comfortable in, multiple paradigms of learning. By emancipating students from a narrow world view about the nature of learning and knowledge, students should be enabled to reflect at a higher level of abstraction, allowing them to make deeper sense and meaning of their own experience during group PBL design courses, and gain the necessary tacit skills and knowledge of creative design.

References

Adams R, Aleong R, Goldstein M, Solis F (2018) Rendering a multi-dimensional problem space as an unfolding collaborative inquiry process. Des Stud

Atman C, Eris O, McDonnell J, Cardella M, Borgford-Parnell J (2015) Chapter 11—engineering design education. In: Cambridge handbook of engineering education research, pp 201–226. Cambridge University Press

Ausubel DP, Robinson FG (1969) School learning: an introduction to educational psychology. Holt, Rinehart and Winston, New York

Boyd EM, Fales AW (1983) Reflective learning: key to learning from experience. J Humanist Psychol 23:99–117

Cambridge Dictionary (2015) Cambridge dictionaries online

Chappell A (2006) Using the 'grieving' process and learning journals to evaluate students' responses to problem-based learning in an undergraduate geography curriculum. J Geogr High Educ 30(1):15–31

Crawley EF, Malmqvist J, Lucas WA, Brodeur DR (2011) The CDIO syllabus v2. 0. An updated statement of goals for engineering education. In: Proceedings of 7th international CDIO conference, Copenhagen, Denmark

Cross N (2007) From a design science to a design discipline: understanding designerly ways of knowing and thinking. In: Design research now, pp 41–54. Birkhäuser Basel

Dorst K (2006) Design problems and design paradoxes. Des Issues 22(3):4–17

Dorst K (2015) Frame innovation: create new thinking by design. MIT Press

Downey G, Lucena JUAN (2003) When students resist: ethnography of a senior design experience in engineering education. Int J Eng Educ 19(1):168–176

Dym CL, Agogino AM, Eris O, Frey DD, Leifer LJ (2005) Engineering design thinking, teaching, and learning. J Eng Educ 94(1):103–120

Eisner EW (1993) Forms of understanding and the future of educational research. Educ Res 22(7):5–11

English MC, Kitsantas A (2013) Supporting student self-regulated learning in problem-and project-based learning. Interdiscip J Prob Based Learn 7(2):6

Edström K, Kolmos A (2014) PBL and CDIO: complementary models for engineering education development. Eur J Eng Educ 39(5):539–555

Findeli A (2001) Rethinking design education for the 21st century: theoretical, methodological, and ethical discussion. Des Issues 17(1):5–17

Galle P (2002) Philosophy of design: an editorial introduction. Des Stud 23(3):211–218

Holland D (2014) Process, precedent and community: new learning environments for engineering design. PhD thesis, University of Dublin

Holland D, Walsh C, Bennett GJ (2012) Troublesome knowledge in engineering design courses. In: 6th annual conference of the national academy for the integration of research, teaching and learning, and the 4th biennial threshold concepts conference

Illeris K (2007) How we learn: learning and non-learning in school and beyond. Routledge

Jaskiewicz T, van der Helm A (2017) Progress cards as a tool for supporting reflection, management and analysis of design studio processes. In: DS 88: proceedings of the 19th international conference on engineering and product design education (E&PDE17), Oslo, Norway, 7 & 8 Sept 2017, pp 008–013

Johnson DW, Johnson RT, Smith KA (1998) Active learning: cooperation in the college classroom. Interaction Book Company, 7208 Cornelia Drive, Edina, MN 55435

Kolb D (1984) Experiential learning as the science of learning and development. Englewood Cliffs NPH:editor1984

McDonnell J, Lloyd P, Valkenburg RC (2004) Developing design expertise through the construction of video stories. Des Stud 25(5):509–525

Moon JA (2013) Reflection in learning and professional development: theory and practice. Routledge

Morgan T (2017) Constructivism, complexity, and design: reflecting on group project design behaviour in engineering design education. PhD thesis, University of Bristol

Morgan T, McMahon C (2017) Understanding group design behaviour in engineering design education. In: DS 88: proceedings of the 19th international conference on engineering and product design education (E&PDE17), Oslo, Norway, 7 & 8 Sept 2017, pp 056–061

Morgan T, Tryfonas T (2011) Adoption of a systematic design process: a study of cognitive and social influences on design. In: DS 68–7: proceedings of the 18th international conference on engineering design (ICED11), vol 7

Savin-Baden M (2003) Facilitating problem-based learning. McGraw-Hill Education (UK)

Schön DA (1984) The reflective practitioner: how professionals think in action. Basic Books

Sheppard SD, Macatangay K, Colby A, Sullivan WM (2008) Educating engineers: designing for the future of the field, vol 2. Jossey-Bass

Simons H (2009) Case study research in practice. SAGE publications

Thorpe K (2004) Reflective learning journals: from concept to practice. Reflective Pract 5(3):327–343

Van Manen M (1991) The tact of teaching: the meaning of pedagogical thoughtfulness. New York Press

Wang T (2010) A new paradigm for design studio education. Int J Art Des Educ 29(2):173–183

Contributions of Academic Makerspaces to Design Education

Vincent Wilczynski

Abstract This chapter provides an overview of academic makerspaces—locations on college campuses for teams of students to design, fabricate, assemble, and test components and complete systems—and explores the impact of these spaces on design education. The spaces are unique in that they are generally open access, much like libraries and other university-wide service-providing facilities, and available for any use. As facilities that promote design activities, academic makers are having a positive effect enhancing design skills in engineering and other disciplines.

Keywords Academic makerspaces · Engineering design · Community · Culture

1 Introduction to Academic Makerspaces

Academic makerspaces—sites where cross sections of a university's creative community assemble to learn, build, and share—are developing as a critical component of campus-based design and innovation ecosystems. Typically, academic makerspaces provide students, faculty, and staff with facilities, resources, and programs for curricular, extracurricular, and personal use (Halverson and Sheridan 2014). With this openness of purpose, academic makerspaces have become popular sites for designers to create concepts, prototypes, and products. Because of this mixed-use within the physical space, the space (through its users) promotes design learning as individuals from unique backgrounds share knowledge and experiences within a common location, thereby promoting peer-to-peer learning. Understanding how the physical characteristics of an academic makerspace promote behavior patterns (referred to as attributes of academic makerspaces) is fundamental for understanding the contributions of academic makerspaces to design education.

For example, one characteristic of academic makerspaces is that they are open in layout, accessibility, and use as illustrated in Fig. 1. Such openness encourages individuals working in academic makerspaces to collaborate, thereby creating a com-

V. Wilczynski (✉)
Yale School of Engineering & Applied Science, 17 Hillhouse Ave., New Haven CT 06511, USA
e-mail: Vincent.Wilczynski@Yale.edu

© Springer Nature Switzerland AG 2019
D. Schaefer et al. (eds.), *Design Education Today*,
https://doi.org/10.1007/978-3-030-17134-6_5

Fig. 1 Open layout for design, fabrication and, assembly within an academic makerspace

munity of users that also promotes design learning. The diversity of users within an academic makerspace creates an environment that supports creative problem-solving.

The most important attributes of academic makerspaces are the supportive community and collaborative culture that form within the space. The resulting making-community and making-culture (i.e., the people and their actions) differentiate these spaces from other learning, fabrication, and assembly locations on college campuses. These characteristics and attributes combine to create environments where design education can flourish because of the accessibility to equipment, training, and resources.

Academic makerspaces first appeared on college campuses in 2010, motivated by grass roots, community-based initiatives where members of the public could meet and have access to traditional and modern digital tools for designing, fabricating, assembling, and testing engineered components and systems. While colleges and universities have always provided design and manufacturing facilities to augment engineering education and research endeavors, the broad access and community-focused aspects of academic makerspaces define their uniqueness when compared to traditional design studios, machine shops, and teaching labs.

This chapter explores the impact on design education within academic makerspaces. The components of academic makerspaces are first detailed as a backdrop to understand their contributions to design education. The differentiation of academic makerspaces is explored to understand how these spaces are unique among institutions of higher education. The history of the making movement and the origins of makerspaces is provided to illustrate how this learning construct was established on college campuses.

Academic makerspace examples from six institutions are presented to highlight the common and unique aspects of these facilities. Makerspaces at Imperial College, University of California—Berkeley, Technical University of Munich, Case Western Reserve University, Seoul National University, and Yale University are reviewed. These profiles examine the space, equipment, staff, and programs at each institution to illustrate how academic makerspaces support and advance design education. Best practices from these institutions are presented and serve as a model for planning and operating academic makerspaces.

Against this backdrop, the contributions of academic makerspaces to design education are then reviewed. Given that these spaces are based on the practice of design, their presence on college campuses increases the number of participants in design activities. These design enthusiasts build skills through hands-on, open-ended problem-solving in academic courses and in extracurricular activities. As such, the presence of academic makerspaces raises awareness and increases the understanding of the value of engineering design while drawing more individuals into design activities.

The field of academic makerspaces developed over the last decade, with its history and impact noted by many authors, including this chapter's author. These ten years of experience establish a foundation for the influence of academic makerspaces in engineering education and beyond, including the impact of these spaces in liberal arts institutions as well as catalyzing student-led entrepreneurship activities. This chapter draws upon that history and summarizes the influence of academic makerspaces as a prelude for describing the contributions of these spaces to engineering design.

2 Components of Academic Makerspaces

Academic makerspaces are the combination of *infrastructure*, *programs*, and *community* that fosters project-oriented, collaborative, and experiential learning through the concept of making (Wilczynski 2015; Andrews and Roberts 2017). In this framework, the physical location, organizational elements, and users create a template for experiential learning. With this broader understanding that academic makerspaces are more than just the space they occupy, academic makerspaces can be characterized as environments that produce knowledge based on experiences. As such, the learning process within academic makerspaces portrayed in Fig. 2 aligns with the theory of constructivism and Confucian scholar Xunzi's synopsis "I hear, and I forget. I see, and I remember. I do, and I understand." (Piaget 1971). This section details the three essential elements of an academic makerspace.

Regarding *infrastructure*, three components are essential: space, equipment, and staff. Given that academic makerspaces are unique to each institution, with the structure of each space created to fulfill a specific local mission, there is no prescription

Fig. 2 Active learning is the defining attribute of an academic makerspace

for the required components of an academic makerspace. There are, however, facets common to most spaces including the fact that a makerspace is a physical location where individuals meet and work. The physical space includes areas where designs are created, components constructed, and individuals meet. In addition, space for support staff is essential within the footprint of an academic makerspace.

Fabrication areas in these spaces are locations where standing machinery and digital fabrication equipment are located, as well as hand tools, computer workstations, and electronics equipment to manufacture components. Large work areas are a standard component of makerspaces, and typically include sturdy and durable benches capable of handling heavy and repeated loads. Support areas such as staff offices, storage, and parts bins are other standard components of these spaces. It is not uncommon to find collaboration rooms within the spaces where teams can jointly design and discuss developments. To facilitate the design process, academic makerspaces include storage for projects as well as basic supplies such as wood/metal stock, fasteners, adhesives, electronic components/microprocessors, and other consumables. Depending on the local purpose of an academic makerspace, instruction areas may also be included in the facility. Finally, a location for the community to gather for presentations is also a common component of many spaces.

Open and flexible layouts are another feature of most spaces. Many include large, open work areas available on a first-come, first-served operating model. These are typically workbenches that a person would use for a specific work session, clearing their material when finished and leaving the space in a condition that another member could immediately use. Space flexibility enhances utilization within aca-

Fig. 3 Hand tools and digital manufacturing equipment are common instruments in academic makerspaces

demic makerspaces. Except for standing machinery, it is common for all furnishings to have wheels to increase the space's flexibility. Spaces that have an open design have clear sightlines, with this aspect fostering attributes related to community and collaboration as everyone can see each other's work.

As illustrated in Fig. 3, the manufacturing equipment in academic makerspaces often includes equipment commonly found in machine shops such as mills, lathes, routers, drill presses, power saws, and other large tools including thermoelectric vacuum formers and injection molding systems. Hand and battery-powered tools complement the standing machinery. It is common to have a high percentage of computer-controlled equipment, such as: laser cutters; 3D printers; vinyl cutters; computer numerical control (CNC) mills, lathes, and routers; plasma cutters; and water jet cutting equipment. Electronics workstations provide a location and tools for circuit design, soldering, and benchtop testing using oscilloscopes, function generators, and power supplies. Tools to measure, cut, and layout paper, cardstock, cloth, foam board, and other materials are also standard equipment. There is generally close correlation between the scheduled courses and events and the equipment housed within each academic makerspace (Freissnig et al. 2016).

To accommodate the diverse community of users and usages, the space's design and fabrication tools and equipment are generally varied and extend across many disciplines. Academic makerspaces do not normally specialize in a single design or manufacturing modality, but rather have a comprehensive inventory of tools and equipment that support a wide spectrum of users. For example, sewing machines are often housed in these spaces as the primary tool to join soft materials. Fume hoods are another common component of academic makerspaces as they provide safe environments for chemical use and painting. Also, some academic makerspaces include wet labs as areas to work with fluid systems, such as in Fig. 4. When academic

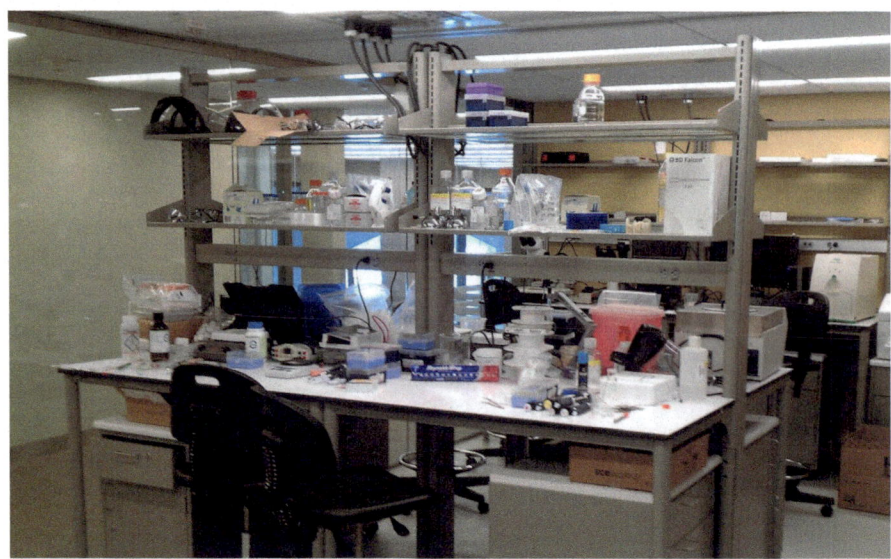

Fig. 4 Wet labs can be a component of academic makerspaces

makerspaces have a special purpose, such as biotechnology, equipment in the space would support that focus (Yao et al. 2017).

Staff members are an essential component of the infrastructure given the complexity of the equipment and operations within academic makerspaces. As these sites support both learning and creation, staff are needed to provide training, oversight, and guidance for the space's users. The staff may be comprised of educators, specialists, and technicians, as well as trained students, with the most common staffing model being a combination of professionals and students (Barrett et al. 2015).

Beyond training and maintaining equipment, staff members play an essential role in establishing and supporting the community aspect of an academic makerspace. A defining attribute is that they are first and foremost educators. This differentiation requires academic makerspace staff members to have strong critical thinking and excellent communication skills. Individuals with engaging personalities and high degrees of empathy are needed to create the welcoming and inviting environment that draws in new members while assisting (and retaining) experienced members.

Serving as a resource is a fundamental responsibility for academic makerspace staff. Given the range of tools, equipment, and processes within a space, combined with the unlimited range of potential needs arising from the diverse backgrounds of the members, academic makerspace staff need to be resourceful. It is also important for staff members to support peer-to-peer learning within the space as there is a limit on how much assistance an individual staff member can provide to a single member. However, the ability for members to learn from each other, with the staff facilitating and assisting with this process, is unbounded.

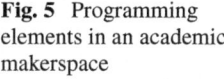

Fig. 5 Programming elements in an academic makerspace

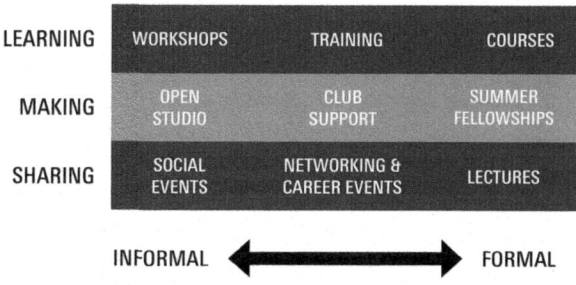

Academic makerspace staff members have many other responsibilities, including: orienting new users to the space; providing tool, equipment, and safety training and oversight; managing operations and logistics to support the space; guiding tours and public outreach activities; teaching courses and workshops; maintaining tools and equipment; establishing and disseminating best practices; measuring impact; and articulating the values of the space by recording and disseminating stories of user successes. This expansive list of skills and responsibilities is much more diverse than those associated with staff members employed in teaching labs and machine shops.

Programming is an important component of academic makerspaces, with the facility's staff overseeing the activities within each space. Programming is the collection of planned and supported activities within an academic makerspace that create opportunities to learn and develop. The programming aspect is a key differentiator from many other learning and working spaces on college and university campuses. The concept of programming is commonly applied by student service organizations such as faith communities and affinity groups. With programming, academic makerspaces become much more than physical locations as the programmed activities within the space connect and engage a community of like-minded individuals.

An approach to understanding the range of programs that might be offered in an academic makerspace is presented in Fig. 5. In this example, activities are organized into those where individuals learn, make, and share. These activities span from informal and loosely structured activities to more formal and tightly structured experiences. Key to each programming activity is a staff member, faculty member, or a member of the makerspace community who directs the activity.

The learn aspects of the presented programming model can be examined to illustrate the span of activities that occur in academic makerspaces. As the most informal activity, workshops are scheduled periods when members gain familiarity with technology such as microprocessors, digital electronics or computer-aided design. These workshops are short duration activities directed by a member of the academic makerspace community. Equipment training is a higher level of formality in this model with the scheduled training delivered by an academic makerspace staff member. This training focuses on teaching members how to use the tools and equipment within the academic makerspace and may be accompanied by proficiency testing. Course instruction delivered by a faculty member is the most formal learning activity within an academic makerspace with many courses combining theory with hands-on learning to complete an open-ended project.

Programming can also support the making dimensions of these spaces, with examples being staffing open studio work periods to provide domain expertise, facilitating student organizations working in the space, and offering summer fellowships to guide student teams through the product design process. Programming can help build community with social events (such as membership parties, movie-nights, and study breaks), career and networking forums, and lectures from industry leaders. A rich and diverse suite of programmed activities creates the vibrancy within and establishes the value of an academic makerspace.

The concept of *community* is essential to academic makerspaces. Individual members, including students, faculty, and staff, form the basis of the academic makerspace community who pursue academic, extracurricular, research-driven, or personal design activities within the space. It is common practice to provide the entire campus access to academic makerspaces, thereby facilitating interactions between diverse groups who are working near and with one another in these spaces. As a community, these users establish culture, build relationships, and create cultural norms (Ali 2016).

The word "members" designate those who use makerspaces, thereby reflecting this group's metaphoric ownership of the space, resources, and programs. As a member of a higher education academic makerspace, an individual is granted privileges to use the space and in turn accepts responsibilities associated with the space (such as enforcing safety). A training session for new members typically reviews the policies and procedures for the space, orients users to the facility, and explains safe operating practices to establish a common knowledge base for all.

As a community, members of the academic makerspace follow expected norms of behavior within the makerspace. These behaviors often include a willingness to share experiences and knowledge with each other as well as to contribute to each other's projects. Such behavioral traits enforce the concepts of teamwork and collaboration that are common in many academic makerspaces. High degrees of peer-to-peer interaction and a willingness of members to share their knowledge with others are commonly found in thriving making communities. Valued members of a makerspace community respect the work of others and recognize the responsibilities and privileges associated with access to such facilities. Academic makerspace communities enforce a culture for safety using training, implementing transparent standards, and empowering users to be responsible for each other (Simon and Haji 2016). A culture for safety is established when the concept of safety is viewed as an inherent component of the design and build process and when the community at large encourages and reinforces safe practices.

A diversity of experiences and backgrounds adds value to an academic makerspace community. The terms density and mixing can be used to promote diversity when applied to problem-solving (Farritor 2017). Density refers to the number of people working near each other within an academic makerspace without barriers between individual projects and people. Mixing refers to interactions (such as social events, pitch sessions, and simply speaking with others) where community members get to know one another. Given the diverse backgrounds of makerspace members and the collaborative culture, combined with the dense packing of individuals working in these spaces and that users know each other from planned mixing encounters, there is an increased probability for discovering innovative solutions within academic makerspaces.

Given this community component, it is important to recognize that academic makerspaces also fulfill social needs on campuses, providing a location and activities that draw in individuals based on common interests. Social programs allow members of the community to get to know each other and share ideas. Bridging the social and functional realms, project and problem pitch events are common methods to recruit problem-solvers, connect individuals, and form teams to investigate a specific cause.

3   Differentiation from Other Campus Resources

A few factors differentiate academic makerspace from traditional shops, project work areas, and teaching labs on college campuses. One view of academic makerspaces is that they are a combination of the normally separated areas for machining (i.e., machine spaces), building (i.e., project spaces), socializing (i.e., community spaces), and teaching (i.e., instruction spaces) (Culpepper 2016). This concept is illustrated in Fig. 6. Each academic makerspace is a combination of the machine, project, community, and instruction spaces that generally support engineering education and related activities. The degree to which each academic makerspace favors machining, building, socializing, and teaching must be aligned with the space's purpose. There is no single template of what an academic makerspace should be or contain, as each academic makerspace is uniquely configured to meet its institutional mission.

This mixed-use model establishes academic makerspaces as spaces that jointly support educational activities, research purposes, and personal pursuits. Subject to each space's purpose, the makerspace may support a mixture of credit-awarding academic courses, personal projects, student design organizations, research projects, and entrepreneurial activities. Few other academic facilities on college campuses

Fig. 6 Hybrid model of academic makerspaces meeting an institution's machine space (MS), project space (PS), community space (CS), and instruction space (IS) needs

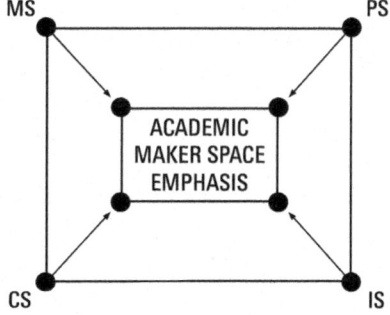

span and support such broad uses. Each personality of a space is unique and results from the space's equipment, programs, and members.

Academic makerspaces are student-focused, and this factor is an additional differentiator from other learning spaces. This differentiation is exemplified with an admittedly narrowed perspective that machine shops are primarily equipment-focused, project spaces are build-focused, educational labs are discipline-focused, and community spaces are event-focused. As a combination space, academic makerspaces address all these areas but do so from the perspective of the student. For example, equipment selection in academic makerspaces is based on student needs regarding safety, training, and use. As another example, access to equipment is best aligned with student schedules, with supervision provided to align with student availability outside of the normal class schedule.

This focus on students promotes a forum for individualized learning accomplished by groups of students. Learning communities and the resulting culture for learning that develops within makerspaces establishes an environment that promotes participatory methods to acquire knowledge (Thomas and Brown 2011). As a student-focused organization, students are empowered as not only members of an academic makerspace but also owners of the space. It is likely that this sense of ownership subtly shifts an individual's perspective, moving from a possible antagonistic "us and them" perspective to a more desirable cooperative "ours" perspective. Benefits from this sense of ownership result in many areas. As owners of the space, individuals help keep the space available for other users by clearing their work areas and returning tools to the appropriate locations at the end of a session. As owners, individuals not only take from the space, but also contribute to the space by assisting others and participating in community functions. Owners are engaged in their community. The concept of ownership is also associated with value. Members value their status in the academic makerspace community and the access to resources to pursue their own projects and interests.

A fundamental characteristic of academic makerspaces is openness, with this concept applicable to many components of each space. Many academic makerspaces are open to the entire campus community, with the members allowed to use the space for any design-centered purpose (Pernia-Espinoza et al. 2017). This open framework often sets the spaces apart from other campus facilities that have more restrictive parameters for access and use. As a result of such open access, academic makerspaces attract students from a wide number of academic majors, with the diversity of the membership contributing to the spaces' effectiveness as a tool to bring students together and advance innovation (Hynes and Hynes 2018; Aidala et al. 2017; Schön et al. 2014). Providing wide temporal access to the space is also important, with many spaces open around the clock. In these cases, access to specific equipment is usually restricted by policies and procedures that ensure user safety. These measures support the notion that academic makerspaces are open to all, for almost any use, at nearly all times.

The collaborative nature of members within an academic makerspace community is also an attribute that differentiates these facilities from others on campus. Community members readily share their knowledge and help others learn, thereby creating a participatory culture based on informal interactions (Roberts and Buckley 2016). The open architecture and visual porosity within an academic makerspace allow users to see all activities within the space, and this awareness promotes interactions and collaboration. The levels of collaboration can be facilitated through informal bench-side, peer-to-peer instruction or more formally in instructional workshops to help members gain knowledge and develop skills. It is important to note that it is the responsibility of the academic makerspace community to create and sustain this collaborative culture (Ali et al. 2016).

Another differentiating feature of academic makerspaces is that they are frequently used for informal learning that is primarily self-directed, social, and accomplished through projects (Jaskiewicz et al. 2018; Tan et al. 2016). Students, either as individuals or in teams, identify problems to solve, create, and test their ideas, and construct solutions. Use may be associated with a course or independent of a student's academic studies. The projects are varied, extending into the domain of student design organizations (such as teams that design and build hybrid vehicles, rockets, and genetic engineering products), entrepreneurial activities, personal design projects, or simply repairing broken objects (such as cell phones). Expertise is developed with the assistance of staff and makerspace members and augmented with self-directed actions based on discovery and practice.

In academic makerspaces, most users are electing to be there, as opposed to being required to be present for a class, lab, or term project. The fact that most users choose to be involved in academic makerspace activities is an important aspect of the community culture that forms around these spaces. As empowered users who pursue their own design interests, makerspace members achieve deeper learning experiences and higher degrees of satisfaction as compared to traditional learning environments (Baleshta et al. 2015).

4   History of Academic Makerspaces

The concepts of hands-on and problem-based learning, as well as laboratory courses, have always been fundamental to engineering design education. Academic makerspaces, however, provide a new framework for these learning environments with increased opportunities for students to use these facilities to guide their own learning and skill development.

Figure 7 illustrates the stages of U.S. engineering education and portrays the disciplinary emphasis as a pendulum swinging between making and calculating. Here, the pendulum starts with an emphasis on *"making"* in 1850 as the first degreed engineers focused on designing and building infrastructure, transportation, and communication solutions. The creation of the American Association for Engineering Education in 1893 amplified the analysis and scientific rigor associated with engineering education programs, moving the pendulum away from the *"engineering as making"* dominance and toward the *"engineering as calculating"* domain. Movement toward the *"calculating"* domain accelerated following World War II and into the Cold War

ENGINEERING EDUCATION TRENDS

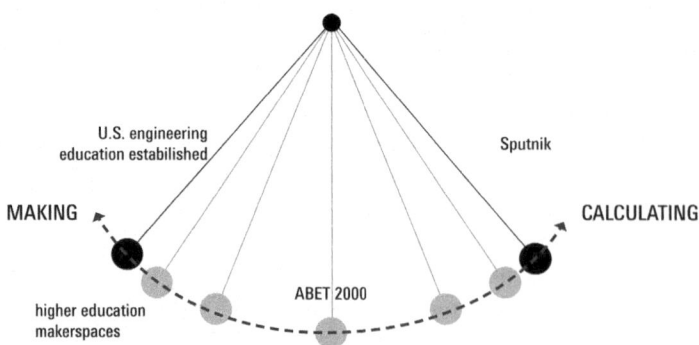

Fig. 7   The swinging pendulum of engineering education

period when science began to be emphasized as the key component of engineering education. Marked by the launch of Sputnik in 1957, a period of *"engineering as calculating"* dominated engineering education.

The role of design and creativity in the engineering profession was emphasized in the 1990s and clearly distinguished the fields of engineering from science. This focus produced an *"engineering as design"* period which was advanced by the Accreditation Board for Engineering and Technology with the publication of accreditation criteria in 2000 that stressed the need for engineering students to have design skills (as well as rigorous grounding in mathematics and the sciences). The revised criteria also promoted teamwork, communications, and other broad skills needed for engineering graduates to make immediate contributions in industry (ABET 1997). These changes helped move the engineering curriculum pendulum away from the *"calculating"* (i.e., analysis-based) dominance and toward the *"making"* domain. The arrival of academic makerspaces on college campuses helped continue this path of the engineering education curriculum pendulum to its present *"engineering as making"* dominance.

The arrival of makerspaces on college campuses was predated by the creation of makerspaces within local communities, a development spawned by two decades of advances in hardware and software products (Anderson 2012; Dougherty 2016; Wilczynski and Adrezin 2016). These products became widely available and moved from industrial settings and academic research labs into consumer spaces, thereby fueling the maker movement. The first advances were in the fields of computer-aided design (CAD), initially applied to electronics and then to three-dimensional (3D) components and systems. The first widely used, open source CAD system was the software product "Eagle" created in 1988 as a tool for printed circuit board design. While two-dimensional CAD systems appeared in the early 1980s, the field of widely available three-dimensional CAD advanced in 1997 with the release of "SolidWorks" for designing mechanical components.

Similar developments also occurred in this time frame for electronics as microprocessors became commonly used in many applications. The microprocessors, mounted on printed circuit boards, accepted digital and analog inputs, manipulated the measured signals, and delivered analog and digital outputs. This development allowed the tools to be incorporated in a variety of sensor-based monitoring and control applications. Their ease of use resulted from the fact that the chips could be readily connected to computers and effortlessly integrated with sensors and actuators. These products included the Parallax Basic Stamp (released in 1995) and the Arduino (released in 2005) microprocessors. The Raspberry Pi microprocessor (released in 2012) extended the utility of microprocessors by including many functions of a stand-alone computer. These products were supported with instructional documentation geared toward the first-time user. Also, advanced users capitalized on developing internet-based communication tools to freely share product information.

These software and microprocessor advancements were soon followed in the hardware domain with machines operating under computer numerical control (CNC). CNC technology led to machines for routing (ShopBot's release in 2000), mechanical cutting (Tormach's release in 2007), laser cutting (Epilog's release in 2008), and 3D printing (Makerbot's release in 2009). These commercial products were developed based on industry/research-grade technologies. They entered the marketplace, in part, as a response to the rise in the number of individuals involved in fabrication at small-scale manufacturing sites (such as in commercial/personal shops) and academic environments, as well as those pursuing personal projects. Like the progression of mainframe computers to personal computers to laptop computers, the software and hardware used to design/manufacture objects moved from tools only available to few (such as at an industrial work site) to many (at any site).

While these software and hardware tools became widely available in 1990–2010, forums to easily exchange information on a global basis also contributed to the creation of making-centered communities. The decade starting in 2000 is associated with a rise of internet-based user groups and blogs devoted to making. These discussion groups (which originally relied on text comments and still images) benefited greatly with the launch of YouTube in 2005, thereby facilitating the use of videos to share design and tool skills. The popular press also contributed to the wide distribution of making skills with the 2005 inaugural publication of "*Make:*" as a consumer magazine devoted to personal creation. According to its first banner-head, the magazine let readers develop "technology on your time," thereby becoming a popular periodical for a developing DIY (Do It Yourself) community.

This history highlights two unique and important components that accelerated the creation of makerspaces. Since 1990, the tools for design and fabrication became more available, affordable, and accessible to a larger number of users. Aided by advancements in computers and fabrication, equipment manufacturers began to offer precise and highly reliable automated machines with prices a tenth of previous versions of the equipment. Advances in online communication by users and manufactures (including product documentation, user-blogs, and training videos), combined with the creation of commercial publications (such as magazines and websites) established platforms for a growing population of users to become familiar, and later proficient, with new technologies.

Community-based facilities, called makerspaces, initially formed in the first decade of this century. These facilities allowed individuals to assemble and share resources, including equipment, knowledge, and skills. The framework for makerspaces was derived from similar loosely affiliated technology-based associations first formed around electronics and personal computers.

Some of the first organized efforts to introduce digital fabrication to wide audiences originated in the United States with the creation of small fabrication labs, termed Fab Labs, at the Massachusetts Institute of Technology (MIT) in 2001. These facilities empowered individuals and increased their creative capacity by providing access to digital fabrication technologies. Just as the personal computer provided computational power to all, Fab Labs provided personal manufacturing capabilities that allowed anyone an ability to design and create new products. In addition to housing machinery and computers, training became a critical component of Fab Labs as they were targeted to introduce, teach, and develop skills for individuals who were unfamiliar with these topics. Supported by academic and government partnerships, over 700 Fab Lab digital machine shops were created across the world in the last 15 years (Gershenfeld 2005).

Against this backdrop of increasingly available design and manufacturing tools and organizational models for engaging local technical communities, individuals with an interest in making began to self-organize. They pooled resources to acquire tools and equipment, creating both locations and organizations where people could access technology and learn new skills. Some of the more widely known organizations include Resistor in New York City, NY (founded in 2005), TechShop in Menlo Park, CA (founded in 2006), and Artisan's Asylum in Cambridge, MA (founded in 2009).

These models influenced grass-roots initiatives in many areas. Finances to equip, maintain, and operate each local makerspace were obtained in ways unique to each space, with many spaces charging members a monthly fee to access the facility. Cultural norms based on collaboration and community developed within these spaces. Members of the makerspace community not only shared resources (such as equipment, tools, electronics, and software), but also shared knowledge by developing and delivering training programs.

The increased affordability of and access to technology, combined with the availability of training tools, advanced the makerspace concept in colleges and universities. Makerspace models in local communities, including their cultural elements of collaboration and open access, accelerated the adoption of these spaces in academia. The arrival of makerspaces in higher education followed long-standing open studio and group-learning practices traditionally applied in architecture, art, and design programs.

The reception for academic makerspaces was facilitated by the hands-on instruction, computer programming, machine/electronics shop training, and project-based learning that has always been a staple of engineering education. Collectively, the adoption of makerspaces at institutions of higher education was catalyzed by the presence of previously existing and similarly serving design and learning spaces on most campuses. The uniqueness of academic makerspaces as a single site for making without disciplinary or end-use bounds differentiated these spaces from others and increased the speed of their adoption within higher education.

5 Case Studies

Six examples are examined to illustrate the scope of academic makerspaces and their range of purposes.

The Imperial College Advanced Hackspace provides a 2,000-member community from across the university access to equipment, events, and expertise related to creation. This academic makerspace is a network of six facilities across two college campuses that support electronic and digital manufacturing, biology-based exploration, as well as metal, plastics and woodworking capabilities. The bio-lab supports investigations in synthetic biology and molecular fabrication. Two faculty members serve as co-directors of the Hackspace from the departments of chemistry and engineering design. The co-director and three other faculty members comprise the Advanced Hackspace Board, with this group providing strategic oversight for the facilities. Seven staff members provide assistance to users and deliver programming. The facilities are generally open 13 hours each weekday and are not used for credit-awarding courses.

As campus-wide facilities, the Imperial College Advanced Hackspace supports education, research, entrepreneurial, and personal interest activities for university members. Members of the public, including industry professionals participate in this making-community by sponsoring challenges and posing questions for teams to solve. The space also supports a public outreach initiative to engage children in the design and creation process. Funding for entrepreneurial endeavors is provided without intellectual property or equity restrictions for both early stage and advanced ventures. The multidisciplinary purpose and varied backgrounds of members have advanced collaborative and community links across the university.

The Jacobs Institute for Design Innovation at the University of California, Berkeley serves as both an academic building (hosting courses from across the university) and as the home for one of UC Berkeley's makerspace communities. The 2,300 m^2 facility includes three teaching studios that service twenty courses each semester (and are open as drop-in space when not used for instruction). The Jacobs Institute had an immediate impact on the number of design-related courses as faculty members developed new courses and revamped existing courses to make use of the space and provide hands-on learning opportunities. These courses included ones in engineering as well as other disciplines such as art, architecture, biology, and management. The Jacobs Institute has also been a catalyst to partner the College of Engineering, College of Environmental Design, the College of Letters and Sciences, and the School of Business to offer a campus-wide undergraduate certificate in design innovation. Demonstrating a similar spirit of collaboration, the Jacobs Institute helped initiate a campus-wide "Fabrication Network" of spaces for prototyping and fabrication.

The facility is led by two faculty members who have the titles of "Director" and "Chief Learning Officer." Programming and operations are directed by four staff members and assisted by student employees. Five design specialists (some full-time and some part-time) provide guidance in eight workspaces at the institute: a metal shop, visualization lab, CAD/CAM lab, project lab, all-purpose makerspace, electronics lab, audiovisual projection lab, and advanced prototyping lab. Access to the workshops is provided to students who have registered, paid an annual fee, and completed a university safety training program. Members of the makerspace community are eligible to attend workshops to explore new skills and projects, schedule consultation time with the design specialists, and have access to a material store that offers common making materials at low costs. Members of the makerspace community are also provided evening and weekend access to the Jacobs Institute.

The Center for Innovative Business and Creation at TUM (Technical University of Munich) is a combination of spaces and services that support the development of technology businesses from ideation to their initial public offering. The *TUM MakerSpace* is one of five major components of the center, with others being an entrepreneurship research institute, start-up incubator, entrepreneurship education area, and a business accelerator. This publically accessible space caters to start-up companies while also supporting individual creators from the university and surrounding community. The 1,500 m^2 MakerSpace is a member of the EU-funded FabLabNet that provides prototyping platforms across Europe and creates a network of users to share experiences and best practices. Access to the facility and courses are fee-based. The MakerSpace staff also provides design and fabrication services, which are also fee-based.

The equipment inventory of the MakerSpace includes over 80 unique machines spanning from consumer to industrial grade equipment. For example, members have access to a 3D printer capable of printing shapes having a volume of 1 m^3. Other large-scale equipment includes automated milling and turning machines, laser sinter systems, water jet cutters, tempering/annealing furnaces, spray and powder coating booths, a curing oven, and welding facilities. The space is organized by tooling areas, with unique areas for welding, wood, metal, textiles, and electronics, in addition to open, multi-team project areas. In addition to supporting start-ups and individuals, the space also supports student team such as the TUM Hyperloop team and includes an instructional area to teach members how to use the space's equipment.

The mission of the *Sears think[box] at Case Western Reserve University* (CWRU) is to serve as a physical and cultural center that provides resources for collaboration, innovation, and creativity, with an emphasis on entrepreneurial growth. Two aspects of the Sears think[box] make it unique compared to other academic makerspaces. The facility is open to the entire university community as well as the surrounding community in Cleveland, Ohio. This "open access" is provided at no charge and has led to strong collaborations beyond the university, including those with nearby art institutes. The second unique aspect of the Sears think[box] is its emphasis on entrepreneurship. Each of the seven floors in the 4,500 m^2 facility has a defined focus that guides an individual along the path of idea generation, rapid prototyping, system improvements, advanced manufacturing, intellectual property and venture capital pursuits, and business incubation. The entrepreneurial mindset is important not only for students who want to be entrepreneurs but is also valued by employers who hire CWRU graduates with Sears think[box] experience.

In addition to the tools and programs commonly found in academic makerspaces, the Sears think[box] provides guidance to students on developing business plans, finding investors, and intellectual property issues. An "IP Venture Clinic" is located within the facility where associates of the CWRU School of Law and graduate students from business and science disciplines help students with all phases of the commercial start-up process including technology protection, developing strategy, establishing a corporate entity, and creating a financial offering.

A faculty director, site manager, business lead, administrator and 6 area managers (covering manufacturing, outreach, and programming) guide this core-facility at CWRU. Sears think[box] staff support courses using the same approach for all members in that they provide access and training. Specialized instruction for course projects is the responsibility of the course instructor. Information on the support the Sears think[box] provides to design and other courses is conveyed with a standard presentation that is incorporated within the courses. Sears think[box] publishes a guide for designing and operating academic makerspaces. This guide includes tutorials on equipment, methodology for certifying each user's ability to operate specific equipment, and operating policies.

The Seoul National University Idea Factory is a combination of space and programs to connect ideas, collaborate, and cultivate creative culture. Its design, which was driven by students, encourages creativity and cooperation through the use of large, open meeting and work areas. Areas are dedicated to ideation, teaming, prototyping, fabrication, and entrepreneurship with the furniture, equipment, and layout in each area supporting these functions. Areas within the Idea Factory can be used for multiple purposes each day (for example as a presentation area for one-time period and as a team meeting area for another time period) with this flexibility maximizing space utilization. The space is open 24 hours each day, 7 days a week. A dedicated staff supports prototyping and equipment training on 3D printers, 3D scanning, laser cutting, CNC milling/routing, machining, and computing.

The Idea Factory was created to provide students with resources to design and develop ideas into commercial products and to improve engineering education. To help achieve those goals, the founding faculty members of the Idea Factory created a new interdisciplinary course for product development that takes advantage of the resources available to students. That original course has also been modified and offered in collaboration with other universities in Hong Kong and China. This development illustrates the space's ability to catalyze collaboration. An additional new course was also created, entitled "Digital Fabrication and Manufacturing," by combining the space's workshops in rapid manufacturing and equipment use into a single academic course that requires students to learn by doing. In addition to the equipment workshops, the Idea Factory also hosts workshops on patent strategy, design processes, and commercialization fundamentals to assist students on creating innovations and start-up companies.

The Yale Center for Engineering Innovation and Design (CEID) was motivated by a faculty-directed strategic plan to increase the culture for engineering at Yale University. Meeting that challenge, the Yale CEID provides 24/7 access to 2,500 members in an 850 m^2 facility. Anyone at Yale is eligible for membership, with its current membership distribution of 60% undergraduates, 30% graduate students, and 10% faculty and staff. A director, two design faculty, two design fellows, and ten undergraduate student aides manage the space.

The facility, which is administered by the Yale School of Engineering & Applied Science, supports approximately ten design courses each year where course instruction is provided in the CEID's classroom and project work is completed in the adjoining work studio. The range of courses reflects the CEID's mission to be a university-wide resource, with courses offered in medical device design, design of museum spaces, musical instrument design, and sustainable design, to name a few. In addition to courses, the Yale CEID supports work associated with research, engineering clubs, entrepreneurial pursuits, and personal creativity. Supporting this scope of work, the Yale CEID is fully embedded in the fabric of the university with its direct engagement in the educational, research, and service components of Yale University.

This collection of examples highlights some common characteristics of these spaces and illustrates the range of infrastructure, programs, and community across a sample set. The variations in each area, for example, the range of space, equipment, and staff, illustrate an important attribute of academic makerspaces: there is not a prescribed formula that defines specific components of an academic makerspace. Rather, each space is created, outfitted, and operated to meet specific goals at each institution. The particulars of the infrastructure, programs, and community are unique to each facility. This concept is illustrated by examining the varied components of the six profiled academic makerspaces.

Floor space is a primary indicator of the size of an academic makerspace (with the number of members also a key metric). For the presented facilities, the smallest footprint is that of the Seoul National University Idea Factory. This facility is comprised of a collection of moderately sized rooms, with each room devoted to a particular topic (ideation, teamwork, prototyping, fabrication, and entrepreneurship). In this facility, the equipment (laser cutter, 3D printers, and hand tools) is located within the prototyping room, with access to industrial grade mills, lathes, and routers provided in a work area that adjoins the makerspace. This arrangement results from the SNU Idea Factory's mission to serve as a site to connect people and ideas. Large-scale manufacturing equipment is not needed to fulfill this mission, thereby keeping the footprint of this academic makerspace small compared to other facilities.

At the other size extreme, the TUM Makerspace has a large footprint and an expanded list of consumer and industrial grade equipment to fulfill its mission supporting start-ups and developing them into established businesses. The equipment portfolio extends beyond that needed to work with wood, metal, and plastics. The space also includes equipment to work with textiles and tools for electronics fabrication. In addition to prototype fabrication, which is common to all academic makerspaces, this facility also supports small batch manufacturing. The larger scale, more varied equipment inventory, and the associated greater number of users are driving factors for a larger size footprint. Similarly, the staff at each facility scales with the complexity of the equipment. While students may have the ability to conduct training, oversee use, and maintain the equipment at the SNU Idea Factory, staff with advanced machining, design, and fabrication skills are needed at the TUM Makerspace.

Programming within a space is created to support the facility's mission. The mission of the Sears [think]box is to provide an ecosystem for venture creation, with that process starting with idea generation, supported with manufacturing and fabrication, and advanced with resources related to intellectual property protection, business planning, and raising capital. Programming supports this mission with workshops devoted to innovation, entrepreneurship, and e-commerce, in addition to the more traditional academic makerspace workshops on CAD, design thinking, microprocessor use, and other making topics. Academic programming can also be offered by an academic makerspace if that is one of the missions of the facility. As many as twenty courses per semester are hosted within the Jacobs Institute for Design Innovation with these courses taught by faculty from the disciplines of engineering, computer science, business, theater, design, and art, for example. To address the large number of students in these courses, as well as other members of the university community who are not taking courses within the space, a staff of five design specialists train users and guide the design process. For these two facilities, the mission drives the investment in each space, with a resulting impact on the level and type of programming.

Each academic makerspace community is also derived from the facility's mission. At the Imperial College Hackerspace, the public at large and industry are drawn to the space with specific events targeting these audiences. For example, hackathons on specific technologies (such as voice-activated actions) have been sponsored by companies looking to integrate developments into their industry. As another example, partnerships with elementary science and technology teachers were established to create hands-on learning activities to increase student engagement in these topics. As a result of these activities, the academic makerspace community users extend beyond the university. At the Yale Center for Engineering Innovation and Design, use is open to the entire campus, with the resulting community of 2,500 users being sixty percent undergraduate students, thirty percent graduate students, and ten percent faculty and staff. For the undergraduate population, fifty percent are from science, engineering and mathematics majors, twenty percent are from social studies and the humanities, and the remaining thirty percent of undergraduates having yet to declare a major. This diverse composition of members meets the center's goal to serve a resource for the entire university.

Collectively, these examples illustrate the variety of ranges of size, equipment, staff, programs and membership that comprise academic makerspaces. There are not prescribed parameters for any of these components, making the field of academic makerspaces itself very diverse. Each space addresses unique needs at each institution, with those specific needs determining the characteristics and attributes of each space. These examples illustrate a range of missions for academic makerspaces that extends beyond engineering education into broad-based design and community building. Design is central to the functions of each space, and the resulting community extends design influence and impact across campus.

6 Contributions to Design Education

The establishment of academic makerspaces on higher education campuses provides new opportunities to enhance design education by increasing the participation in, demand for, and impact of design activities. Specific contributions to an institution's design environment from academic makerspaces result from the following factors:

1. Academic makerspaces fuel the pipeline of design-interested students.
2. Academic makerspaces accommodate design interests in an accessible and central location with tools, training, and support for design pursuits.
3. Academic makerspaces increase the number of curricular, co-curricular, and extracurricular design activities.
4. Academic makerspaces create social communities centered on design.
5. Academic makerspaces showcase the design process and achievements.
6. Academic makerspaces increase design proficiencies.

These contributions illustrate a cycle of removing barriers, driving interest, creating structure, establishing communities, and celebrating success that applies to many areas beyond design. Based on this model, it is claimed that the presence of academic makerspaces at an institution of higher education exposes more students to design practices, with a potential to establish a self-sustaining, student-centered, interest-driven interest in design.

The design interest growth cycle starts with a recognition that academic makerspaces catalyze interests in making, with design an integral part of the idea to realization pathway. Establishing an academic makerspace is a visible confirmation that the process of creating is valued at an institution of higher education. Experience shows that individuals will be drawn to that opportunity, and especially so when the space supports sequenced courses that progress from cornerstone to capstone topics (Bill and Fayard 2016; Wilczynski et al. 2016). For the campus community at large, academic makerspaces also provide an alternative learning opportunity outside of the classroom, thereby increasing the number of students interested in design principles and applications (Burke 2015). Within engineering programs, academic makerspaces provide a location for applying theory to create solutions, with proficiency in the hands-on aspects of the engineering profession essential to an individual's success as an engineer.

The accessibility to resources, including tools, training, and guidance, removes roadblocks as a person develops an idea into a physical object. The availability of these resources and a systematic methodology to access them increases the likelihood that the resources will be effectively used. Regarding design, these resources include instruction on design processes as well as the application of engineering fundamentals to solve problems. Learning how to design and create engineered systems is an inherent responsibility of academic makerspaces, with specific programs crafted to establish and improve design skills. One study of fifty participants quantified the value of an academic makerspace for enhancing design skills with nearly ninety percent of the makerspace participants having a positive impact while working in these facilities (Forest et al. 2014).

While design projects can be pursued as solo activities, they are more commonly a component of team-based curricular, co-curricular, and extracurricular activities. Academic makerspaces commonly host these activities, thereby promoting their growth on university and college campuses. Because of an access to centralized resources, student teams have a higher probability for a successful design experience, thereby increasing their confidence as designers and fueling interest in other projects. Responding to this increased demand, even more design courses and projects are spawned by such success (Cooke et al. 2018).

Inherent in this cycle is the creation of design communities within academic makerspaces. The spaces serve as a magnet for individuals interested in design and provide a forum for individuals to meet each other. A combination of low barriers for initial participation and the integration of users through social activities and projects promote connections and collaborations within academic makerspaces (Smith 2017). By collaborating, the number and uniqueness of possible solutions increases, thereby increasing the likelihood that innovative solutions to problems can be found.

Academic makerspaces serve as a stage to display design results and artifacts, with those displays including work in progress and completed projects. Increasing visibility of any product is essential to drive demand for that product, and that fact also applies to design education. Students benefit from seeing the work of others and by realizing that they too can be successful in design activities. With the open layout of academic makerspaces, work is always on display and individuals can see the design process in action. Academic makerspaces also frequently host presentations where students share their results from courses and extracurricular activities.

These factors combine to enhance design skills at institutions hosting academic makerspaces. Students who are active in academic makerspaces improve their design skills by practicing elements of their profession—be that in computer-aided design, electronics, fabrication, or testing—through hands-on work. Over the course of four years, undergraduate participants can sharpen their skills and become proficient fabricators using a variety of traditional and digital tools. Increases in design self-efficacy and creative confidence for students active in academic makerspaces have been quantified by a collection of researchers (Saorín et al. 2017; Morocz et al. 2016). Hands-on construction of physical artifacts in academic makerspaces serves as a mechanism to practice the iterative process of problem-solving (Rosenbaum and Hartmann 2018). As a mechanism for integrating design into courses, academic makerspaces have been effective at developing design competencies as well as increasing student skills with design tools (Kim and Ruters 2016). Enrollment in academic makerspace courses has shown to have a significant increase in confidence, motivation, and success solving design challenges (Hilton et al. 2018).

The research on the impact of academic makerspaces on increasing design skills aligns with other studies that document their positive influence on developing product development and entrepreneurial skills. Networking opportunities, combined with easy access to hardware and software, and the support of the makerspace community, accelerate the product development process (Freissnig et al. 2018). It is proposed that the democratization of resources and the social structure of academic makerspaces promote empowerment and develop entrepreneurial skills as members progress from apprentices to practitioners (Hui and Gerber 2017).

In summary, academic makerspaces follow a traditional product growth process where their very existence generates demand, the access to resources ensures success, and the resulting positive experience drives addition interest (from the individuals achieving success and others who observe the success). As a result, the number of students interested in design topics increases, thereby creating additional demand for design experiences on college and university campuses, and enhancing the design skills of the participating students.

Acknowledgements This work is partially supported by Yale Center for Engineering Innovation and Design and the Yale School for Engineering & Applied Science.

References

ABET (1997) Engineering Criteria 2000. ABET

Aidala KE, Baker N, Fledman R, Klemperer PF, Mensing S, St John A (2017) Types of academic makerspaces. In: Proceedings of the 2nd international symposium on academic makerspaces, Cleveland, OH, USA

Ali PZ (2016) The power of investing in building relationships, collaboration and ownership. In: Proceedings of the 1st international symposium on academic makerspaces, Cambridge, MA, USA

Ali PJ, Cooke M, Culpepper ML, Forest CR, Hartmann B, Kohn M, Wilczynski V (2016) The value of campus collaboration for higher education makerspaces. In: Proceedings of the 1st international symposium on academic makerspaces, Cambridge, MA, USA

Anderson C (2012) Makers: the new industrial revolution. Crown Business Publisher

Andrews D, Roberts D (2017) Academic makerspaces: contexts for research on interdisciplinary collaborative communities. In: SIDOG'17, Halifax, NS, Canada

Baleshta J, Teertstra P, Luo B (2015) Closing the loop: integrating 3D printing with engineering design graphics for large class sizes. In: Proceedings of the Canadian engineering association conference, Hamilton, ON, Canada

Barrett TW, Pizzico MC, Levy B, Nagel RL, Linsey JS, Talley KG, Forest CR, Newstetter WC (2015) A review of university maker spaces. In: American society for engineering education annual conference proceedings, Seattle, WA, USA

Bill VG, Fayard AL (2016) Building a makerspace to nurture the innovation culture at the NYU Tandon School of Engineering. In: Proceedings of the 1st international symposium on academic makerspaces, Cambridge, MA, USA

Burke J (2015) Making sense: can makerspaces work in academic libraries? In: Proceedings of the association of college and research libraries conference, Portland, OR, USA

Cooke M, Forest CR, Hartman B, Hoover AM, Hunt J, Kohn M, Culpepper ML, Wilczynski V (2018) Models for curricular integration of higher education makerspaces. In: Proceedings of the 3rd international symposium on academic makerspaces, Palo Alto, CA, USA

Culpepper ML (2016) Types of academic makerspaces. In: Proceedings of the 1st international symposium on academic makerspaces, Cambridge, MA USA

Dougherty D (2016) Free to make: how the maker movement is changing our schools, our jobs, and our minds. North Atlantic Books

Farritor S (2017) University-based makerspaces: a source of innovation. Technol Innov 19:389–395

Forest CR, Moore RA, Jariwala AS, Fasse BA, Linsey J, Newstetter W, Ngo P, Quintero C (2014) The invention studio: a university maker space. Adv Eng Educ, Summer 2014

Freissnig M, Karre H, Schnöll HP, Ramsaeuer C (2016) Development of an educational program using capabilities of academic makerspaces. In: Proceedings of the 1st international symposium on academic makerspaces, Cambridge, MA

Freissnig M, Handy L, Schnöll HP, Ramsaeuer C (2018) The role of makerspaces in product development of hardware start-ups. In: Proceedings of the 3rd international symposium on academic makerspaces, Palo Alto, CA, USA

Gershenfeld N (2005) Fab, the coming revolution of your desktop—from personal computers to personal fabrication. Basic Books

Halverson ER, Sheridan KM (2014) The maker movement in education. Harv Educ Rev 84(4):495–504

Hilton EC, Forest CR, Linsey JS (2018) Slaying dragons: an empirical look at the impact of academic makerspaces. In: Proceedings of the 3rd international symposium on academic makerspaces, Palo Alto, CA, USA

Hui JS, Gerber EM (2017) Developing makerspaces as sites of entrepreneurship. In: Proceedings of the 2017 ACM conference on computer-supported cooperative work and social computing, Portland, OR, USA

Hynes MM, Hynes WJ (2018) If you build it, will they come? student preferences for makerspace environments in higher education. Int J Technol Des Educ 28:867–883

Jaskiewicz T, Mulder I, Verburg S, Verhu B (2018) Leveraging prototypes to support self-directed social learning in makerspaces. In: International conference on engineering and product design education, London, United Kingdom

Kim MS, Ruters J (2016) Competency-based curriculum for digital fabrication and makerspaces. In: Proceedings of edmedia: world conference on educational media and technology, Vancouver, BC, Canada

Morocz RJ, Levy B, Forest C, Nagel RL, Newstetter WC, Talley KG, Linsey JS (2016) Relating student participation in university maker spaces to their engineering design self-efficacy. In: American society for engineering education annual conference proceedings, New Orleans, LA, USA

Pernia-Espinoza A, Sodupe-Ortega E, Pecina-Marqueta S, Martinez-Banares S, Sanz Garcia A, Blanco-Fenrnandez A (2017) Makerspaces in education. In: Proceedings of the 3rd international conference on higher education advances, Valencia, Spain

Piaget J (1971) Psychology and epistemology: towards a theory of knowledge. Grossman

Roberts D, Buckley J (2016) The role of a design studio in a mechanical engineering department. In: Proceedings of the 1st international symposium on academic makerspaces, Cambridge, MA, USA

Rosenbaum LF, Hartmann B (2018) Making connections: project courses improve design self-efficacy and interdisciplinary awareness. In: Proceedings of the 3rd international symposium on academic makerspaces, Palo Alto, CA, USA

Saorín JL, Melian-Díaz D, Bonnet A, Carrera CC, Meier C, De La Torre-Cantero J (2017) Makerspace teaching-learning environments to enhance creative competence in engineering students. Teach Skills Creat 23:188–198

Schön S, Ebner M, Kumar S (2014) The maker movement. Implications of new digital gadgets, fabrication tools and spaces for creative learning and teaching. eLearning Pap 39

Simon K, Haji MN (2016) Building a safety-based culture for a student-run makerspace. In: Proceedings of the 1st international symposium on academic makerspaces, Cambridge, MA, USA

Smith A (2017) Social innovation, democracy and makerspaces. Science policy research unit working paper series, University of Sussex

Tan M, Yang Y, Yu P (2016) The influence of the maker movement on engineering and technology education. World Trans Eng Technol Educ 14(1)

Thomas D, Brown JS (2011) A new culture of learning: cultivating the imagination for world of constant change. CreateSpace Publishing

Wilczynski V (2015) Academic makerspaces and engineering design. In: American society for engineering education annual conference proceedings, Seattle, WA, USA

Wilczynski V, Adrezin R (2016) Higher education makerspaces and engineering education. In: Proceedings of the ASME 2016 IMECE, Phoenix, AZ, USA

Wilczynski V, Wilen L, Zinter J (2016) Teaching engineering design in a higher education makerspace. In: American society for engineering education annual conference proceedings, New Orleans, LA, USA

Yao AI, Lucero S, Facciotti MT (2017) Prototyping biomolecules to machines: a case study of launching and sustaining an academic biomarker lab. In: Proceedings of the 2nd international symposium on academic makerspaces, Cleveland, OH, USA

Engineering Capstone Design Education: Current Practices, Emerging Trends, and Successful Strategies

Susannah Howe and Jay Goldberg

Abstract Capstone design courses play an important role in the preparation of engineering students for professional practice and career success. They allow students to apply what they have learned prior to their senior year along with newly developed skills relating to the solution of real-world design problems, often in a team environment with external clients. Team-based capstone design courses provide opportunities for students to develop teamwork and communication skills. In the United States, accredited engineering programs are required to offer a major design experience; most programs fulfill this requirement through a capstone design course. Differences in course learning outcomes, management, structure, duration, projects, student teams, and required deliverables exist among institutions. This chapter presents the current state of engineering capstone design education and highlights changes to capstone design practices in the U.S. over the past 25 years, based on extensive, seminal surveys of capstone design programs. It discusses current practices and successful strategies in engineering capstone design education. The chapter includes perspectives and feedback from hundreds of engineering capstone design faculty regarding their personal experiences with capstone design programs. This chapter also provides recommendations for supporting engineering capstone design experiences based on the authors' vast experience teaching and managing capstone design courses and engaging with the capstone design community. These recommendations include scaffolding the design curriculum, fostering industry involvement in capstone design courses, keeping courses and faculty up-to-date with current design practices, obtaining organizational support for capstone design courses, sourcing capstone design projects, and preparing students for professional practice.

Keywords Engineering capstone design · Career preparation · Design pedagogy · Design projects · Professional practice

S. Howe (✉)
Smith College, Picker Engineering Program, 151 Ford Hall, Northampton, MA 01063, USA
e-mail: showe@smith.edu

J. Goldberg
Department of Biomedical Engineering, Marquette University and the Medical College of Wisconsin, P.O. Box 1881, Milwaukee, WI 53201-1881, USA
e-mail: jay.goldberg@mu.edu

© Springer Nature Switzerland AG 2019
D. Schaefer et al. (eds.), *Design Education Today*,
https://doi.org/10.1007/978-3-030-17134-6_6

1   Introduction and History of Engineering Capstone Design Education

The concept of an engineering capstone design course was pioneered in the U.S. at Harvey Mudd College (HMC) in the early 1960s in response to perceived deficiencies in engineering curricula at the time, including inexperience with open-ended problems, limited project skills, and little exposure to professional practice (Bright and Phillips 1999). As such, the HMC Engineering Clinic provides a clinical experience for engineering students, similar to that for medical students. "But in the Engineering Clinic setting, services are furnished for clients from industry, commerce, or public service agencies, while the students are introduced to the responsibilities of professional engineering work—under the restrictions of time, budget, and interpersonal relationships" (Bright and Phillips 1999, p. 190).

Engineering capstone design courses serve many purposes. A primary function is to help prepare students for the real world, enabling smooth transition to professional practice. In addition, capstone design courses provide students with opportunities to solve complex open-ended problems, manage a project, and create value for a customer, either external or internal. When structured with a team component, capstone design courses expose students to the joys and challenges of collaboration, diversity of perspectives, communication, and task delegation. Capstone design courses strike a balance between application of material students have learned in previous courses and acquisition of new knowledge and skills. Indeed, the capstone design experience can change the way engineering students think about technology, themselves, society, and the world around them. Capstone design can make students aware of their potential to make a positive contribution to the world and generate excitement for and pride in the engineering profession. As Pembridge and Paretti (2010) note, even when other design courses occur throughout the curriculum, "capstone courses form an important bridge between school and work as students bring their classroom learning to bear on the process of conducting and managing a complex extended project" (Pembridge and Paretti 2010, p. 15.1217.2).

Capstone design courses and/or the experiences they provide to students are both highly valued and required by accrediting organizations such as ABET (the Accreditation Board for Engineering and Technology). ABET defines engineering design as "the process of devising a system, component, or process to meet desired needs. It is a decision-making process (often iterative), in which the basic science and mathematics and engineering sciences are applied to convert resources optimally to meet these stated needs" (ABET 2018, p. 5). ABET General Criterion 3, student outcome (c), requires that students demonstrate "an ability to design a system, component, or process to meet desired needs within realistic constraints such as economic, environmental, social, political, ethical, health and safety, manufacturability, and sustainability" (ABET 2018, p. 4). ABET General Criterion 5 states that "Students must be prepared for engineering practice through a curriculum culminating in a major design experience based on the knowledge and skills acquired in earlier course work and incorporating appropriate engineering standards and multiple realistic constraints" (ABET 2018, p. 6). Many engineering programs rely heavily on the capstone design course to provide the "major design experience" required by ABET, and evidence demonstrating that student outcome (c) has been achieved. Capstone design can often be the only course in the curriculum that addresses many, if not all, of the items mentioned in outcome (c) and the ABET definition of design.

Organizations dedicated to the improvement of engineering education, such as ASEE (American Society for Engineering Education), also recognize the value of capstone design experiences. A recent ASEE report (ASEE 2012) states that for many years engineering education emphasized theory, equations, and modeling more than hands-on experience. However, engineering practice has changed and currently involves team-based, cross-disciplinary projects requiring additional skill sets and personal characteristics. For this reason, it is important for engineering students to experience real-world engineering as part of their professional formation. The report also states that the content of a more relevant learning experience in U.S. engineering curricula should include among other things, "an emphasis on creativity, critical thinking, design, and leadership" and should "reflect upon the experiential aspects" of engineering education (ASEE 2012, p. 22). Capstone design courses are uniquely positioned to enhance engineering education and better prepare our students to meet the new demands of a changing engineering profession (Goldberg 2012a).

The work presented in this chapter draws extensively on the authors' previously published scholarship in the field of capstone design education. By synthesizing information regarding current practices and recommended strategies, this chapter provides a more complete picture of capstone design experiences and their value to engineering design education today.

The focus of this chapter is the engineering capstone design experience in particular, but the student experience and recommended strategies may also apply to other non-engineering disciplines. The (1) emphasis of capstone courses on active, student-centered, and experiential learning, (2) opportunity for students to apply what they have learned in previous courses, and (3) exposure of students to real-world problems of importance to members of their particular profession are common benefits provided by capstone courses and experiences in any discipline (Hauhart and Grahe 2015).

2 Current State of Engineering Capstone Design Education and Emerging Trends

Although capstone design courses are common across engineering programs, they vary substantially in the way they are implemented. The first United States survey of capstone courses was conducted in 1994 in an effort to better understand current practices at the time (Todd et al. 1995). This was followed in both 2005 and 2015 with subsequent nationwide surveys (Howe 2010; Howe et al. 2017) using many of the same questions and including some new categorical and open-ended response questions.

Additional surveys across multiple institutions and capstone programs have been conducted by a variety of researchers on topics such as assessment (McKenzie et al. 2004), teaching load and funding (Howe 2008), content in capstone design courses (Howe et al. 2009), faculty experiences with capstone design pedagogy (Pembridge and Paretti 2010), capstone design problem statements (Dixon 2012), and technical design reviews (Dixon 2014), for example. Other researchers have focused their surveys on specific engineering disciplines (Silverstein et al. 2013; Rao et al. 2015). The 2015 U.S. survey was also extended to Australia and New Zealand (Howe and Rosenbauer 2017) to begin exploring how capstone design education is implemented in other countries.

This section presents the current state of engineering capstone design education, drawing comparisons across disciplines, and highlighting changes within capstone design practices in the past 25 years, based on the extensive, seminal surveys of capstone design programs in the U.S. The 2015 survey dataset, in particular, includes 522 respondents representing 464 distinct departments at 256 institutions, including a wide swath of engineering disciplines: biomedical, chemical, civil/environmental, electrical/computer, general, industrial, mechanical/aerospace, and multidisciplinary. Although the data are focused on U.S. programs, results from the Australia and New Zealand surveys (Howe and Rosenbauer 2017) plus a decade of conversations at the biennial Capstone Design Conferences (www.capstoneconf.org) suggest that the essence of capstone design education is quite similar: there are variations in implementation details, but these may be greater across programs or institutions than they are across nations. Focus areas include course logistics and management, faculty involvement, pedagogy, projects, teams, expenses and funding, and faculty experience and opinion.

2.1 Course Logistics and Management

Capstone courses vary widely in size, depending largely on the number of students in a given department or institution. Figure 1 shows the number of students per capstone design course cycle from both the 2005 and 2015 survey data. Some classes have fewer than 10 students and others have more than 200 at a time. Of particular interest, however, is the fact that student numbers appear to be increasing: the median bracket in 2015 is higher than that in 2005. Capstone design is aptly also known as "senior design": according to 2015 data, 88% of the 463 respondents noted their capstone design students were undergraduate seniors, whereas 7% noted a mix of undergraduate seniors and juniors, and only 3% of respondents had a mix of undergraduate seniors and graduate students.

Fig. 1 Number of students per capstone design course cycle (Longitudinal Data)

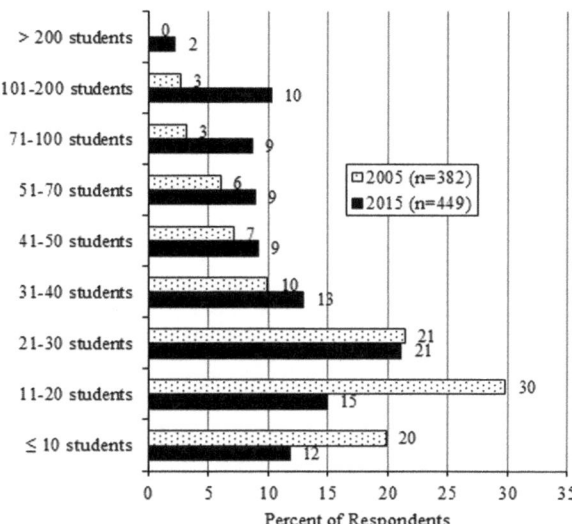

Table 1 Student/faculty ratio (2015 Data)

Student/faculty ratio	% of respondents (n = 440)
<1	3
1–10	33
11–20	22
21–40	27
41–60	9
61+	6

Student/faculty ratio is also variable, depending on student enrollment and how much teaching power is devoted to capstone design instruction. Table 1 shows the average student/faculty ratio for the 2015 survey respondents. While student/faculty ratios of 20:1 or less were most common, some programs had ratios exceeding 60, with one program topping out at 170. The key takeaway is that capstone design has sufficient flexibility to be taught in a variety of different formats for student/faculty interaction.

Capstone design courses exist within and between engineering disciplines and sometimes in conjunction with non-engineering disciplines. The 2015 survey asked respondents "What departments (faculty and/or students) are part of your capstone design course? Select all that apply." The 500 responses included more than 20 different engineering disciplines (ranging from 1 to 36% of respondents for each), as well as multiple non-engineering disciplines (<4% of respondents for each): art/architecture/design, business/marketing, communication, health/medical/nursing, humanities, mathematics/statistics, natural sciences, and social sciences. Of the 500 respondents, 262 (52%) included faculty and/or stu-

Fig. 2 Capstone design
course structure and
sequence (Longitudinal
Data)

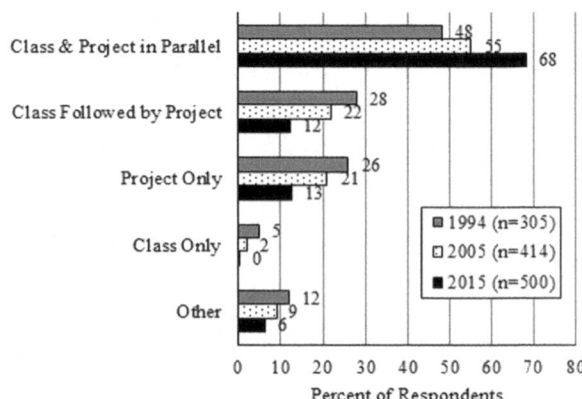

Fig. 3 Capstone design
course duration
(Longitudinal Data)

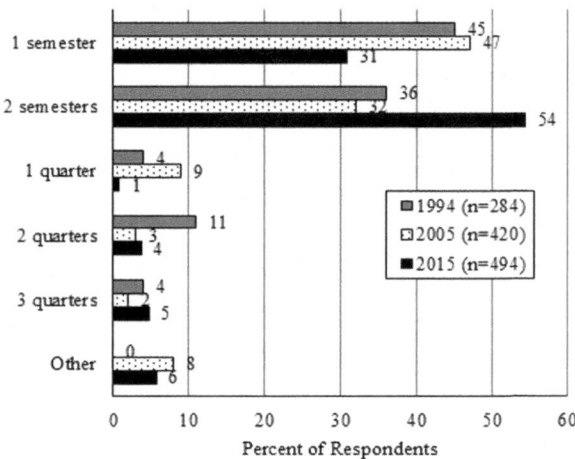

dents from at least two different disciplines in their capstone courses; 57 (11%) had
at least five different disciplines represented.

The increasingly common structure of capstone design courses is a class in parallel
with a design project, as shown in Fig. 2.

While capstone courses can be implemented in a variety of timeframes, longitudinal survey data suggest that the course duration is increasing, as shown in Fig. 3.
More than half of the 2015 respondents had a 2-semester capstone design course,
which is a sizable increase from previous years. In addition, the 2015 data show
a drop in both 1-semester and 1-quarter durations. The "Other" responses in 2015
mostly reflected even longer durations, including 2–3 trimesters, 4 quarters, and even
3–4 semesters.

A question on both the 2005 and 2015 U.S. surveys asked about how the capstone
program ensures that student teams are able to meet; the data are shown in Fig. 4.
Although responses were more evenly split between several options in 2005, the

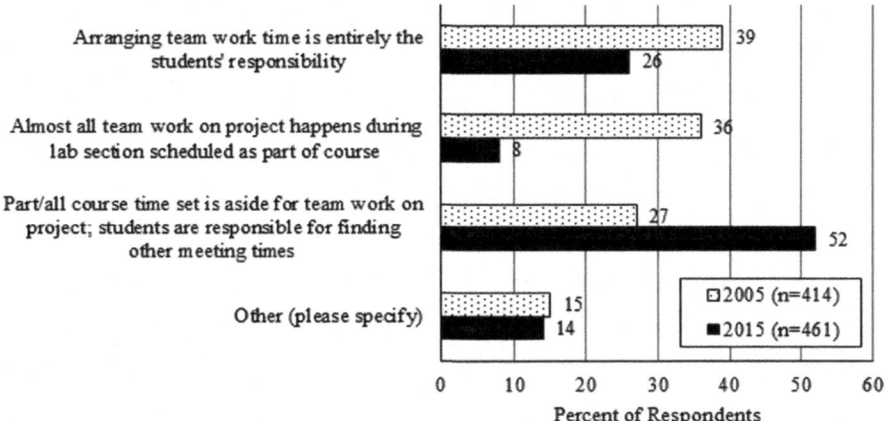

Fig. 4 Approaches to ensure student work time (Longitudinal Data)

Table 2 Faculty position type held by respondents (2015 Data)

Position type	% of respondents (n = 441)
Tenure-track, tenured	53
Tenure-track, untenured	8
Non-tenure track, permanent	24
Non-tenure track, temporary	7
Other	8

majority of 2015 respondents employed the hybrid model where some/all class time was provided to start but students were responsible for finding other times outside of class. The most common write-in response—in both 2015 and 2005—from respondents who selected "Other" was that student teams had weekly meetings with their faculty coach.

2.2 Faculty Involvement

The majority of faculty members teaching capstone design courses in the U.S. are tenured at their institutions, as shown in Table 2, though permanent non-tenure-track positions are not uncommon. The "Other" responses included adjunct positions, emeriti, and combinations of multiple options.

Capstone design instructors typically bring substantial experience from engineering industry and government work to their teaching. More than half of respondents in 2015 had at least six years of work experience outside academia, whereas only 15% had none or less than one year. Clarifying comments provided by some respondents suggested that the length of professional work experience far exceeded 11 years; respondents included multiple write-ins for 25 and 30 years of experience. Furthermore, 85% of respondents' non-academic work involved design, further strengthening their preparation to teach capstone design.

2.3 Capstone Pedagogy

Capstone design courses cover a lot of topics, most commonly as part of the team project or in lecture. Table 3 shows the results from the 2015 U.S. survey, in which respondents selected topics from a predetermined list and noted whether topics were covered specifically in lecture (L), in an individual assignment (IA), as part of the team project (TP), or not covered (NC). Beyond the list of topics provided, 184 respondents also provided more than 100 distinct write-in topics. Most common were engineering economics/financial analysis, design for X, professional preparation and licensure, and safety/liability.

Table 3 Topics covered in capstone design courses (2015 Data, n = number of respondents)

Category	% of respondents				
	N	L	IA	TP	NC
Analysis tools	450	45	25	83	9
CAD design/layout	440	23	18	67	28
Concept generation	453	62	23	80	7
Concept selection	451	61	24	81	7
Creativity/problem solving	462	53	24	80	6
Decision making	458	58	19	80	7
Engineering ethics	455	69	30	45	12
Functional specifications	455	56	24	77	12
Intellectual property	440	51	12	33	37
Leadership	443	47	14	64	19
Optimization	430	36	12	57	32
Oral communication	469	57	29	89	1
Project management	468	67	26	89	2
Prototyping, testing	445	41	16	71	22
Sketching	422	18	14	46	44
Standards/regulation	448	59	17	70	10
Sustainability	434	44	13	53	27
Teamwork	463	57	21	81	5
Written communication	472	56	42	91	1

Table 4 Top five topics covered in capstone design courses (Longitudinal Data)

1994	2005	2015 Lecture	2015 Overall
Oral communication	Written communication	Engineering ethics	Written communication
Concept generation	Oral communication	Project planning and scheduling	Project planning and scheduling
Teamwork	Engineering ethics	Concept generation/selection	Oral communication
Planning/scheduling	Project planning	Standards and regulations	Concept generation/selection
Engineering ethics	Decision making	Decision making	Team building/teamwork

Table 4 displays the top five topics covered throughout the 1994, 2005, and 2015 surveys. The data, which have changed very little over the years, reveal a notable emphasis on professional skills.

What is and can be taught in capstone design courses depends in part on what is covered earlier in the curriculum, especially in a design context. The 2015 survey asked respondents "What design courses do you require as prerequisites for capstone design?" The responses were coded into nine primary categories, each of which had additional content themes, as shown in Table 5. Specific design courses and related engineering and design-based topics are common precursors to capstone design courses, though nearly a fifth of respondents noted no specific design-based prerequisites.

An often discussed topic at the biennial capstone design conferences is that of product versus process in capstone design. As such, the 2015 survey asked respondents "How do you balance product versus process in your capstone design projects?" Responses were coded into seven distinct categories based on numerical value provided (51–74% = "emphasis," 75–94% = "heavy emphasis," 95–100% = "all") or interpretation of the response by the researchers based on wording and adjectives. While more than 208 responses were received, only those 208 that could be definitively coded were included in the analysis. Figure 5 shows the results of the coding, with representative quotes. Although there are capstone programs that focused solely on product or solely on process, the majority of respondents either weighed the two equally or emphasized process.

Evaluation of student performance in capstone design is commonly informed by many people and based on many different types of work. As shown in Table 6, course instructors and project coaches had the highest level of input on grades in 2015, whereas departmental advisory board members and competition judges had a limited role if any. The final report, final oral presentation, and final product each had the biggest role in evaluation among 2015 respondents, as depicted in Table 7. The design process, interim work and design reviews were of similar importance in assigning grades. A large majority of respondents (81%) also indicated that peer

Table 5 Categories and content themes regarding design prerequisites (2015 Data)

Category	# Resp. (n = 312)	Content themes (in descending order of frequency)
Specific Courses	132	Specific elective courses and labs; department-specific courses; sequence of design courses; other design course(s)
Specific engineering topics	91	Machine design; design components in other courses; CAD; mechanical design; thermal design; software engineering, software design; design and manufacturing; simulation/testing; design theory/methods; experimental methods; construction management; component design; system design; modeling with architectural drawings; product design
None	61	None/nothing specific
Specific years	52	Freshman design/intro to design; junior design; sophomore design; senior design
Criteria-based	31	Senior standing or minimum # of credits
Other topics	17	Economics; project management; technical communications, technical writing; math; physics
Varies	11	Varies by department/major
Most/all	11	Most or all elective courses; all core courses through 300 level; all core courses
General/yes	8	(No themes—response affirms that there are design prerequisites)

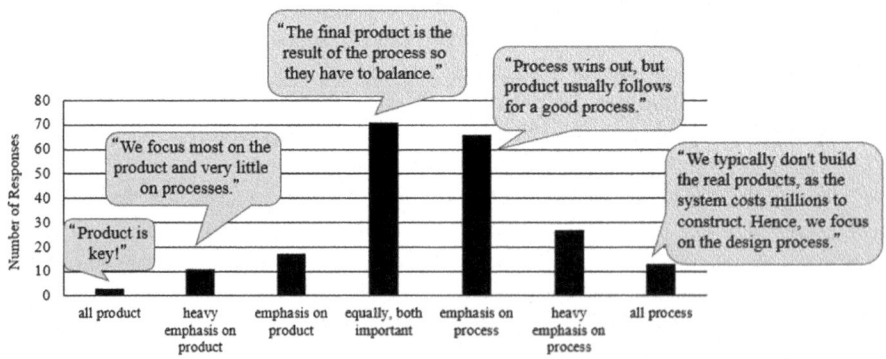

Fig. 5 Balance of product versus process (2015 Data)

Table 6 Evaluators of students' work (2015 Data)

Evaluator	Percent of respondents				
	High input on course grade	Moderate input on course grade	Limited input on course grade	Evaluate projects, don't contribute to course grade	No role at all
Course instructors (n = 467)	83	8	5	2	2
Project advisors/coaches (n = 462)	36	26	14	11	14
Industry liaisons (n = 460)	8	17	24	28	24
Other department faculty (n = 457)	5	12	25	21	37
Students (n = 455)	3	14	36	22	25
Department Advisory Board (n = 454)	2	6	12	20	61
National competition judges (n = 450)	1	1	2	12	83

feedback played at least a minor role. When asked how grades were assigned in capstone design, 90% of 469 respondents selected "Individually assigned based on both individual and team performance."

2.4 Design Projects

A key component of a capstone design experience is the design project itself. Projects can come from many different sources, but as shown in Fig. 6, the most popular sources are industry and government, followed by faculty research. That the numbers for any given year sum to well over 100% indicates that most respondents obtained projects from different types of sources. The 2015 data indicate an increase in entrepreneurial projects, as well as the emergence of service learning projects.

Table 7 Evaluation of deliverables (2015 Data)

Deliverable	Percent of respondents			
	Major role in evaluation	Moderate role in evaluation	Minor role in evaluation	Not considered in evaluation
Final written report (n = 469)	77	20	2	0
Final oral presentation (n = 468)	63	29	7	1
Final product (n = 460)	62	24	5	8
Design process (n = 464)	41	44	13	3
Mid-course reports (n = 462)	23	48	21	8
Mid-course presentations (n = 463)	18	45	25	12
Design reviews (n = 464)	17	41	29	13
Individual assignments (n = 458)	10	24	37	29
Logbooks (n = 458)	9	17	27	48
Peer feedback (n = 459)	8	33	40	18
Hours spent on project (n = 458)	6	18	22	54
Quizzes/exams (n = 455)	6	12	19	62
Business plan (n = 453)	4	9	25	62

The "Service Learning" option was provided on only the 2015 survey, so there is no longitudinal comparison. Sources in the "Other" category for 2015 included clinicians and instructor ideas. Note: the 1994 data were reported only as "Industry," "Internally," and "Other," so they are shown in a separate box.

Analyzing the 2015 survey further reveals some differences by discipline, as shown in Table 8 (the numbers in the table indicate the percentage of respondents who indicated having at least one project from the project source category). Industry and government were the most common project source for most departments, in particular for nearly all industrial engineering and multidisciplinary engineering respondents. Projects based on faculty research were especially prominent in biomed-

Fig. 6 Sources of capstone design projects (Longitudinal Data)

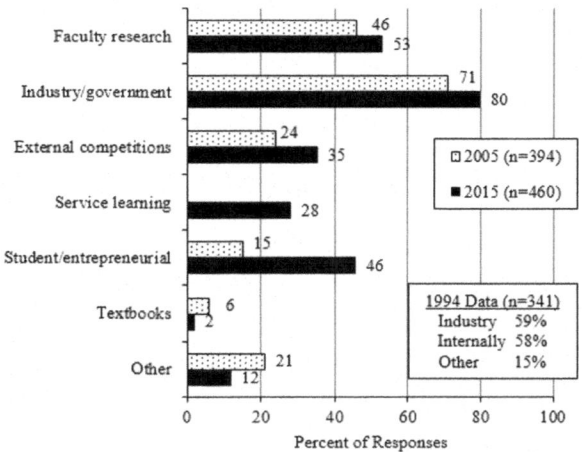

Table 8 Source of capstone design projects by discipline (2015 Data)

Project sources	Percent of respondents							
	BME (n = 34)	ChE (n = 31)	CEE (n = 74)	EECS (n = 77)	IE (n = 22)	MAE (n = 109)	MULTI (n = 27)	OTHER (n = 86)
Industry/government	71	71	84	68	95	87	96	78
Faculty research	74	39	15	75	14	64	52	60
External competitions	35	55	18	34	5	63	26	22
Service learning	35	10	34	23	5	36	30	28
Student/entrepreneurial	47	35	16	77	9	58	41	43
Textbooks	0	16	1	3	0	2	0	0
Other	32	19	16	4	0	9	7	8

BME biomedical engineering, *ChE* chemical engineering, *CEE* civil and environmental engineering, *EECS* electrical engineering and computer science, *IE* industrial engineering, *MAE* mechanical and aerospace engineering, *MULTI* respondents including more than one listed disciplinary grouping, *OTHER* other engineering disciplines not listed

ical, electrical/computer, and mechanical/aerospace disciplines. Projects from competitions were most often found in chemical and mechanical/aerospace disciplines, and entrepreneurial projects were common in electrical/computer engineering disciplines specifically. Nearly a third of biomedical, civil/environmental, and mechanical/aerospace disciplines sourced at least some of their projects from service learning opportunities. It is also worth noting that 32% of biomedical respondents indicated that some of their projects were from other categories, including clinicians.

Finding capstone design projects is an ongoing task, and can often be a time-consuming challenge for capstone design instructors. The 2015 survey asked respondents "What strategies do you use for finding capstone design projects?" The

Table 9 Categories and content themes for finding capstone design projects (2015 Data)

Category	# Resp. (n = 321)	Content themes (in descending order of frequency)
External contacts	173	Local and regional industries; alumni; industrial advisory board; previous sponsors; connections in general; personal contacts of capstone instructor; faculty and department contacts; development office; word of mouth; student contacts; co-op and internship contacts; clinicians; other university's capstone project sponsors
Internal sources	92	Student-proposed; faculty research and ideas; brainstorming; on-campus projects
Marketing	85	Solicitation and networking; advertising; internet searches
Prefabricated	28	Competitions; repeat previous projects; textbooks
Criteria-based	21	Global trends and industry needs; multidisciplinary groups
Magnet	16	Approached externally; reputation
Who finds	14	Dedicated capstone personnel; leave to faculty mentor
Extreme	10	No coordinated strategy; anything and everything
Events-based	7	Demo day or project day; attend career day; conferences

responses clustered into nine categories, each of which had additional content themes, as shown in Table 9. More than half of respondents used external contacts for at least one method of finding projects.

2.5 Student Teams

Capstone projects can be conducted individually or in teams; the number of students on a team is a common topic of discussion among capstone instructors. Figure 7 depicts reported average team size data in general across three survey years in the graph, plus 2015 data in more resolution in the table. Note that more than 80% of 2015 respondents had between 3 and 5 students per team, but a handful had team sizes exceeding 10 students per team. Moderate team size has remained fairly consistent over time.

Respondents also reported on the number of teams assigned per project; in all three surveys, the majority response was one team per project (73% of n = 458 in 2015). However, some of these 2015 respondents also noted that occasionally they assigned two teams per project, especially when enrollment numbers warranted the change. Another outcome of increasing student enrollments (see Fig. 1) with relatively constant team size is an increase in number of projects per capstone course cycle. Indeed, in 2005, the mean and median number of projects per capstone course

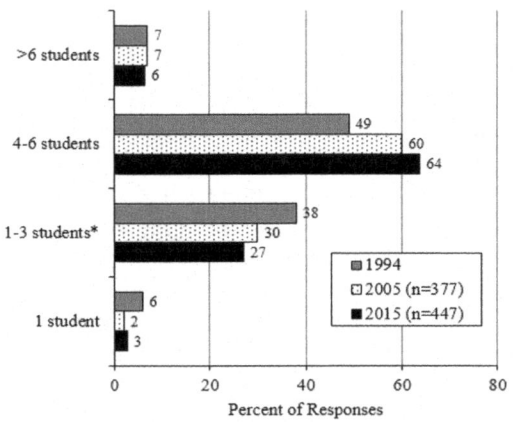

2015 U.S. Data	
Average Team Size	Percent of Respondents (n=447)
1 student	3
2 students	3
3 students	24
4 students	40
5 students	18
6-7 students	6
8-10 students	3
>10 students	3

* In 1994, "1-3 students" was a specific choice
In 2005, "1-3 students" refers to all responses > 1.5 and <3.5 students
In 2015, "1-3 students" refers to an average of 2 or 3 students

Fig. 7 Number of students per capstone design team (Left: Longitudinal Data; Right: 2015 Data)

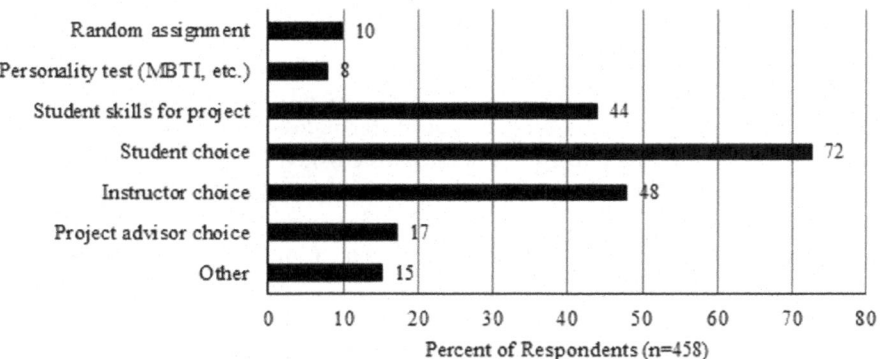

Fig. 8 Methods for assigning students to capstone design teams (2015 Data)

cycle were 8.1 and 5, respectively; in 2015 these numbers increased to 12.4 and 9, respectively.

Another topic regarding capstone design teams is the method for assigning students to teams. As shown in Fig. 8, the most common way to assign students to teams was student choice, followed by instructor choice and student skills. Many 2015 respondents utilized a combination of options, however, as evidenced by the fact that the sum of the data far exceeds 100%. The category labeled "Other" includes write-in responses such as GPA, schedules, and CATME software (Layton et al. 2010).

2.6   Expenses and Funding

Capstone design courses and projects have a number of different associated expenses, as shown in Fig. 9. Project supplies, hardware, and software were the most commonly report expenses, noted by more than two-thirds of respondents in 2015. Some of the more common "Other" responses included external fabrication/analysis, personnel and summer salary, and respondents who noted that they had no expenses at all.

The sources of funding for these expenses included a wide range of options, as shown in Fig. 10. Note that the values sum to far more than 100%, indicating that many respondents selected more than one option. Department and industry were sources for more than half of the respondents. The few "Other" responses included approaches such as crowdfunding or self-funding by the capstone instructor.

A comparison over time, as shown in Fig. 11, illustrates that institutions and sponsors have remained the most common funding sources in the past 20 years. (Note, so as to match the broader categories from the earlier surveys, "Department" and "Institution" were combined to be "Institution," "Industry" and "Government/foundations" were combined to be "Sponsor," and "Individuals" and "Other" were combined to be "Other".) Students were less likely to fund their own project in 2015 than they were previously, and individuals such as alumni were nearly as likely as current students to fund capstone projects in 2015.

The breakeven costs for capstone design projects span a broad range depending on type of project, but the vast majority of projects break even financially at a relatively low level. On the 2015 survey, respondents were asked to provide the minimum, average, and maximum breakeven cost per project (though the survey did not formally define "breakeven cost," so respondents may have interpreted it differently). Figure 12 shows the average values for the 325 respondents from 2015 who provided such

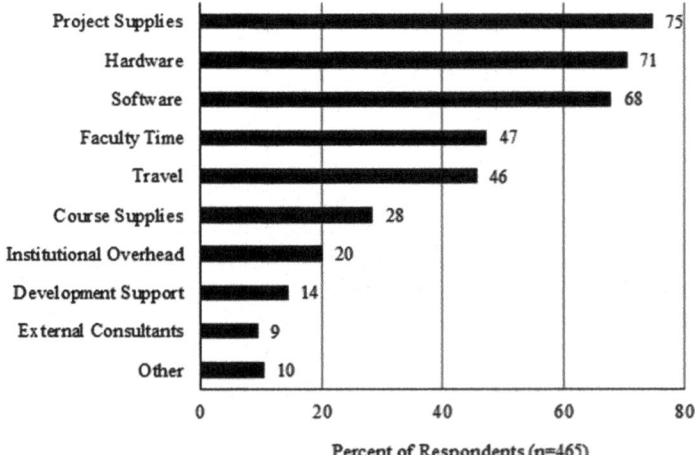

Fig. 9 Types of expenses associated with capstone design projects and courses (2015 Data)

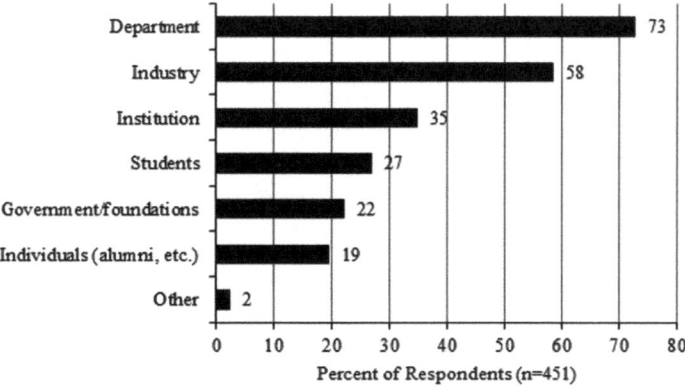

Fig. 10 Funding sources (2015 Data)

Fig. 11 Funding sources
(Longitudinal Data)

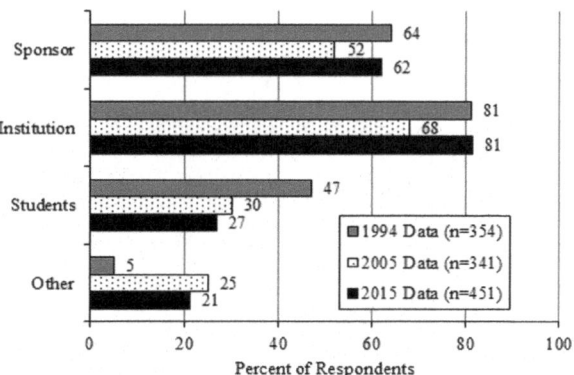

data, with each point representing one respondent. The maximum reported average breakeven cost was $50,000, but the vast majority of respondents had values much lower. As is evident from the inset, 300 of the 325 respondents had breakeven costs less than $5000, 200 were less than $1000, and 50 had no costs at all.

Analyzing the data by discipline revealed that chemical, civil, and industrial engineering respondents had the lowest average breakeven costs, likely because their capstone projects do not typically involve building physical prototypes. Multidisciplinary capstone programs reported the largest mean and median values of average breakeven cost within their populations, but they also had the widest data spread.

The amount of funding that capstone programs request from external sponsors also varies widely, both across capstone programs and even within them for different sponsors. Respondents to the 2015 survey who had external sponsors were asked to provide the average, minimum, and maximum amount of financial support that the external sponsors provided. Table 10 shows the average level of support from external sponsors. For just over half of the respondents, external sponsors provided an average of $2000 or less per project, whereas for just 5% of respondents, external sponsors

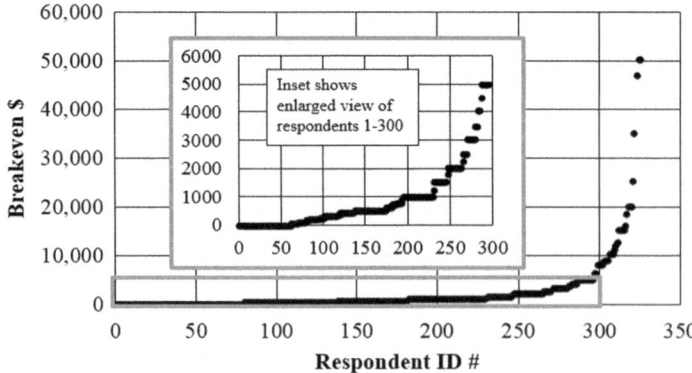

Fig. 12 Average breakeven cost per project (2015 Data)

Table 10 Average financial support provided by external sponsors (2015 Data)	Average financial support (USD)	% of respondents (n = 266)
	$0	16
	$1–500	13
	$501–1000	12
	$1001–2000	11
	$2001–3000	11
	$3001–4000	5
	$4001–5000	11
	$5001–8000	6
	$8001–12,000	6
	$12,000–20,000	4
	$20,001–30,000	3
	>$30,000	2

provided average funding exceeding $20,000 per project. The minimum level of external funding was $0 for all funding ranges, suggesting that even programs with sizable average funding levels had some projects without any funding from external sponsors. The maximum external funding level was $250,000 for one particular capstone program, but the median maximum was only $5,000, indicating that the largest maximum was quite an outlier. Variation was accentuated by discipline: nearly half of chemical engineering capstone programs averaged zero external funding for their projects, whereas nearly a quarter of multidisciplinary capstone programs averaged funding greater than $20,000 per project.

2.7 Faculty Reflection

Capstone faculty members are drawn to teach capstone for many different, but overlapping, reasons. According to the 2015 survey, respondents most enjoy being involved with capstone because of student growth and accomplishment, application of students' previous learning, student pride and excitement, interaction with students and clients and other faculty, the transformation of students to professionals, broad topics beyond just technical skills, and the variety and variability of projects. Sample responses included *"Seeing the student succeed and overcome their struggles during the semester. Coming together as a team to produce real, quality project work"*, *"Seeing students mature in confidence"*, and *"I enjoy the variety of the projects and the organizations that I work with. It interests me to be aware of the issues/problems our sponsors face."*

Given the complex and demanding nature of capstone courses, however, capstone faculty members also face their share of challenges. The biennial Capstone Design Conferences feature regular formal and informal conversations about various issues and difficulties with running a capstone design program. Based on responses to the 2015 survey, capstone faculty's biggest challenges include workload/time, finding and supporting projects, encouraging student commitment and ownership, lack of institutional or departmental support, and unhealthy team dynamics. Representative comments included *"The short amount of time involved and the time it takes to complete a project"*, *"Finding projects that are appropriately challenging from sponsors that are willing to contribute financially"*, and *"Motivating students to their full potential."*

2.8 Summary

Capstone design courses vary widely in implementation, demonstrating the versatility of the educational model, but the essence of the capstone design experience in the United States is surprisingly similar across programs and institutions.

- *Course Logistics and Management*: Capstone design courses are typically for final year undergraduates and can be offered to few students or many students simultaneously, by a single faculty member or with a team of faculty. Capstone design courses offer the opportunity to include multiple engineering disciplines as well as non-engineering disciplines. Typical format for capstone design courses is a class in parallel with a design project, most commonly spanning two semesters. Students are often given time to work in class but must also coordinate other work times.
- *Faculty Involvement*: The majority of U.S. engineering faculty members teaching capstone design courses are tenured at their institutions but bring substantial design experience from employment in engineering industry and government.

- _Capstone Pedagogy_: Capstone design courses typically cover many topics, the most common of which include written and oral communication, project planning, conceptual design, and teamwork. Prerequisites for capstone design vary across institution, but often include specific courses or engineering topics taught prior to the final year. Most capstone programs strike a balance between process and product, recognizing that both are important outcomes. The evaluation of capstone design students is typically informed by many people (course instructors/coaches, industry liaisons, peers, and others) based on different types of work (reports, presentations, design reviews, final product, etc.) but grades are usually assigned individually based on individual and team performance.
- _Design Projects_: Capstone design projects are most commonly sourced from industry and government, but can come from many different sources including the students themselves. Finding suitable projects is typically an ongoing task that relies heavily on external contacts such as local industries, alumni, past employers, and personal connections.
- _Student Teams_: Students typically collaborate in 3–5 person teams for their capstone design projects, usually with each team working on its own distinct project. The most common approaches to assign students to teams are student choice, instructor choice, and student skills, or a combination thereof.
- _Expenses and Funding_: Typical expenses in capstone design courses include project supplies, hardware, and software, among others. Funding sources for these expenses vary, but commonly include the institution and industry sponsors. While the range of expenses may vary significantly, many capstone design projects have breakeven costs less than $1000. Likewise, sponsor funding varies widely—from $0 to a reported high of $250k per project—but the majority of capstone programs receive less than $5000 (and many receive less than $2000) per project from sponsors.
- _Faculty Reflection_: Capstone design faculty find joy in teaching capstone for many reasons, including student growth and accomplishment, opportunities to interact with clients and other faculty, and the variety of projects. At the same time, capstone design faculty commonly lament their heavy workloads, the challenge of finding and supporting projects, a lack of institutional support, and difficult team dynamics.

3 Strategies for Supporting and Enhancing Engineering Capstone Design Experiences

Capstone design course instructors need to find ways to support and maintain their courses to provide an effective learning environment for their students. This section presents multiple strategies for supporting capstone design courses and enhancing the students' experience in these courses: scaffolding the design curriculum, fostering industry involvement in capstone design courses, keeping courses and faculty up-to-date with current design practices, obtaining organizational support for capstone design courses, sourcing capstone design projects, and preparing students for professional practice.

Since each institution has different constraints, requirements, structures, and cultures, what works for one institution may not be best for another. There is no single best approach to capstone design projects, pedagogy, or course administration. The strategies presented here are based on the authors' extensive experience teaching capstone design courses and their active and continuous involvement with the broader capstone design community.

3.1 Scaffolding the Design Curriculum

Students often begin their capstone design experience with several deficiencies and lack the following:

- previous team project experiences and the necessary teamwork skills needed for success,
- ability to depend on other students (teammates) for their grade and willingness to relinquish control over tasks that may affect their grade,
- project management skills and the ability to use tools such as work breakdown structures, risk management methods, or project schedules to manage projects,
- time management skills and knowledge on how to run effective team meetings,
- experience in leading a project team and motivating fellow team members,
- knowledge of the product development process and the standards that specify how new products are designed, developed, and tested, and
- familiarity with finance (time value of money, methods for determining return on investment) and marketing.

Given that most capstone design courses are taught in the senior year (Howe et al. 2017), to best prepare students for capstone design, it is helpful to ensure that students have as many team-based design experiences as possible prior to their senior year. For an optimal experience, the capstone design project should not be the first design activity of students' undergraduate careers. Curricular and co-curricular design components can be helpful in scaffolding the design curriculum. They provide students with additional design and team experiences to help prepare them for their capstone design course and project.

3.1.1 Curricular Components

A design map is a useful tool to determine where in the undergraduate curriculum students will have design experiences or learn about design concepts (design spine) (Lulay 2015). These experiences may be provided through lower level design and non-design courses in the form of learning modules (a few class periods on a specific design topic), formal dedicated design courses, or design projects that are part of design or non-design courses. A design map can be created by surveying engineering faculty as to their inclusion of various design components or topics in their courses, through either a design project or lectures or reading assignments. The survey can inquire about items such as the following:

- individual or team design projects
- in-class design competitions
- written and oral reports discussing design projects
- use of schedules to manage projects
- use of engineering drawings to document design
- use of analytical methods to predict performance and evaluate design
- search for applicable standards and prior art
- identification of customer needs
- development of design specifications
- concept generation, evaluation, and selection
- prototyping
- design verification and validation
- intellectual property
- legal and regulatory requirements
- ISO and other standards (design controls, risk management, etc.)
- project management
- opportunity statements, value propositions
- commercialization
- financial, marketing, and other business issues

Providing information on which courses include design projects and relevant design topics allows faculty to perform a gap analysis to identify deficiencies in the design curriculum. Once this information is available, gaps in the design curriculum can be addressed by modifying some existing courses to provide additional opportunities for students to solve open-ended problems and develop their communication, interpersonal, and design skills in courses throughout the four-year curriculum instead of only the senior capstone design course. Course modifications could include adding assignments or classroom activities that would require students to (1) complete group projects, (2) speak in front of groups, (3) write technical reports and other documents, (4) complete design projects (team or individual), and (5) participate in design competitions contained within a particular course. If only one of each of these modifications is made to each of a few courses in the engineering curriculum, students would benefit from the additional opportunities to further develop their skills and would be better prepared for the senior capstone course and beyond. By including a design project in courses offered during the freshman, sophomore, and junior years, engineering design programs can create a four-year design curriculum.

Including some of the components listed above is a good way to incorporate design into non-design courses. Design projects can enhance learning by providing hands-on, active learning opportunities for students to learn about practical applications of the course material. Team projects can provide opportunities to develop teamwork skills, and oral presentations and written reports can help develop oral and written communication skills.

Many engineering schools require first-year introduction to engineering courses during the freshman year. These serve to expose new students to different disciplines within engineering, the profession, career opportunities, and ethics, and they often include design challenges to show students how engineering methods can be applied to solve real-world problems. They also allow students to experience engineering and design early in the curriculum, creating excitement and interest in the field of engineering and in being an engineering student. The prevalence of these courses prompted the American Society for Engineering Education (ASEE) to create the First Year Programs Division (https://sites.asee.org/fpd/) to address the needs of instructors of these freshman courses. Some schools are creating new design courses to cover important design topics (prior to the capstone design course) in the sophomore or junior year. This allows students to learn about design earlier in the curriculum and focus on their design project during the capstone course.

3.1.2 Co-curricular Components

Cooperative education (co-op) and internship opportunities in industry serve as field apprenticeships for undergraduate engineering students (College Affordability Guide 2016) and allow them to see, firsthand, what it will be like to work as an engineer, learn what engineers do on a daily basis, and learn what is needed to be successful on the job. During their experiences, many co-op and internship students become very familiar with standards and regulations that impact how products and processes are developed and tested, which can be helpful to their capstone design projects.

Co-ops and internships provide students with hands-on, experiential learning opportunities that can supplement and/or reinforce what they learn in their capstone design courses, as discussed in Goldberg 2016. Anecdotally, many capstone design course instructors have found that students who have been exposed to the real world through these or other relevant work experiences tend to appreciate the value of their capstone design courses more than those students without such experiences. Prior to the capstone design course, many co-op students are already familiar with some of the topics covered in the course, and often have had some design and testing experience as part of their co-op job. These students often indicate that they are looking forward to the course and the opportunity to work with fellow students to complete an entire design project. During their co-op jobs, they may have been involved with only a few phases of a project and were prevented from seeing it completely through to completion. They now want to learn how to manage a project through more phases of the design process and feel that the capstone design course will allow them to do this.

Another co-curricular option is design competitions. Many national student engineering organizations focus on a design project in which individual student chapters form design teams and compete in national design competitions. Examples include the American Society of Civil Engineers (ASCE) Concrete Canoe, the American Society of Mechanical Engineers (ASME) Human Powered Vehicle, the Society of Automotive Engineers (SAE) Mini Baja Car, and the Engineering World Health (EWH) Design Competitions. These design projects may be part of a capstone design course, but often they are not and team membership is not limited to seniors. Students participate due to their interest in the subject and desire to be part of a team of fellow engineering students from freshman to seniors. National student design competitions increase students' interest in becoming engineers, and help focus student energy, enthusiasm, and talent on a solution to a real-world problem. They help develop students' design and problem-solving skills and can be helpful in preparing underclass students for their capstone design projects in the senior year (Goldberg 2012a).

3.2 Fostering Industry Involvement in Capstone Design Courses

Industry involvement in capstone design courses can provide many benefits to students, faculty, and academic institutions. Industry representatives can support capstone design courses by serving as guest speakers, project sponsors and technical advisors, curriculum advisors, and members of industrial advisory boards.

3.2.1 Guest Speakers

As guest speakers, practicing engineers provide a relevant, practical real-world perspective of their topic, reinforcing its importance to the practice of engineering. Students (and design faculty) appreciate the up-to-date treatment of the topic provided by guest lecturers from industry. Practicing engineers in industry are usually willing to speak to students about topics in their areas of expertise; the opportunity allows them to share their experiences with students and may also be considered as professional development. Being asked to be a guest speaker is flattering to many engineers, as it suggests that they are thought of as experts in a particular topic. In areas with local industry, finding engineers to speak, especially alumni, should not be difficult. Otherwise, travel expenses for guest speakers and honoraria can be offered as an incentive, or guest speakers can participate remotely via videoconference.

3.2.2 Project Sponsors and Technical Advisors

As project sponsors of and technical advisors to capstone design projects, industry representatives provide opportunities for students to work on real-world problems of importance to industry, gain exposure to specific industry and market segments, gain experience with project management and the product development process, and become familiar with various design requirements and constraints.

Academic institutions benefit from industry sponsorship of capstone design projects through the building of relationships with industry (which can lead to research collaborations and grants), maintenance of a high-quality capstone design course and project experiences, and addition of resources available to students to complete their design projects. Many capstone courses require students to construct and test prototypes for design verification and validation. Students often have access to their school's computer network, libraries, makerspaces, and laboratories. Construction and testing of prototypes can be costly and may require processes or specialized test equipment not typically available in the academic setting. In these situations, the industry sponsor may be able to provide the team with the resources needed for prototyping and testing, as well as funds for materials and supplies, and for travel to meet with the sponsor if appropriate.

3.2.3 Industrial Advisory Board Members and Curriculum Advisors

Members of industrial advisory boards can serve as curriculum advisors to ensure that the design curriculum is up-to-date, and the content and objectives of capstone design courses are relevant to professional practice and are helping to prepare students for careers in engineering. Advisory board members can provide a periodic review of course objectives, lecture topics, and required deliverables to help fine-tune the capstone design course curriculum and ensure that students develop the skills needed for successful careers.

3.3 Keeping Course and Faculty Up-to-Date with Current Design Practices

New standards, regulations, and project management and design tools are constantly changing and evolving. To ensure that design courses are (1) up-to-date, (2) relevant to professional practice, and (3) preparing students for the real-world challenges they will face in industry, it is important for capstone design faculty to keep current with these developing trends. In addition to seeking feedback from members of industrial advisory boards on curricular issues, and inviting engineers working in industry to serve as guest speakers as discussed above, there are other ways for faculty to keep themselves and their courses up-to-date with current design practices.

3.3.1 Industry Trade Publications

There are many industry trade publications to which faculty can subscribe (often at no charge) to learn about the latest trends in a particular industry. For example, Medical Device and Diagnostic Industry (MDDI) magazine reports on the latest trends in the medical device industry involving new product development, product design, technology, manufacturing, quality assurance, project management, legal and regulatory requirements, and marketing. Examples in other industries include Engineering News Record, DesignNews, and PE Magazine.

3.3.2 Workshops, Seminars, and Professional Society Meetings

Many professional societies offer workshops, seminars, and courses on various topics of importance to their members. These continuing education and professional development opportunities help keep faculty up-to-date on the latest trends and changes in their fields. Some of these activities may count toward professional development hours (PDH's) needed to maintain professional engineering licensure.

3.3.3 Faculty Internships

A faculty internship or sabbatical experience involving working in industry (Goldberg 2014; Howe 2015) can be extremely helpful in maintaining an up-to-date, relevant capstone design course, especially for faculty with no industry experience. For faculty with industry experience, if it has been several years since working in industry, these experiences can serve as a refresher course to keep current.

3.3.4 Relevant Lecture Topics

Although the focus of the capstone design course is typically on the project, lectures can be an important part of the course. They can provide information not included elsewhere in the curriculum that students need to properly execute their projects and/or prepare them for their careers. Lecture topics should be chosen carefully to make the best use of class time. Topics that are needed by students to effectively manage their projects or prepare them for work as professional engineers should be given priority. Depending upon the engineering discipline, these could include the following:

- project management
- teamwork
- communication skills
- design process/product development process
- design constraints and requirements

- risk management
- business issues
- value creation and commercialization
- career management
- professional legal issues

As shown in Table 4, in 2015, the top five topics most commonly covered in capstone courses were written communication, project planning and scheduling, oral communication, concept generation/selection, and team building/teamwork. Other topics included in capstone design courses are shown in Table 3.

Since few faculty are familiar enough with all of these topics, other faculty or working engineers with expertise in these areas can be asked to serve as guest speakers. This helps spread the load of lecturing among more than one instructor, provides students with good role models (particularly if guest speakers comprise a diverse group), and ensures that students are presented with the most accurate, up-to-date information from experts who deal with it on a daily basis.

3.4 Obtaining Organizational Support for Capstone Courses

To provide a positive experience for students and meet the goals of the capstone design course, organizational support is needed from several areas to provide the resources required for a successful capstone design course, including personnel, equipment, tools, space, funding, policies, and attitudes. These are all part of creating an inclusive culture of design within the academic environment.

3.4.1 Capstone Design Course Instructor

A dedicated instructor is needed who understands and has experience with engineering design, project management, and the product development process. Ideally, the instructor would have some industry experience to share with students and use to guide course design and teaching activities (Goldberg 2007).

3.4.2 Department Chairs, Faculty Project Advisors

In courses with many projects and project teams, the instructor may be unable to effectively advise all project teams single-handedly, depending on his/her other responsibilities and area of expertise. Teams often require a mix of engineering disciplines to solve the problem. Few, if any, faculty are experts in all areas of engineering. In this case, other faculty members with the expertise required for a particular project can be recruited to advise these teams.

To be effective, the department chair and faculty project advisors must (1) recognize the importance of the capstone design experience in helping prepare students for careers in engineering, (2) support the goals of the course, (3) become familiar with the design process, and (4) be willing to commit their time to advising student project teams. In some departments it is assumed that all faculty members will advise a project team. Faculty consider project advising to be part of their jobs, and volunteer advisors are easy to find. In other departments, the department chair might need to provide incentives, such as additional salary or credit toward a reduced teaching load, to incentivize faculty members to participate. Junior faculty may not want to spend time advising project teams if this activity does not count toward their requirements for tenure. In this case, allowing time spent advising project teams to count toward the requirements for tenure could help get more junior faculty involved.

Department chairs should also provide the financial resources needed to manage the capstone design course. This could include salaries for support staff to assist in soliciting projects, honoraria for guest speakers, and money for teams to build and test prototypes, especially for teams without industry sponsors that would normally provide resources for these activities. The creation and maintenance of a library containing up-to-date industry standards would also be helpful to capstone design teams (Goldberg 2007).

3.4.3 Deans of Colleges of Engineering

If individual departments are unable to provide necessary resources, then the College of Engineering may need to provide support. For example, additional funds could be made available for project teams as well as resources (such as a machine shop, prototyping facilities, and test equipment) needed to construct and test prototypes. The Dean of the College must be supportive of the goals of the design program and the capstone design course. For capstone design courses involving students from several engineering disciplines (biomedical, electrical, mechanical, etc.), the College of Engineering should also provide adequate staff for managing the course as well as funding for guest speaker honoraria.

3.4.4 University Administration

The University administration can support design projects by (1) providing a responsive institutional review board to expedite projects involving human subjects, (2) establishing a faculty- and industry-friendly intellectual property policy that helps facilitate industry sponsorship of senior capstone design projects, and (3) providing support from grants/contracts personnel to assist with externally sponsored projects.

3.4.5 Project Funding

As shown in Figs. 9 and 12, expenses in capstone design projects are multifaceted, and can range substantially in magnitude. As shown in Fig. 10 and Table 10, possible sources and levels of funding may also vary significantly. The most important consideration in soliciting and securing funding for capstone design projects is to ensure that it meets or exceeds expenses so that the course remains financially sustainable; note that this may well be an ongoing task for the capstone design instructor or other course support staff.

3.4.6 Entrepreneurial Ecosystem

For schools with an established entrepreneurial ecosystem, support from other programs that are part of the ecosystem can be helpful in creating truly multidisciplinary capstone design teams. To engage students and faculty from other programs (such as business, natural sciences, social sciences, arts, or medicine), incentives must be provided to encourage them to collaborate on capstone design projects. Offering credit for participation in the capstone design course can help get more students from other disciplines involved in design projects.

3.4.7 Creating a Culture of Design

Creating a culture of design throughout all four years of the engineering curriculum is an excellent way to prepare students for a successful capstone design experience and support the course. Implementing some of the methods for scaffolding the design curriculum (discussed previously) also serves to create a culture of design throughout the curriculum. Sponsoring departmental or college of engineering design competitions, creating a senior (and other) design project public showcase event, and providing resources for non-project–related design activities (that would allow students to "tinker" with their own ideas not associated with any course or course-related project) are additional ways to help create a culture of design. Support from college of engineering administration and faculty is vital to creating a vision and lasting culture of design.

3.5 Sourcing Capstone Design Projects

There are several types of design projects often found in the project portfolios of capstone design courses. The goal of each is to develop a design solution to a specific problem. Each type provides a different learning experience for the students. Capstone design projects typically come from industry, government, faculty, design competitions, community or national organizations, clients from the local community, and students. Ideally, projects are based on problems and unmet needs that present opportunities for new products or processes that can provide value to customers and users.

3.5.1 Industry-Sponsored Projects

Industry-sponsored projects are typically based on a defined problem that a company (or government agency) needs a team of engineering (and other) students to solve for them. This type of problem may involve the design of a new or improved product (new feature, easier to use, lower cost, new-to-the-world product), production process, or specialized test equipment. Students get the opportunity to work on real-world problems of importance to industry/government and learn about the needs of a particular market and the operations of a company. Students learn firsthand about the requirements and constraints of design. Experience gained from industry-sponsored projects can help prepare students for careers in industry or government.

3.5.2 Faculty-Sponsored Projects

Faculty-sponsored projects typically involve the design of a tool, apparatus, or piece of equipment to be used to conduct research in the laboratory of the faculty sponsor. To provide an adequate design experience to students, these projects should contain significant design content, including identification of customer needs, development of potential designs, and construction and testing of prototypes, and should not require students to conduct research for the faculty sponsor and/or write research papers for publication.

3.5.3 Design Competitions

Projects based on design competitions, such as those mentioned previously, often involve a problem presented by the sponsoring organization. Students interested in solving the stated problem are encouraged to form teams. Design constraints and requirements may be defined by the sponsoring organization as well as rules involving team size, composition, and required documentation (in addition to what is required by the capstone design course).

3.5.4 Community Organizations

Local community or international organizations can be sources of projects that address problems supported by the organization's mission. These may include service learning projects that address community needs or the needs of developing nations. For example, physicians or family support groups can help identify local clients with disabilities who are in need of assistive technologies and devices. Humanitarian groups might propose the design of a water treatment system or community building.

3.5.5 Student Ideas

In some disciplines graduates will eventually work for companies and become involved in identifying unmet needs that can lead to opportunities for new products. Providing students with a list of design projects from which to choose prevents them from gaining experience in needs-finding activities. Several approaches to enabling student-generated projects include (1) soliciting student project ideas in the summer before the capstone design course starts, (2) providing a few weeks at the beginning of the semester for students to generate project ideas, or (3) offering a course prior to the senior year that focuses on need finding to identify a problem to pursue during capstone (Goldberg 2012b).

3.6 Preparing Students for Professional Practice

The capstone design course is perhaps the most important course of the undergraduate engineering curriculum. It provides a valuable opportunity for students to apply what they learned up until (and including) their senior year and get a preview of what it will be like to work as an engineer. Opportunities to develop the communication, teamwork, and project management skills needed for successful careers are embedded in most capstone design courses and are discussed in previous sections. Additional skills and experiences to enhance the course and better prepare students for professional practice include the ability to create value for a customer, experience working on a multidisciplinary team, and experience with a virtual team.

3.6.1 Value Creation

Engineering students should understand that design is not about creating "cool" technologies, but solving a problem in a way that provides value to a customer or client. Requiring students to understand the customer's needs, associated problems, existing solutions, and potential improvements will help students provide value to the customer. The Value Proposition Canvas is a useful tool that can be used to graphically define and display customer jobs, pains, gains, pain relievers, gain creators, and other information that helps the team create a value proposition for their project (Osterwalder et al. 2014). This value proposition can serve to guide a team toward a solution that creates value for their customer.

3.6.2 Multidisciplinary Teams

Many capstone design courses include students from more than one engineering discipline (Howe et al. 2017). Although project teams in these courses may consist of students from different disciplines, they are all engineering students who tend to approach problems in the same way. When engineering graduates enter the workforce, they will be expected to work on truly multidisciplinary teams. In industry, these teams typically consist of members of research and development, marketing, production, finance, regulatory affairs, and other departments. Engineering graduates will work with people with different areas of expertise, levels of education, and social and cultural backgrounds who tend to look at problems in different ways. This multidisciplinary approach often leads to better solutions to a problem. For example, engineers and industrial designers often collaborate in industry on the design of new products; their approaches to problem solving are different and they emphasize different aspects of design, but both provide valuable contributions that one alone could not (Goldberg and Malassigne 2017). Providing capstone design students with the opportunity to work with students from other disciplines can help prepare them for work on multidisciplinary teams in industry. Recommendations for effectively implementing multidisciplinary teams in capstone design courses have been documented elsewhere (Goldberg et al. 2014).

3.6.3 Virtual Teams

Global innovation requires collaboration between groups of people located in different parts of the world, and is a growing trend in industry. Virtual teams are often used to manage new product development projects. These teams are similar to traditional teams but are geographically separated and rely heavily on virtual methods of communication (email, video conferencing, teleconferencing, etc.) instead of regular face-to-face meetings. Experience working as a member of a virtual capstone design team can help prepare students for this growing trend, but requires careful and thoughtful implementation to maximize learning. Several recommendations for including virtual teams in capstone design courses have been documented elsewhere (Goldberg and Howe 2015).

4 Summary

This chapter presented the current state of engineering capstone design education and highlighted changes to capstone design practices in the U.S. over the past 25 years, based on extensive, seminal surveys of capstone design instructors. It discussed current practices and successful strategies in engineering capstone design education. The chapter included perspectives and feedback from hundreds of engineering capstone design faculty regarding their personal experiences with capstone design programs.

This chapter also provided recommendations based on the authors' vast experience teaching and managing engineering capstone design courses and engaging with the engineering capstone design community. It included multiple strategies for supporting capstone design courses and enhancing student experiences in these courses: scaffolding the design curriculum, fostering industry involvement in capstone design courses, keeping courses and faculty up-to-date with current design practices, obtaining organizational support for capstone design courses, sourcing capstone design projects, and preparing students for professional practice. These strategies can be helpful to current and future engineering capstone design instructors and may be applicable to non-engineering capstone design courses as well.

References

ABET (2018) Criteria for accrediting engineering programs 2018–2019. http://www.abet.org/wp-content/uploads/2018/02/E001-18-19-EAC-Criteria-11-29-17.pdf. Accessed 11 July 2018

ASEE (2012) Innovation with impact: creating a culture for scholarly and systematic innovation in engineering education. American Society for Engineering Education, Washington, DC

Bright A, Phillips JR (1999) The Harvey Mudd engineering clinic: past, present, future. J Eng Educ 88(2):189–194

College Affordability Guide (2016) Degree prospects: course credit plus a paycheck. http://www.collegeaffordabilityguide.org/co-op-programs. Accessed 11 July 2018

Dixon, D (2012) Experiencing capstone design problem statements. In: Proceedings of the American Society for Engineering Education, San Antonio, TX

Dixon, D (2014) Technical design reviews in capstone. In: Proceedings of the American Society for Engineering Education, Indianapolis, IN

Goldberg J (2007) Capstone design courses: producing industry-ready biomedical engineers. Morgan and Claypool, Boulder CO

Goldberg J (2012a) Capstone design courses: part II: preparing biomedical engineers for the real world. Morgan and Claypool, Boulder CO

Goldberg J (2012b) Learning to identify unmet needs and new product opportunities. Int J Eng Educ 28(2):349–354

Goldberg J (2014) Faculty internships for capstone design instructors: maintaining an up-to-date capstone design course. IEEE Pulse

Goldberg J (2016) Co-ops and capstone design: are they interchangeable? IEEE Pulse

Goldberg J, Cariapa V, Corliss G, Kaiser K (2014) Benefits of industry involvement in multidisciplinary capstone design courses. Int J Eng Educ 30(1):6–13

Goldberg J, Howe S (2015) Virtual capstone design teams: preparing for global innovation. Int J Eng Educ 31(6B):1773–1779

Goldberg J, Malassigne P (2017) Lessons learned from a 10-year collaboration between biomedical engineering and industrial design students in capstone design projects. Int J Eng Educ 33(5):1513–1520

Hauhart R, Grahe J (2015) Designing and teaching undergraduate capstone courses. Jossey-Bass, San Francisco, CA

Howe S (2008) Focused follow-up to 2005 national survey. In: Proceedings of American Society for Engineering Education, Pittsburgh, PA

Howe S (2010) Where are we now: statistics on capstone courses nationwide. Adv Eng Educ 2(1):1–27

Howe S (2015) Industry immersion: the impacts of a sabbatical deep-dive. In: Proceedings of American Society for Engineering Education, Seattle, WA

Howe S, Lasser R, Su K, Pedicini S (2009) Content in capstone design courses: pilot survey results from faculty, students, and industry. In: Proceedings of American Society for Engineering Education, Austin, TX

Howe S, Rosenbauer L (2017) Extending the 2015 capstone design survey: data from Australia and New Zealand. In: Proceedings of American Society for Engineering Education, New Orleans, LA

Howe S, Rosenbauer L, Poulos S (2017) The 2015 capstone design survey results: current practices and changes over time. Int J Eng Educ 33(5):1393–1421

Layton R, Loughry M, Ohland M, Ricco G (2010) Design and validation of a web-based system for assigning members to teams using instructor-specified criteria. Adv Eng Educ 2(1):1–28

Lulay K (2015) Implementation of a design spine for a mechanical engineering curriculum. In: Proceedings of American Society for Engineering Education, Seattle, WA

McKenzie, L, Trevisan M, Davis D, Beyerlein S (2004) Capstone design courses and assessment: a national study. In: Proceedings of American Society for Engineering Education, Salt Lake City, UT

Osterwalder A, Pigneur Y, Bernarda G, Smith A (2014) Value proposition handbook. Wiley, Hoboken, NJ

Pembridge J, Paretti M (2010) The current state of capstone design pedagogy. In: Proceedings of American Society for Engineering Education, Louisville, KY

Rao M, Hayes C, Lynch K (2015) Survey of mechanical engineering capstone design courses—summary of results. ASME International Mechanical Engineering Education Leadership Summit, Newport Beach, CA

Silverstein D, Bullard L, Seider W, Vigeant M (2013) How we teach: capstone design. In: Proceedings of American Society for Engineering Education, Atlanta, GA

Todd R, Magleby S, Sorensen C, Swan B, Anthony D (1995) A survey of capstone engineering courses in North America. J Eng Educ 84(2):165–174

Bio-inspired Design Pedagogy in Engineering

Jacquelyn K. S. Nagel, Christopher Rose, Cheri Beverly
and Ramana Pidaparti

Abstract Every day, engineers are confronted with complex challenges that range from personal to municipal to national needs. The ability for future engineers to work in cross-disciplinary environments will be an essential competency. One approach to achieving this essential competency is teaching biomimicry or bio-inspired design in an engineering curriculum. Bio-inspired design encourages learning from nature to generate innovative designs for man-made technical challenges that are more economic, efficient and sustainable than ones conceived entirely from first principles. This chapter reviews current teaching practices and courses in engineering curricula for training students in multidisciplinary design innovation through bio-inspired design. Emphasis is placed on theory-based and evidenced-based approaches that have demonstrated learning impact. The significance and implications of teaching bio-inspired design in an engineering curriculum are discussed, and connections to how the essential competencies of future engineers are fostered is addressed. Teaching bio-inspired design in an engineering curriculum using cross-disciplinary approaches will not only develop essential competencies of tomorrow's engineer, but also enable students to become change agents and promote a sustainable future.

Keywords Bio-inspired design · Biomimicry · Innovation · Cross-disciplinary

J. K. S. Nagel (✉)
Department of Engineering, James Madison University, Harrisonburg, VA 22807, USA
e-mail: nageljk@jmu.edu

C. Beverly
Learning, Technology & Leadership Education Department, James Madison University,
Harrisonburg, VA 22807, USA
e-mail: beverlcl@jmu.edu

R. Pidaparti
College of Engineering, University of Georgia, Athens, GA 30602, USA
e-mail: mparti@uga.edu

C. Rose
Department of Biology, James Madison University, Harrisonburg, VA 22807, USA
e-mail: rosecs@jmu.edu

© Springer Nature Switzerland AG 2019
D. Schaefer et al. (eds.), *Design Education Today*,
https://doi.org/10.1007/978-3-030-17134-6_7

1   Introduction

It is well known that engineering involves integrating broad knowledge towards some purpose, generally to address a need or solve a problem. As we move into a global future, engineers can no longer isolate themselves and must be prepared to work across disciplinary, cultural, political, and economic boundaries. Every day, engineers are confronted with complex challenges that range from personal to municipal to national needs. The ability for future engineers to work in cross-disciplinary environments will be an essential competency. Furthermore, with greater emphasis being placed on understanding social, economic and environmental impacts of engineered solutions, another essential competency is the cognitive flexibility to think about the whole system at different levels of fidelity and in different time scales. Undergraduate education must train students to not only solve engineering challenges that transcend disciplinary boundaries, but also communicate, transfer knowledge, and collaborate across technical and non-technical boundaries. One approach to achieving this goal is teaching biomimicry or bio-inspired design in an engineering curriculum. Bio-inspired design encourages learning from nature to generate innovative designs for man-made technical challenges that are more economic, efficient and sustainable than ones conceived entirely from first principles.

Cross-disciplinary instruction in bio-inspired design can increase engineering students' creativity and adaptive problem solving skills. Having to retrieve and transfer knowledge from domains outside of engineering forces students to adapt to unfamiliar languages and content formats (which addresses non-technical skills) in order to apply the biological information intelligently to engineering problems (which addresses technical skills). Additionally, biomimicry touches on many areas of engineering including electrical, mechanical, materials, biomedical, chemical, manufacturing and systems, which makes it applicable in a wide range of engineering programs, from discipline-specific to general ones. Multidisciplinary design through bio-inspired design will be vital to promoting future innovation in engineering design.

Showing engineering students the significance and utility of bio-inspired design is easy. Teaching them how to do bio-inspired design without requiring them to also be fully trained as biologists is much more difficult. This chapter reviews current teaching practices and courses (undergraduate and graduate) in engineering curricula for training students in design innovation through bio-inspired design. Emphasis is placed on theory-based and evidenced-based approaches that have demonstrated learning impact. The significance and implications of teaching bio-inspired design in an engineering curriculum are discussed, and connections to how the essential competencies of future engineers are fostered is addressed. Teaching bio-inspired design in an engineering curriculum using cross-disciplinary approaches will not only develop essential competencies of tomorrow's engineer, but also enable students to become change agents and promote a sustainable future.

2 Bio-inspired Design Overview

The natural world provides numerous cases for inspiration in engineering design. Biological organisms, phenomena, and strategies, which can be grouped and referred to as biological systems, are exemplary systems that provide insight into sustainable and adaptable design. Biological systems offer engineers billions of years of valuable experience, which can be used to inspire engineering designers. Biomimetic design "offers enormous potential for inspiring new capabilities for exciting future technologies" (Bar-Cohen 2006a) and encourages engineering innovation (Bar-Cohen 2006b; Lindemann and Gramann 2004). Furthermore, engineers can not only mimic what is found in the natural world, but also learn from those natural systems to create reliable, smart and sustainable designs.

The name *biomimetics* was coined by the biophysicist Otto Schmitt during the 1950s (Schmitt 1969). During that same time period U.S. Air Force flight surgeon and psychiatrist Major Jack Steele coined the term *bionics* (Steele 1960), which is widely used in Europe. It was not until the 1990s that the term *biomimicry* was popularized by Benyus (1997) who co-founded the Biomimicry Institute. However, *biologically-inspired design*, or *bio-inspired design*, is preferred in engineering, especially in the USA, to avoid confusion with cybernetics and medical bionics, but also to include solutions that were inspired by biology and do not directly copy a specific feature. Internationally, however, the terms of *bionics* and *biomimetics*, are preferred. Fundamentally, all these names refer to the same goal, which is to study and learn from nature to solve human problems. Bio-inspired design results in the discovery of non-conventional solutions to problems that are often more efficient, economic and elegant. Furthermore, it is a problem solving lens which has resulted in technical innovation.

A handful of engineering innovations resulting from studying and mimicking nature have become so integrated into our society that they have become commonplace. Velcro©, aircraft, and pacemakers, are all based on bio-logical inspiration, and are engineering breakthroughs for materials, aeronautics and medicine, respectfully. Within the last thirty years several breakthroughs in fluid dynamics, sensors, materials, computational algorithms, alternative energy, and sustainable architecture have had the commonality of taking inspiration from nature (Nachtigall 2000; Brebbia et al. 2002; Brebbia and Collins 2004; Brebbia 2006, 2008; Brebbia and Carpi 2010; Forbes 2006; Bar-Cohen 2006b).

Bio-inspired designs are often considered novel and innovative, and inspiration can occur one of three ways: chance observation, systematic exploration, or through dedicated study. These three reveal that a spectrum of inspiration approaches exists, which further alludes to the struggle of how to navigate the vast amount of biological information available. With bio-inspired design emerging as its own field, engineering design research has begun to investigate methods and techniques to systematically transfer biological knowledge to the engineering domain. The main goal of these research efforts is to create tools, methods, and knowledge to facilitate design activities for engineers with a variable background in biology, reduce the element of

Fig. 1 Generic bio-inspired
design process

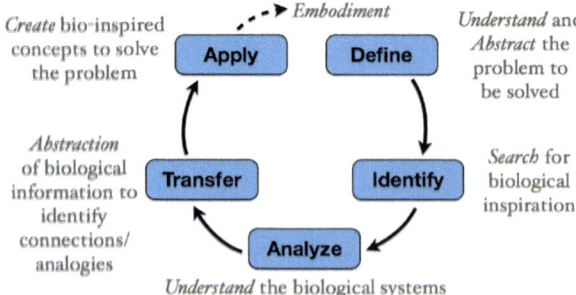

chance, and reduce the amount of time and effort required for developing solutions. Prominent research in bio-inspired design to support systematic exploration to find biological inspiration has led to development of inspiration facilitators, representation methods, information transfer methods, and concept generation techniques. Consequently, studies and surveys have been conducted on the tools and methods to reveal strengths as well as gaps in the literature with respect to the process of bio-inspired design (Fu et al. 2014; Wanieck et al. 2017; Fayemi et al. 2017; Kruiper et al. 2018).

Multiple process models for bio-inspired design exist and vary depending on the tools used, level of automation involved, theory basis, or the intended audience. For example, a process model based in function-based design theory intended for engineers will include specific steps for creating or using functional models, whereas a process model for a multidisciplinary team will include a step for biologizing the design challenge, or reframing the challenge into a biological context. A generic process for bio-inspired design, irrespective of theory basis and audience, is given in Fig. 1. This generic process model includes the fundamental steps that are necessary to start from a problem and arrive at a conceptual design that meets the needs of the problem. Once the generic process is completed, it is expected that the designer would transition to the later phases of a traditional engineering design process to vet the bio-inspired solution.

3   Bio-inspired Design Teaching Practices in Engineering Curricula

This section reviews a variety of teaching practices that have been employed to teach bio-inspired design with an emphasis on design theory-based and evidenced-based approaches. Each teaching practice is described as well as explained in terms of how it can be implemented for teaching bio-inspired design.

3.1   Design Theory Based Practices

Using engineering design theory-based tools and methods offers an opportunity to bridge theory and practice in the classroom. Design theory-based instruction in bio-inspired design has occurred through the practices of function and concept-knowledge theory and TRIZ.

3.1.1   Function Theory Based Design

Function-based design is a theory that utilizes solution-neutral, qualitative models (abstractions) in early design to reduce fixation on a solution prematurely while broadening the design solution space. Although these models are intended to bridge the system requirements to conceptual design, they can also be used to capture functionality of existing engineered or biological systems. Inclusion or analysis of biological systems into function theory for engineering design has taken three avenues: flow based functional modeling, structure-behavior-function (SBF) modeling, and SAPPhIRE modeling.

A flow based functional model is a description of a product in terms of the elementary functions and flows that are required to achieve the product's overall function or purpose. Function is the solution-neutral, verb statement of what something does, while a flow is the solution-neutral, noun object being operated upon by the function. A graphical form of a flow based functional model is represented by a collection of sub-functions connected by the flows on which they operate or transform. This structure is an easy way for a designer to see what type of functions are performed without being distracted by any particular form the product may take. The technique is widely used in early design and can be found in many texts on engineering design methodologies (Ulrich and Eppinger 2004; Cutherell 1996; Fenves 2001; Hundal 1990; Miles 1972; Pahl and Beitz 1996; Otto and Wood 2001; Dieter 1991; Ullman 2002). To aid with prescribing functionality during the modeling process, a taxonomy of function (verb) and flow (noun) terms called the Functional Basis is used (Stone and Wood 2000; Hirtz et al. 2002). The Functional Basis spans all engineering domains while retaining independence of terms. The function set of the Functional Basis is broken down into eight categories termed the primary classes. These classes have further divisions at secondary and tertiary levels that offer increasing degrees of specialization. The primary class represents the broadest definition of distinct function while the tertiary class provides a very specific description of function. The flow set of the Functional Basis allows for the associated function's input and output flows to be described. Similarly to the function set, there are three distinct classes within the flow set of the functional lexicon. Within the primary class of the flow set, there are three main categories used to describe flow: material, signal and energy, as popularized by (Pahl et al. 2007). Each of these categories has the capability to represent the input or output of a function. It is the secondary classes of functions and flows of the Functional Basis that are primarily used when describing a product.

At Oregon State University and James Madison University flow based functional models are used to facilitate bio-inspired solutions, starting from a problem or curiosity, following a systematic methodology supported by a design tool framework (Nagel et al. 2010b, 2014; Nagel and Stone 2011a, b). When starting from a problem to solve, the models assist with identification of functions that are used to search for biological inspiration. When starting from a curiosity, functional models of biological systems are created to directly compare biological and engineering system functional models to formulate analogies and identify engineering components that can solve the biological system functions (Nagel et al. 2008, 2010a). Functional modeling and abstraction converts the biological information into an engineering context to facilitate understanding, reduce fixation on physical features, and archive design information. Integration of bio-inspired design with flow-based functional modeling aims to leverage existing function-based design tools and methods, while also reducing the amount of time and effort required for intentionally generating bio-inspired engineering solutions.

Structure-behavior-function (SBF) modeling is a technique that uses abstractions for understanding the causal behaviors of physical systems (Goel et al. 2009). Structure specifies the physical description for components and the connections among them (what). Function refers to the pre- and post- conditions for the behavior of the system (why). Behavior specifies the causal processes or mechanisms occurring in the system including the transition between states (how). The graphical form of an SBF model is represented by independent behavior and structure models as well as a hierarchal function diagram with structure and behavior information.

At the Georgia Institute of Technology SBF models are used to populate a searchable database to support searching for inspiration with the intention of supporting a computational approach to bio-inspired design through analogy (Yen et al. 2010; Vattam et al. 2010; Goel et al. 2012). Additionally, the models aid in understanding the biological system (Vattam et al. 2011). Specifically they are used to organize the information learned during a found object exercise where students are asked to go outside and observe an interesting phenomena in nature (Yen et al. 2014).

The third functional modeling technique is SAPPhiRE, which is comprised of seven elementary constructs: State-Action-Part-Phenomenon-Input-oRgan-Effect (Chakrabarti et al. 2005). The SAPPhiRE model of causality was developed to explain the functioning of an entity that uses physical phenomena for attaining its functions. These models provide a rich description of behavior through multiple views of function and behavior that include properties of the system and its context or environment. They also link function, behavior, and structure of a system to each other in a way that is common for natural and technical systems at various levels of abstraction.

At the Indian Institute of Science, SAPPhiRE models are used to support bio-inspired design (Sarkar et al. 2008). They are used for ideation through a software package entitled Idea-Inspire, which interfaces with a database of natural and complex artificial mechanical systems indexed by the SAPPhiRE model (Srinivasan and Chakrabarti 2009a, b). With the problem to be solved described in terms of the SAPPhiRE model constructs, the software allows the search for analogical entries in the database to stimulate ideation of bio-inspired solutions.

3.1.2 C-K Theory Based Design

Concept-Knowledge (C-K) theory, introduced by Hatchuel and Weil, integrates creative thinking and innovation by utilizing two spaces: (1) The knowledge space (K)—a space containing propositions that are known; and (2) The concepts space (C)—a space containing concepts that are propositions or groups of propositions that are undetermined in K (Hatchuel et al. 2004, 2011; Shai et al. 2009; Hatchuel and Weil 2003, 2009). Concepts generate questions and research to answer those questions will generate new knowledge that will provide new attributes for new concepts. The design path is defined as a process that generates concepts from an existing concept or transforms a concept into knowledge. Although specific tools are not embedded, C-K theory has shown to reduce fixation and improve the knowledge and creativity of the user (Hatchuel et al. 2004, 2011; Shai et al. 2009; Hatchuel and Weil 2003, 2009).

There are four operations allowed: expansion of each space ($C \rightarrow C, K \rightarrow K$), conjunction meaning when a concept proposition is tested and leads to new knowledge ($C \rightarrow K$), and disjunction meaning when a new concept is generated from existing knowledge ($K \rightarrow C$). Concepts can be partitioned or included, but not searched or explored in the C space. Adding new properties to a concept results in the concept being partitioned into sets or subsets of concepts. The reverse, subtracting properties from a concept, results in subsets being included into the parent set. After partitioning or inclusion, concepts still remain concepts ($C \rightarrow C$), but they can also lead to the creation of new propositions in K ($C \rightarrow K$). The combination of knowledge and addition of new discoveries expands the knowledge space ($K \rightarrow K$) and can result in new concepts ($K \rightarrow C$).

Innovation is the direct result of the two operations that move between the spaces: using the addition of new and existing concepts to expand knowledge, and using knowledge to expand concepts. Knowledge is therefore not restricted to being a space of solutions, but rather it can be leveraged to improve the chance of innovative designs. C-K theory provides a framework for a designer to navigate the unknown, to build and test connections between the knowledge and concept spaces (analogies), and to converge on a solution grounded in theory combined with new knowledge. It is adaptive and generalizable across scientific and engineering domains, which makes it amenable to bio-inspired design (Hatchuel and Salgueiredo 2016).

As C-K theory emphasizes connection building as well as exploration and expansion of both spaces to iterate to a better solution, thus inherently modeling cross-domain linkages, it was used to create instructional resources for teaching bio-inspired design at the undergraduate level (Nagel et al. 2016; Pidaparti and Nagel 2018). Because C-K theory is a visual approach to structuring the discovery process of learning from the knowledge and concept spaces, a C-K mapping template was created. The template is an adaptable instructional resource that can facilitate the knowledge transfer processes of bio-inspired design going from biology to engineering (biology-driven) as well as engineering to biology (problem-driven) if starting from the knowledge or concept side, respectively. An accompanying set of guidelines for filling out the template was created to assist novice learners. As an adaptable resource, the template can be used at multiple learning levels (e.g., novice, interme-diate, expert) by adding or subtracting supplemental information and by choice of design path. This approach is further detailed in Sect. 6.2.

3.1.3 TRIZ Based Design

The Theory of Inventive Problem Solving (TRIZ) is a method used to trigger creativ-ity in problem solving (Altshuller 1984). The foundation of TRIZ is the compilation and organization of technical knowledge through the analysis of patents. Recurring conflicts and their solutions were identified and categorized into a set of 39 technical characteristics and 40 inventive principles for overcoming contradictions. Design problems are formulated as technical contradictions in terms of the 39 character-istics, with one characteristic worsening and one characteristic improving. These two characteristics are found within the TRIZ matrix to identify which of the 40 principles has potential to solve the contradiction.

To extend TRIZ to facilitate bio-inspired design, Vincent and Mann abstracted and categorized biological information and principles by engineering trade-offs to create BioTRIZ (Vincent and Mann 2002). There are six biological characteristics (fields) that the 40 inventive principles are mapped to for both technical and biological approaches to the resolution of the trade-offs, and are termed the PRIZM matrices. These matrices allow translation between BioTRIZ and TRIZ. Although problems in engineering and biology can be similar, the inventive principles that nature and technologies use to solve the problems can be very different. The BioTRIZ matrix is used simultaneously with the TRIZ matrix to develop complete concepts.

At the University of Detroit Mercy, the combination of TRIZ and biomimicry is used to teach innovation and entrepreneurship to undergraduate engineers in an engineering design course (Weaver et al. 2012). In their approach, a designer utilizing the TRIZ methodology is referred to the relevant principles with man-made and biological examples that illustrate the application of the principles. This expands the scope of the design space and facilitates the concept development phase.

3.2 Evidence Based Practices

Instructional practices shown empirically to enhance learning are frequently used in engineering classrooms. Evidence based instruction in bio-inspired design has occurred through the practices of project based and experiential learning.

3.2.1 Project Based Learning

Project-based learning is a well-established, active learning teaching method in which learners engage with an authentic, ill-structured problem (Jonassen and Hung 2008).

It provides students with an opportunity to gain knowledge and skills by addressing a complex question or problem through investigation, research, and application. In practice, project-based learning is student centered and inquiry driven, and allows for students to learn deeply through their own discovery and mastery.

Bio-inspired design concepts and examples have been used by several institutions to educate students on design innovation and as another source of design inspiration. Institutions include Oregon State University, University of Georgia, James Madison University, Purdue University, Clemson University, Penn State University-Erie, University of Maryland, Indian Institute of Science, University of Toronto and Ecole Centrale Paris to name a few. Often the instruction is across less than four lectures, which reduces the burden of integration into existing courses. These institutions also require engineering students to complete assignments or a project involving bio-inspired design to practice the technique and demonstrate its value. Integration occurs at the freshman through senior levels, in a variety of departments, and primarily depends on when engineering design is offered in the curriculum. Consequently, varying levels of instruction and support are provided to the students, and many rely on the resources provided by the Biomimicry Institute, such as the database AskNature.org.

At Kettering University, in the Industrial and Manufacturing Department, biomimicry is integrated into an ergonomics course through project-based learning (Lynch-Caris et al. 2002). Students work individually on projects using the Biomimicry Innovation Tool, which blends aspects of problem-based learning, innovation, biomimicry, and ergonomics into a single student experience. They present their bio-inspired concept at the end of the course.

3.2.2 Experiential Learning

Experiential education was once considered as experiences beyond the classroom and curriculum such as internships, study abroad, service-learning and field experiences. Kolb, however, defines experiential learning as incorporating learning through experience as well as the intentional role of reflection (Kolb 1984), and does not explicitly state that the experience needs to be outside the classroom. Successful experiential learning in the classroom can be achieved through the Kolb learning cycle, which describes experiential learning in four stages: concrete experience, reflective observation, abstract conceptualization, and active experimentation, which flows back to concrete experience. It is rooted in constructivist theory of learning (Kolb 2014) and has been used successfully in undergraduate engineering, science, and nursing education.

An example is the first-year engineering design and communication course at the University of Calgary (Eggermont et al. 2017). The course is a requirement for all incoming engineering students before declaration of an engineering program in year two. The course introduces students to teamwork, hands on design projects and individual paper-based design projects. Biomimicry is introduced as a design process with the expectation that students will discover multidisciplinary solutions that are motivated by nature and that incorporate elements of natural principles such as feedback loops, multifunctionality, water-based chemistry. The course emphasizes communication through flow-chart based challenges and strategy diagrams as well as hand sketching of design principles, biological strategies, and final concepts. Students are exposed to experience and reflection through execution of the bio-inspired design process and the many visual modes of communication.

A second example of experiential learning applied to bio-inspired design is the course at James Madison University described in Sect. 7.1.

3.3   Other Practices

In addition to theory-based and evidence-based practices, instruction in bio-inspired design has also been conveyed in the context of sustainability and a multidisciplinary environment.

3.3.1   Sustainability-Focused Instruction

Sustainability can be defined in terms of three dimensions: environmental, social and economic (also known as people, planet, and profit) (Pappas et al. 2013; Moore 2005). It is not only concerned with environmental aspects and consequences of our actions, but also making sure that society's needs are fulfilled while businesses remain profitable. A whole systems perspective is necessary to understand the integration and interaction of these different dimensions (Fiksel 2012; Clayton and Radcliffe 1996). Systems thinking supports the construction of dynamic models that enable integrated assessment of the costs and benefits of possible solutions.

At Villanova University, bio-inspired design is taught in the context of sustainability as part of the sustainable engineering graduate program (Cross-cutting Sustainable Engineering Courses 2016). This course starts with assessing current, non-biomimetic solutions to sustainability issues and their associated benefits. Following the introduction to sustainability, students receive instruction on nature's principles and capabilities related to the issues. The last third of the course focuses on creating improved solutions by applying what was learned from nature to reduce sustainability issues and achieve greater societal benefits. A key outcome of the course is a whole system understanding of sustainability and how to assess the impact of technological solutions.

3.3.2   Multidisciplinary Instruction

Bio-inspired design offers an exciting and relevant approach to multidisciplinary education. Multidisciplinary courses are co-taught by at least two faculty from different disciplines, which aims to promote cross-disciplinary learning, foster teamwork among the student teams, and improve student engagement (Ludwig et al. 2017). The benefits for students taking courses that are taught by a multidisciplinary teaching team include improved student–teacher relationships, more engaging course format, greater diversity of perspectives, learning to incorporate information from another discipline into their own discipline of study, critical thinking skills, and improved communication skills (Letterman and Dugan 2004).

At Virginia Tech a multidisciplinary course in bio-inspired design is offered that provides training in the methods and techniques of bio-inspired design that facilitate the use of biological functions and forms in solutions for human challenges (Kennedy et al. 2014). The pedagogy used in the course is collaborative teaching and follows a team teaching approach. Course planning and instruction of all students is a shared responsibility. All instructors play an active role in the course both in and out of the classroom, and leadership is rotated among the instructors based on the topic of the week. The course consists of two halves, the first of which presented relevant materials about bio-inspiration in lecture format. The second half of the course switched to a project format wherein students tackled a design challenge centered around solving a "wicked problem" (i.e., Water Stewardship). The output of the course is a conceptual design and prototype.

4   Best Practices

When incorporating bio-inspired design into a course or developing a course dedicated to the topic there are multiple best practices to consider. Each of the following paragraphs explains a single best practice.

For students to fully appreciate the time and effort required to arrive at a bio-inspired solution it should be taught as a process rather than a single step within the engineering design process. There are multiple methodologies accessible through research journals and conferences, some grounded in engineering design theory, as well as through websites. Additionally, there are tools that can aid with specific steps within the bio-inspired design process.

Students should be trained to recognize the difference between biological inspiration and biological imitation so they can work toward innovative solutions. Imitation occurs when visual or behavioral characteristics of a biological system are directly copied, such as in the creation of an artificial organ. Inspiration occurs when learning from the biological system and making connections (e.g., first principles, analogies) to knowledge within the domain of the problem.

Students should be taught that inspiration can be derived from a diverse set of biological system physical and non-physical characteristics. Novice designers performing bio-inspired design tend to focus heavily on observable aspects or physical characteristics of a biological system (form, shape, material, structure, visual patterns). Non-physical characteristics such as function, process, behavior, and system level attributes are more likely to result in design inspiration than imitation. Thus, students should be encouraged to consider non-physical biological characteristics, or to consider them in combination with physical biological characteristics.

To lower the hurdle to learning biological knowledge as well as applying that knowledge, students should be taught that it is not necessary to search for an uncommon, unique, or "right" biological system. A single or well-known biological system can provide multiple different engineering solutions. Recognizing that inspiration can come from physical and non-physical characteristics, a designer can consider familiar biological systems that are well documented from different analysis perspectives and the less obvious information can be found.

Exposing students to the abundance of design examples that can be found in nature demonstrates the breadth of applicability. Bio-inspired design touches on many areas of engineering including electrical, mechanical, materials, biomedical, chemical, manufacturing and systems, which makes it applicable in a wide range of engineering programs, from discipline-specific to general ones. Broad exposure also helps students to recognize that bio-inspired design results in multidisciplinary design innovation.

Allowing students to find and present examples of bio-inspired design increases their curiosity and engagement with the topic. Not only does daily or weekly presentation of examples from literature broaden students' knowledge about bio-inspired design, but it also allows them to practice presentation skills. Starting class with a brief example gets the class into the mindset of bio-inspired design and could be aligned specifically with the lecture topic of the day.

Inviting guest speakers from industry to share their bio-inspired design experiences signifies that the topic is relevant and respected. Reading and learning about bio-inspired design applications follows the typical model of learning, but does not demonstrate they could do this as a professional. Inviting external speakers to share their work and interact with the students, either through video conferencing or in person, demonstrates industry acceptance and helps the students to see how they might potentially perform bio-inspired design in the workplace.

Providing a multidisciplinary team experience with non-engineers prepares students for future communication and teamwork challenges in the workplace. Engineering students find working with non-engineers extremely valuable and enriching to their education. Finding non-engineering faculty to co-teach with can assist with course curriculum design, provide a support network for implementation, and model cross-disciplinary collaboration for the students. And could assist faculty with getting support from university administration.

Introduction of a new course in bio-inspired design is easier at the graduate level as those courses tend to be less structured and much of the material for such a course (e.g., tools, methods, case studies) is found in research journals. However, successful introduction at the undergraduate level occurs through student-centered approaches such as experiential learning, or piloting through honors courses which offer a smaller class size and multiple disciplines.

Resources for curriculum development are primarily found in research journals as outlined in Sect. 3 and ASEE conference papers. Published resources are vetted through a variety of means and offer frameworks for creating new resources.

Other sources include the Biomimicry Educators Network through AskNature.org and teachEngineering.org which offer resources for teaching college and pre-college students.

5 Current Status of Bio-inspired Design in Engineering Curricula

In response to the increased emphasis on cross-disciplinary thinking skills and adaptive and sustainable designs by professional societies, industry and today's global marketplace, engineering colleges in the United States and abroad are increasingly expanding the scope and focus of their curricula to include bio-inspired design topics and projects that expand systems thinking skills, and has been integrated at the module, project, or course levels (Bruck et al. 2004, 2006, 2007; Nelson 2008; Thomas and Breitenberg 2007; Lynch-Caris et al. 2002; Weissburg et al. 2010; Nagel and Stone 2011b; Nagel et al. 2013; Glier et al. 2011a, b; Hsiao and Chou 2007; Nelson et al. 2009; Seipel 2011; Farel and Yannou 2013; Jenkins 2011a, b; Cattano et al. 2011). Multiple institutions offer full-term courses in bio-inspired design within engineering or multidisciplinary courses that bring together students from across the science, technology, engineering, art, and math (STEAM) disciplines. While instruction in bio-inspired design is more common in engineering programs at the graduate level, as it is often considered an advance design topic, it is exciting to note that bio-inspired design instruction is also being incorporated into curricula at the undergraduate level from first-year through senior. Table 1 gives an overview comparison of the bio-inspired design courses at different academic levels and what they tend to focus on.

The following sub-sections detail how bio-inspired design instruction has been integrated into undergraduate and graduate engineering curricula as both single and multidisciplinary courses. More information about courses can be found in Nagel and Pidaparti (2016), Gleich et al. (2010). The third sub-section gives an overview of graduate programs within engineering are affiliated with engineering programs.

Table 1 Comparison of courses at different academic levels

First-year	Sophomore	Junior	Senior	Graduate
Introductory course that introduces the topic and its philosophy—Tend to be multidisciplinary		Technical elective focused on a specific topic (e.g., materials, robotics, design process, ergonomics, infrastructure)—Tend to be single discipline		Focused on the theory of the topic and based on research—Mix of single and multi discipline

5.1 Single-Discipline Courses

In this category are the courses that are introducing bio-inspired design, but in only one discipline (e.g. engineering or architecture). The courses tend to teach an introduction to the topic and then focus on a specific aspect of bio-inspired design, such as the biological analysis step or bio-inspired materials, to give depth of knowledge through detailed research. These courses are also generally focused on the subject matter of the discipline they are taught out of, and can include the use of software for simulation, computational tools, data analysis, working with an industrial partner, or even simulating a cross-domain collaboration, but always with students of the same discipline. Thus, it is expected that students have a certain foundation of knowledge in their discipline before taking the course. Additionally, these courses are taught by a single instructor from the same discipline as the students. Projects are often used as the culminating experience and serve as a means for students to demonstrate their understanding.

An example course is the senior level technical elective on bio-inspired engineering offered in the undergraduate mechanical engineering department at Montana State University (Jenkins 2011b). The course covers relevant bio-inspired design and engineering design processes with a focus on structures and materials from both nature and engineering. Another example is the graduate level course at Oregon State University (Stone 2013). In this course, students receive instruction on design theory in the areas of analogy, function-based design, TRIZ, and optimization as they are applied to bio-inspired design.

5.2 Multidisciplinary Courses

In this category are the courses that are introducing bio-inspired design and simulating a real collaboration with students from two or more disciplines at the same institution. These courses are often taught by multiple faculty from different disciplines to mimic the composition of the students in the course. Instruction is often focused on the fundamentals of the bio-inspired design philosophy as well as processes and tools that are accessible to a mixed audience. Students work in mixed teams both in and out of class to learn from each other and build respect for what the other disciplines contribute to the process. While these courses can be placed at any point in the curriculum, they tend to be in the first years and, in some cases, coupled with general education requirements. Thus, these courses do not typically have course-based prerequisites.

An undergraduate multidisciplinary course example is the one offered through the Georgia Tech Center for Bio-inspired Design (Goel 2007; Yen et al. 2011, 2014; Helms and Goel 2014). The course is co-taught by faculty from biology and engineering and admits junior and senior level students from all fields of engineering and biology. Students work in multidisciplinary teams on assignments and projects throughout the course which teaches the process of bio-inspired design.

5.3 Graduate Programs

In addition to single courses at the graduate level are full graduate programs dedicated to preparing students for professional work. These programs aim to provide multidisciplinary training in biology, design, engineering, and, in some programs, business to meet the growing demand for biomimicry practitioners in industry. Similar to traditional degree programs, the curriculum for the graduate programs focused on biomimicry have required and elective courses as well as a minimum number of practicum or research credit hours. It is expected that students entering these graduate programs have completed an undergraduate course of study in biology, engineering, design, or business from which to build upon.

HTW SAAR in Saarbrucken Germany offers in six semesters a Master of Engineering or in four semesters a university certificate in Construction Bionics (Weber 2018). The program is designed for engineers in industry with backgrounds in mechanical engineering, mechatronics, material sciences or comparable fields of study who desire to acquire the tools and skills necessary to solve engineering problems and development tasks through inspiration from nature. Graduates are prepared to bring products and processes to market faster, develop more effective solutions to problems, and conserve resources (Gleich et al. 2010). The program is only offered in German and taught by engineering faculty.

Within the United States, there are two prominent graduate programs. Arizona State University (ASU) offers a Master of Science in biomimicry (Master of Science (MS) in Biomimicry—Online 2018) and a certificate in biomimicry (Graduate Certificate in Biomimicry—Online 2018). All coursework is delivered online and in conjunction with the Biomimicry Institute. The ASU program and certificate view biomimicry as a methodology for sustainable design, systems thinking, creativity, and multidisciplinary collaboration that will prepare graduates for a twenty-first century career. The University of Akron (UA) offers a five-year biomimicry fellowship program that is associated with the Doctor of Philosophy in integrated bioscience (Biomimicry Research and Innovation Center 2018). The fellowship program is supported by The University of Akron's Biomimicry Research & Innovation Center, in collaboration with Great Lakes Biomimicry. A unique aspect of this program is that each biomimicry fellow works closely with an industrial sponsor, while undertaking biomimicry-focused dissertation research. Both the ASU and UA programs are taught by faculty from a variety of disciplines.

6 Significance of Bio-inspired Design in Engineering Curricula

Engineers in the twenty-first century can no longer isolate themselves and must be prepared to work across disciplinary, cultural, political, and economic boundaries to meet challenges facing the US and the world. Innovative solutions will be necessary to address the complex challenges of the future. Following the discussion on the link to innovation, this section reviews the literature pertaining to the essential competencies of the twenty-first century engineer and offers a synthesis that can be connected to the learning outcomes of bio-inspired design. The implications of teaching bio-inspired design in an engineering curriculum are also discussed in this section.

6.1 Innovation

Innovative engineering design tools and methods are essential to creating new and better products and industries, and are also important for the US to maintain and sustain its global economic leadership. "Design Quality" is the main factor that differentiates one competing product from another. Companies such as Toyota, Apple, and Samsung are pioneers in positioning design as a key contributor to innovation. "Design Innovation" has been identified as an important learning approach for students in science, technology, and engineering disciplines by national organizations, like the National Science Foundation and the National Academy of Engineering, among others.

Undergraduate engineering programs that focus exclusively on engineering principles throughout the curriculum will not be able to train students to recognize interrelationships or be adaptive problem solvers. The connections within engineering may be obvious to students, or students might take it for granted that some aspects are similar. With bio-inspired design, however, many of the connections to be made between biology and engineering will not be obvious and making these connections will require critical thinking and investigation from multiple system levels and viewpoints, thus emphasizing systems thinking.

The knowledge base in biology has proven to be a useful resource for engineers searching for novel approaches to solving complex design problems. Biological systems provide insight into sustainable and adaptable design, which has been used to inspire engineering innovation. Bio-inspired designs are viewed as creative and novel solutions to human problems and are often efficient, economic, elegant and sustainable. Moreover, some bio-inspired designs, such as Velcro, have become so commonplace that it is hard to image life without them. Other imitations of nature now on the cusp of practical usefulness, such as artificial photosynthesis, could lead to enormous societal benefits including regional energy independence and reduced greenhouse emissions. The overarching motivation is not just to train students to explore the biological domain for solutions, but to have the cognitive flexibility, creativity, and adaptive problem solving skills to explore any contextual domain from which they might find solutions to complex, cross-disciplinary engineering problems. Bio-inspired design is a cross-disciplinary field and requires diversity of thought and perspective, which leads to innovation. Thus, teaching students about bio-inspired design improves their cross-disciplinary thinking skill, which is among the essential competencies outlined in the Engineer 2020 Report and by other organizations and researchers as discussed in the following section.

6.2 Essential Competencies of the Twenty-first Century Engineer

The Engineer 2020 report, ABET, and engineering design education researchers have identified the essential competencies for engineers to be prepared to work across disciplinary, cultural, political, and economic boundaries to solve complex design challenges. Undergraduate education must train students to not only solve engineering challenges that transcend disciplinary boundaries, but also communicate, transfer knowledge, and collaborate across technical and non-technical boundaries. The competencies given below can be divided into two groups, task-specific, and meta competencies. Task-specific competencies are skill sets that define how well-prepared graduates are to meet the workforce challenges that lie ahead based on their level of attainment (Earnest and Hills 2005). Meta competencies are skill sets that enable graduates to function globally while meeting technical demands, have the cognitive flexibility to think about the whole system at different levels of fidelity and in different time scales, and transfer task-specific skills to new challenges or tasks they have not encountered before (Radcliffe 2005).

While not an exhaustive inventory of the literature on engineering competencies, the essential competencies for the twenty-first century engineer were analyzed from the following sources: Engineer 2020 report (National Academy of Engineering (NAE) 2005), Accreditation Board for Engineering and Technology (ABET) (ABET 2011), and the competency-based approach to personalized education for twenty-first century engineers (Hawthorne et al. 2012; Siddique et al. 2012; Bertus et al. 2012; Panchal et al. 2012). A synthesis of the three different perspectives on engineering competencies reveals the following common themes:

- Holistic, Critical thinking
- Complex, Multidisciplinary problem solving
- Creativity
- Communication across disciplines
- Understand impact in global, economic, environmental, and societal contexts
- Collaboration in a multidisciplinary team
- Self-regulated learning
- Flexibility and agility
- Global Awareness

Students learning through bio-inspired design have the potential to meet all of these competencies. Bio-inspired design is inherently multidisciplinary. It blends information from biology, engineering, physics, mathematics, architecture, and design and consequently fosters the first four competencies. With communication also including scientific literacy (Chae et al. 2010). Understanding impact is fostered through comparing natural to engineered solutions and recognizing that all materials, forms, and processes of natural systems have a purpose (a function) and sometimes multiple purposes as well as projects that require designing for those outcomes. Bringing together teams of people across disciplinary boundaries, within and outside engineering, fosters innovation through the diversity of thought and communication. While multidisciplinary teams are the strongest, there may be institutional limitations that prevent such teams. Learning, flexibility, and agility are fostered through open-ended questions and projects that require the student to define the problem and inspiring biological system. The final competency can be fostered through considering the biological systems in regional areas across the globe or working with teams abroad.

In the current context, it is widely recognized that most students that go into engineering have high school level training in biology. Adding biomimicry into the engineering curriculum encourages students to utilize and build off their prior knowledge, which fosters making connections and recognizing interrelationships across STEM disciplines (Weissburg et al. 2010; Nagel et al. 2013). Moreover, requiring knowledge transfer across domains as well as organizing that knowledge into logical constructs helps to develop future flexibility and adaptive expertise that will facilitate innovation and efficiency (McKenna 2007; Bransford 2007). These competencies have also been identified as critical key skill areas for engineers by the Partnership for Twenty-first Century Skills (Partnership for 21st Century Skills 2005) and the Assessment and Teaching of Twenty-first Century Skills working group (Binkley et al. 2012). Instruction on bio-inspired design concepts will help engineering colleges achieve a number of the recommendations made by the National Academy of Engineering in reference to educating the engineer of 2020 (National Academy of Engineering (NAE) 2005), as well as ABET student outcomes c, d, e and h (ABET 2011) and foster competencies that support engineering innovation.

A specific example of the synthesized engineering competencies connected to the learning outcomes of bio-inspired design is given in Sect. 7.2.

6.3 Implications

Implementing biomimicry concepts into engineering curricula presents a unique opportunity to incorporate fundamental curiosity-driven and technology perspectives and involve collaborations from multiple disciplines. In addition, faculty from various disciplines will have the opportunity to engage in a collaborative teaching environment and share valuable experiences and insights. Moreover, anecdotal evidence suggests that students find bio-inspired design exciting, as it offers relevance to professional practice as well as an effective hook to frame complex, cross-disciplinary problems. Courses incorporating bio-inspired design into engineering curricula can help students to think innovatively in a multidisciplinary fashion. Another advantage to including bio-inspired design courses in undergraduate engineering curricula is that bio-inspired design touches on many areas of engineering including electrical, mechanical, materials, biomedical, chemical, manufacturing and systems, which makes it applicable in a wide range of engineering programs, from discipline specific to general ones. Thus, there are several opportunities to foster the competencies of twenty-first century engineers in a variety of institutional settings through bio-inspired design with engineering design courses being the most advantageous.

Many of the current offerings are at the senior level, which provides the advantages of students being able to apply complex engineering theories, work efficiently in teams, communicate well, perform research, and think abstractly and holistically. It is expected that students are meeting the ABET outcomes by their senior year, which can allow for a richer course experience, but may not carry over into their professional work. On the other hand, introductory level courses in the first year expose students to a new way of thinking that could be reinforced throughout their college coursework thus embedding the approach in their problem solving process and will foster some of the competencies. Ideally, students would apply bio-inspired design throughout an engineering curriculum to ensure the competencies are met.

7 Examples

This section offers two explicit examples of bio-inspired design pedagogy in engineering. The first is a detailed view of a multidisciplinary course offered at the undergraduate level, while the second gives an overview of modular implementation in existing design courses with connection to engineering competencies addressed.

7.1 Example Multidisciplinary Course at JMU

As bio-inspired design is a design methodology that looks to nature's examples for impactful, responsible, and innovative solutions to humankind's problems, the semester-long course is open to all disciplines. Disciplines that have enrolled in the course include engineering, biotechnology, and industrial design. It is co-taught with an engineering faculty member and a biology faculty member. The course explores ways that biological insight can play a significant role in solving technical challenges with an emphasis on the process. Rather than focus on a single bio-inspired design process, students are introduced to the generic process outlined in Sect. 2 to gain a fundamental understanding before practicing four methods from published literature. Class time is dedicated to learning the fundamentals about bio-inspired design, approaches to bio-inspired design (both problem-driven and biology-driven), explor-

Table 2 Course activities aligned with the Kolb learning cycle

Concrete experience	Students work individually, in small groups and in multidisciplinary teams to learn course content as well as knowledge from each other. Students interact with and get feedback consistently and frequently with members of their small groups, team and the instructor throughout the course
Reflective observation	Students engage in reflective writing after each guest speaker as well as at the mid- and end-of-term where they respond to prompts about their evolving perspective of bio-inspired design. There is a daily student presented example of bio-inspired design that sparks interesting group discussions and expansion on prior topics
Abstract conceptualization	Students learn in ways that are different from their typical experience. Engineering students comment on how much they enjoyed the opportunity to work with people from outside their major and share their knowledge of the design process. Non-engineering students comment on increasing their creativity and learning processes for problem solving and contributing to the design process
Active experimentation	Throughout the course students generate and present ideas based on biological inspiration from a biology-driven and problem-driven approach. During the end of the course when teams are working solely on the course project they are required to present at multiple points in the project cycle to get feedback. They culminate by presenting a final idea and the bio-inspired design process followed to stakeholders

ing biological systems for solution principles, and techniques to transfer knowledge from biology to the domain of the problem to provide a solution.

The course is designed using experiential learning, thus students engage in activities during class as well as a semester-long project. Course activities mapped to the Kolb learning cycle are described in Table 2. Students study the multidisciplinary topic of biomimicry though team and individual work. Student teams complete a project for an application area of interest or one of the suggested student design competitions. Class activities include: a student chosen and presented example of biomimicry each class period, presentation of bio-inspired concepts through digital means or drawn on the whiteboard, gallery walks, whole class discussions, practice with tools and methods from the literature for search, abstraction, and transfer, and guest speakers from industry and academia focused on biomimicry work/experiences. All students are evaluated using the same course objectives and assessments.

During the first half of the semester, students practice the steps of the generic bio-inspired design process in reverse to help ease the transition into the new way of thinking. The second half of the course is focused on application of four specific methodologies of bio-inspired design found in the literature followed by project teams choosing the methodology they thought best fit their project problem or defining their own process as a mix of those experienced. Guest speakers with biomimicry experience from industry and academia occur throughout the semester to reinforce that the topic is more than an academic exercise. Following the guest speaker, students must make a forum post explaining what they learned and how it informs their perspective on bio-inspired design.

7.2   Example of JMU/UGA Approach for Integration and Competencies Addressed

At James Madison University (JMU) and University of Georgia (UGA) bio-inspired design instruction is integrated into the second-year engineering design course (Nagel and Pidaparti 2016; Pidaparti and Nagel 2018). Instructional resources that can help to effectively scaffold students to transfer knowledge across disciplinary boundaries and train engineers in cross-disciplinary thinking are developed with support from the National Science Foundation. Our approach in developing instructional resources uses Concept-Knowledge (C-K) Theory (Shai et al. 2009; Hatchuel et al. 2011; Hatchuel and Weil 2003, 2009) as it is a well-established approach for integrating multiple domains of information and facilitating innovation through connection building. The instructional resources are designed to foster the competencies of holistic, critical thinking; complex, multidisciplinary problem solving; creativity; communication across disciplines; understand impact in global, economic, environmental, and societal contexts; self-regulated learning; and flexibility and agility.

During a pilot study at James Madison University, the C-K theory instructional approach was adopted for teaching a sophomore engineering design class to specifically address how the twenty-first century competencies can be targeted and achieved.

To implement the C-K theory instructional approach a bio-inspired design teaching module, learning activity, and assignment that incorporated a C-K mapping template with guidelines was created and integrated into the course during the topic of concept generation and introduced as a creative method for design. All assignments in the sophomore design course tie to a year-long course project of developing a human powered vehicle for a client in the community who has cerebral palsy, including the bio-inspired design assignment. To integrate bio-inspired design into the human powered vehicle design project each member of a team applied bio-inspired design to a different sub-system (e.g., propulsion, steering, braking) of their design to showcase a variety of design problems and analogies that enable bio-inspired design. All students completed the C-K mapping template three times, twice in class as part of a learning activity to understand the process of discovery and again in their assignment to scaffold application to the human powered vehicle.

The developed teaching module introduces bio-inspired design as a design philosophy and provides several examples of how biological systems were used as inspiration for innovative solutions. Students learn about the two major paths to a bio-inspired de-sign, biology-driven and problem-driven, as well as how analogies

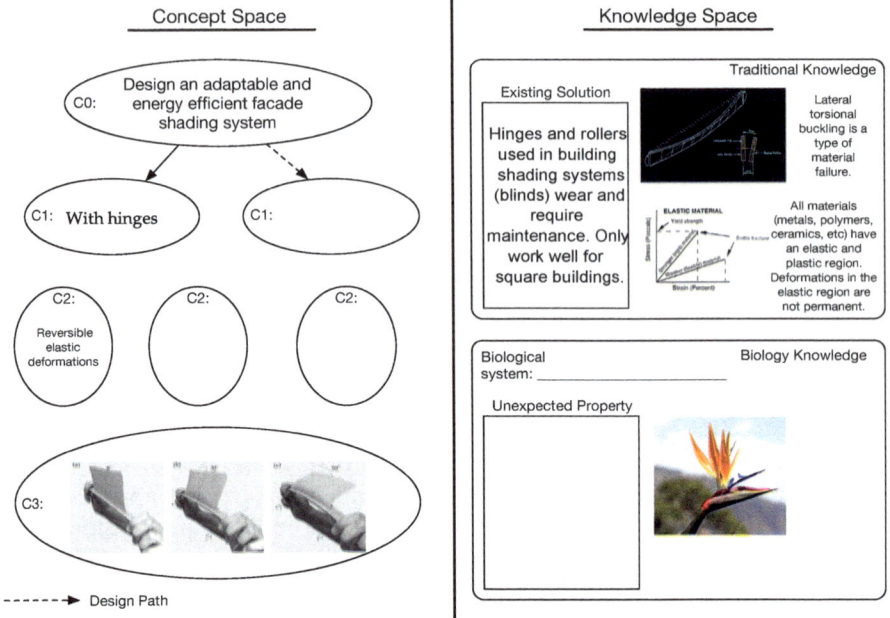

Fig. 2 C-K template for hingeless facade shading mechanism learning activity

are used to assist with transferring the knowledge from biology to engineering. For the purposes of scaffolding the sophomore engineering de-sign students in their application of bio-inspired de-sign, two problem-driven examples using C-K theory were provided with accompanying learning activities. One problem-driven example and learning activity focused on the hingeless facade shading mechanism, Flectofin®, inspired by the bird-of-paradise flower (Salgueiredo 2013). Shading buildings with irregular geometries is very difficult since most sun protection systems were developed for planar façades and include the use of hinges. The pollination mechanism of the bird-of-paradise flower offers inspiration based on the elastic kinematics of plant movements. After the initial problem is explained, students are provided the partially filled in template shown in Fig. 2 to complete during the explanation of the example. This scaffolds the students through the C-K theory mapping process. Students are walked through the thought processes and analogies of the discovery process for arriving at a bio-inspired solution using the C-K theory framework.

The developed assignment that compliments the teaching module and learning activities includes three parts: (1) complete the C-K mapping template for a human-powered vehicle sub-system, (2) use the sketches in the C3 level of the template along with the team generated morphological matrix to create a full human powered vehicle concept, and (3) a W/H/W reflection essay answering three questions about the content and process. The W/H/W reflections require learners to reflect on and respond to three questions: What did I learn?, How did I learn it?, and What will I do with it? These three prompts structure reflection so that learners focus on concepts, knowledge and skills, processes, and utilization/generalization/sustaining of learning. The W/H/W reflections provide formative snap-shots of learning and application that the learners are making as they progress through the material.

Analysis of the W/H/W reflection questions aims to identify which twenty-first century competencies were achieved by incorporating bio-inspired design education in an engineering design course. Fifteen (65%) students consented to participate in the research. Transcriptions of the reflection questions for consenting participants were de-identified and analyzed using qualitative content analysis. Qualitative content analysis identifies themes in the student reflections. This method involves reducing participants' comments to their smallest meaningful unit, coding these units, identifying categories for these codes, and then finally identifying themes from the categories (Patton 2002). The reflection statements resulted in 206 (108 for content questions and 98 for process questions) unique/coded meaningful units. Multiple themes and categories emerged for each question based on coded meaningful units. Themes for each reflection question including the number of student responses that support that theme, and the distilled competencies that the instructional resources were intended to foster are given in Table 3.

Focusing on the content (biology knowledge), students learned that biological systems are surprising complex but have attributes that can easily be applied to design problems. Recognition that nature has a lot to offer resulted in valuing what can be learned from biological systems. Conversely, a few students concluded that biological knowledge is not always applicable to the design problem. Both of these themes as well as forming cross-domain linkages link to the competency of critical thinking as students had to analyze the information they were finding and manage their own thinking about the information. The competencies of self-regulated learning and communication across disciplines link strongly to the themes of students learning about biological systems by engaging with scholarly resources through independent research, and forming cross-domain linkages as students had to dive deeper into the literature than just looking at the pictures to understand how the biological system relates to their chosen problem. Similar trends were observed for what students learned about the process of bio-inspired design. Critical thinking was exhibited in recognizing the value (or not) of including biological inspiration in a design process and that the process facilitates knowledge transfer between the domains that results in solutions to engineering problems. Communication across the disciplines is also evident in the recognition of knowledge transfer across domains. Understanding impact in a broader context was evident as looking to nature for inspiration resulted in students finding possible solutions that they thought were more sustainable than the existing engineering solution.

The competencies of self-regulated learning and communication across disciplines directly relate to how students learned the content and process. In both cases, internal resources (the instructional materials) and external resources (scholarly works and websites) were used to learn the content and process. It is not surprising that external resources were heavily used for learning the biology knowledge as it was required for students to identify and learn about the inspiring biological system

Table 3 Mapping of reflection questions to themes and competencies

Reflection question	Themes (n = supportive categories)	Competencies addressed
What did I learn about the content?	• Valued what can be learned from nature and biology (17) • In-depth understanding of chosen biological system (14) • Cross-domain linkages (11) • Biology is not always applicable (4)	• holistic, critical thinking • self-regulated learning • communication across disciplines
How did I learn the content?	• Scholarly or external resources (31) • Course learning resources (4)	• self-regulated learning • communication across disciplines
What am I going to do with the content?	• Apply to immediate problem—course project (16) • Facilitate a future design path (11)	• flexibility and agility • complex, multidisciplinary problem solving • creativity
What did I learn about the process?	• Valued the inclusion of biology in engineering design (22) • Recognized knowledge transfer between domains for problem solving is possible (17) • Bio-inspired design is not always applicable (3)	• holistic, critical thinking • communication across disciplines • understand impact in global, economic, environmental, and societal contexts
How did I learn the process?	• Course learning resources (20) • External or other resources (13)	• self-regulated learning • communication across disciplines
What am I going to do with the process?	• Facilitate a future design path (20) • Apply to immediate problem—course project (3)	• flexibility and agility • complex, multidisciplinary problem solving • creativity

independently. Whereas all the resources for learning the process were modeled in class and provided for the assignment.

With respect to what the students are going to do with the content and process, it was not surprising to see the main trends of application to the course project and future opportunities. Creativity as well as flexibility and agility are expressed in the application of the analogically distant information (biology) to the target problem in engineering, and generally gaining a new perspective when designing. The competency of complex, multidisciplinary problem solving is embedded in the application of bio-inspired design to a specified problem, and it is encouraging to see the trend of wanting to use bio-inspired design when designing or solving engineering and non-engineering problems in the future. Future implementation has occurred at JMU through capstone and research projects. Some students (about 10%) have proposed capstone projects that explicitly use bio-inspired design as the problem solving process, have applied bio-inspired design during their capstone project, or have joined a bio-inspired design research project.

One student expressed that learning about biology helped in gaining further knowledge about a specific sub-system of the human-powered vehicle. Similarly, two students expressed that they would use existing biology knowledge to help understand engineered components and systems, meaning the students learned more about engineering through biology. This unanticipated result points toward the significance of teaching bio-inspired design in an engineering curriculum. Teaching bio-inspired design in an engineering curriculum using multidisciplinary approaches will not only develop competencies of the twenty-first century engineer but also enable undergraduate students to become change agents and promote a sustainable future.

8 Conclusion

This chapter introduced the current state of bio-inspired design inclusion in engineering curricula and the pedagogical methods used to deliver the topic. Bio-inspired design is gaining acceptance in engineering programs as both graduate and undergraduate programs across the globe are introducing the topic as a course module, independent course, and fully integrated program curricula. Best practices for teaching the topic within an existing course or a course dedicated to the topic are provided.

Although much research has been conducted in the application of bio-inspired design to solve engineering challenges, little has been done to create resources for teaching the topic. One approach to addressing this gap is the work using C-K theory to develop instructional resources for teaching engineers how to perform bio-inspired design. This approach was shown to promote multiple competencies for the twenty-first century engineer which highlights the significance of incorporating bio-inspired design into engineering curricula. Teaching bio-inspired design in an engineering curriculum will not only develop essential competencies of tomorrow's engineer, but also enable students to become change agents and promote a sustainable future.

Acknowledgements This work is partially supported by the National Science Foundation under Grant No. 1504612 and 1504614. Any opinions, findings, and conclusions or recommendations expressed in this material are those of the author(s) and do not necessarily reflect the views of the National Science Foundation. The authors also gratefully acknowledge the helpful comments and suggestions of the reviewers, which have improved the presentation.

References

ABET (2011) Criteria for accrediting engineering programs. Engineering Accreditation Commission

Altshuller G (1984) Creativity as an exact science. Gordon and Breach, Luxembourg

Bar-Cohen Y (2006a) Biomimetics—using nature to inspire human innovation. J Bioinspiration Biomim 1:P1–P12

Bar-Cohen Y (2006b) Biomimetics biologically inspired technologies. CRC/Taylor & Francis, Boca Raton, FL

Benyus JM (1997) Biomimicry innovation inspired by nature. Morrow, New York

Bertus C, Khosrojerdi A, Panchal JH, Allen JK, Mistree F (2012) Identifying dilemmas embodied in 21st century engineering. In: ASME international conference on engineering design, DETC2012-71163, Chicago, IL, 2012

Binkley M, Erstad O, Hermna J, Raizen S, Ripley M, Miller-Ricci M, Rumble M (2012) Defining twenty-first century skills. In: Griffin P, Care E, McGaw B (eds) Assessment and teaching of 21st century skills. Springer, Dordrecht

Biomimicry Research and Innovation Center (2018) The University of Akron. http://blogs.uakron.edu/biomimicry/. Accessed 28 June 2018

Bransford J (2007) Preparing people for rapidly changing environments. J Eng Educ 96(1):1–3

Brebbia CA (2006) Design and nature III: comparing design in nature with science and engineering, vol 3. WIT, Southampton

Brebbia CA (2008) Design & nature IV: comparing design in nature with science and engineering. WIT, Southampton, Boston

Brebbia CA, Carpi A (2010) Design & nature V: comparing design in nature with science and engineering. WIT, Southampton, Boston

Brebbia CA, Collins MW (2004) Design and nature II: comparing design in nature with science and engineering, vol 3. WIT, Southampton

Brebbia CA, Sucharov LJ, Pascolo P (2002) Design and nature: comparing design in nature with science and engineering, vol 1. WIT, Southampton, Boston

Bruck HA, Gershon AL, Gupta SK (2004) Enhancement of mechanical engineering curriculum to introduce manufacturing techniques and principles for bio-inspired product development. Paper presented at the ASME international mechanical engineering congress and RD&D Expo, Anaheim, CA

Bruck HA, Gershon AL, Golden I, Gupta SK, Gyger LS Jr, Magrab EB, Spranklin BW (2006) New educational tools and curriculum enhancements for motivating engineering students to design and realize bio-inspired products. Paper presented at the Design and nature 2006, Southampton, UK

Bruck HA, Gershon AL, Golden I, Gupta SK, Gyger LS, Magrab EB, Spranklin BW (2007) Training mechanical engineering students to utilize biological inspiration during product development. Bioinspiration Biomim 2(4):S198–S209

Cattano C, Nikou T, Klotz L (2011) Teaching systems thinking and biomimciry to civil engineering students. J Prof Issues Eng Educ Pract 137(4):176–182

Chae Y, Purzer S, Cardella M (2010) Core concepts for engineering literacy: the interrelationships among STEM disciplines. In: American Society for Engineering Education annual conference and exposition, Louisville, KY, 2010

Chakrabarti A, Sarkar P, Leelavathamma B, Nataraju BS (2005) A functional representation for aiding biomimetic and artificial inspiration of new ideas. Artif Intell Eng Des Anal Manuf 19:113–132

Clayton T, Radcliffe N (1996) Sustainability—a systems approach. Routledge, London

Cross-Cutting Sustainable Engineering Courses (2016) Villanova University. Accessed 30 Aug 2018

Cutherell D (1996) Chapter 16: product architecture. In: Rosenau M Jr (ed) The PDMA handbook of new product development. Wiley and Sons

Dieter G (1991) Engineering design: a materials and processing approach. McGraw-Hill Series in Mechanical Engineering, 2nd edn. McGraw-Hill, New York

Earnest J, Hills S (2005) Abet engineering technology criteria and competency based engineering education. In: 35th ASEE/IEEE frontiers in education conference, Indianapolis, IN, 2005, pp 7–12

Eggermont M, Hepp AF, Shyam V (2017) Pigs in space: a bio-inspired design and space challenges cornerstone project. Paper presented at the ASEE, Columbus, OH, USA

Farel R, Yannou B (2013) Bio-inspired ideation: lessons from teaching design to engineering students. In: International conference on engineering design (ICED), Seoul, Korea, 2013

Fayemi P-E, Wanieck K, Zollfrank C, Marnzana N, Aoussat A (2017) Biomimetics: process, tools and practice. Bioinspiration Biomim 12(1):011002

Fenves S (2001) A core product model for representing design information. National Institute of Standards and Technology, Gaithersburg, MD

Fiksel J (2012) A systems view of sustainability: the triple value model. Environ Dev 2:138–141. https://doi.org/10.1016/j.envdev.2012.03.015

Forbes P (2006) The gecko's foot: bio-inspiration: engineering new materials from nature. W. W. Norton & Co., New York

Fu K, Moreno D, Yang M, Wood K (2014) Bio-inspired design: an overview investigating open questions from the broader field of design-by-analogy. J Mech Des 136(11):111102

Gleich A, Pade C, Petschow U, Pissarskoi E (2010) Biomimetics in education and training. In: Potentials and trends in biomimetics. Springer, Berlin

Glier MW, McAdams DA, Linsey JS (2011a) Concepts in biomimetic design: methods and tools to incorporate into a biomimetic design course. In: ASME 2011 International design engineering technical conferences & computers and information in engineering conference, Washington, DC, 2011

Glier MW, Tsenn J, Linsey JS, McAdams DA (2011b) Methods for supporting bioinspired design. In: ASME 2011 International mechanical engineering congress and exposition, Denver, CO, 2011

Goel A (2007) Center for Biological Inspired Design. http://www.cbid.gatech.edu/

Goel AK, Rugaber S, Vattam S (2009) Structure, behavior, and function of complex systems: The structure, behavior, and function modeling language. Artif Intell Eng Des Anal Manuf 23(1):23–35

Goel AK, Vattam S, Wiltgen B, Helms M (2012) Cognitive, collaborative, conceptual and creative—four characteristics of the next generation of knowledge-based CAD systems: a study in biologically inspired design. Comput Aided Des 44:879–900

Graduate Certificate in Biomimicry—Online (2018) Arizona State University. http://biomimicry. asu.edu/education/asu-online-certificate-program/. Accessed 28 June 2018

Hatchuel A, Salgueiredo CF (2016) Beyond analogy: a model of bioinspiration for creative design. Artif Intell Eng Des Anal Manuf 30:159–170. https://doi.org/10.1017/S0890060416000044

Hatchuel A, Weil B (2003) A new approach of innovative design: an introduction to C-K theory. In: International conference on engineering design (ICED), Stockholm, 2003

Hatchuel A, Weil B (2009) C-K design theory: an advanced formulation. Res Eng Des 19(4):181–192

Hatchuel A, Masson PL, Weil B (2004) C-K Theory in practice: lessons from industrial applications. Paper presented at the International design conference, Dubrovnik

Hatchuel A, Masson PL, Weil B (2011) Teaching innovative design reasoning: how concept–knowledge theory can help overcome fixation effects. Artif Intell Eng Des Anal Manuf 25(1):77–92

Hawthorne B, Sha Z, Panchal JH, Mistree F (2012) Developing competencies for the 21st century engineer. In: ASME international conference on design education, DETC2012-71153, Chicago, IL, 2012

Helms M, Goel A (2014) The four-box method of analogy evaluation in biologically inspired design. In: ASME 2014 International design engineering technical conferences & computers and information in engineering conference, Buffalo, NY, 2014

Hirtz J, Stone R, McAdams D, Szykman S, Wood K (2002) A functional basis for engineering design: reconciling and evolving previous efforts. Res Eng Design 13(2):65–82

Hsiao H-C, Chou W-C (2007) Using biomimetic design in a product design course. World Trans Eng Technol Educ 6(1):31–35

Hundal M (1990) A systematic method for developing function structures, solutions and concept variants. Mech Mach Theory 25(3):243–256

Jenkins CH (2011a) Bio-inspired engineering. New York. Momentum Press

Jenkins CH (2011b) Doing BiE: lessons learned from teaching bio-inspired engineering. In: ASME 2011 International mechanical engineering congress and exposition, Denver, CO, 2011

Jonassen DH, Hung W (2008) All problems are not equal: implications for problem-based learning. Interdiscip J Probl-Based Learn 2(2):4

Kennedy B, Buikema A, Nagel JKS (2014) Integrating biology, design, and engineering for sustainable innovation. Paper presented at the IEEE integrated STEM education conference, Princeton University

Kolb D (1984) Experiential learning: experience as the source of learning and development. Prentice-Hall, New York

Kolb DA (2014) Experiential learning: experience as the source of learning and development. FT Press

Kruiper R, Vincent J, Abraham E, Soar R, Konstas I, Chen-Burger J, Desmulliez M (2018) Towards a design process for computer-aided biomimetics. Biomimetics 3(3):14. https://doi.org/10.3390/biomimetics3030014

Letterman MR, Dugan KB (2004) Team teaching a cross-disciplinary honors course. Coll Teach 52(2):76

Lindemann U, Gramann J (2004) Engineering design using biological principles. In: International design conference—DESIGN 2004, Dubrovnik, 2004

Ludwig PM, Nagel JK, Lewis EJ (2017) Student learning outcomes from a pilot medical innovations course with nursing, engineering, and biology undergraduate students. Int J STEM Educ 4:33. https://doi.org/10.1186/s40594-017-0095-y

Lynch-Caris TM, Waever J, Kleinke DK (2002) Biomimicry innovation as a tool for design. In: American Society for Engineering Education annual conference and exposition, San Antonio, TX, 2012

Master of Science (MS) in Biomimicry—Online (2018) Arizona State University. http://biomimicry.asu.edu/education/asu-online-masters-degree/. Accessed 28 June 2018

McKenna AF (2007) An investigation of adaptive expertise and transfer of design process knowledge. J Mech Des 129(7):730–734

Miles L (1972) Techniques of value analysis engineering. McGraw-Hill, New York

Moore J (2005) Seven recommendations for creating sustainability education at the university level. Int J Sustain High Educ 6:326–339

Nachtigall W (2000) The big book of bionics: new technologies, following the example of nature. German Verlags-Anstalt

Nagel JKS, Pidaparti R (2016) Significance, prevalence and implications for bio-inspired design courses in the undergraduate engineering curriculum. Paper presented at the ASME IDETC/CIE DEC-59661, Charlotte, NC

Nagel JKS, Stone RB (2011a) A systematic approach to biologically-inspired engineering design. Paper presented at the ASME IDETC/CIE 2011, DTM-47398, Washington, DC, USA

Nagel JKS, Stone RB (2011b) Teaching biomimicry in the context of engineering design. In: Biomimicry in higher education webinar. The Biomimicry Institute

Nagel R, Tinsley A, Midha P, McAdams D, Stone R, Shu L (2008) Exploring the use of functional models in biomimetic design. J Mech Des 130(12):11–23

Nagel JKS, Nagel RL, Stone RB, McAdams DA (2010a) Function-Based Biologically-Inspired Concept Generation. Artificial Intelligence for Engineering Design, Analysis and Manufacturing 24 (4):521–535

Nagel JKS, Stone RB, McAdams DA (2010b) Function-based biology inspired concept generation. In: Biomimetics, learning from nature. In-Tech, Croatia

Nagel JKS, Nagel R, Eggermont M (2013) Teaching biomimicry with an engineering-to-biology thesaurus. Paper presented at the ASME IDETC/CIE, Portland, OR

Nagel JKS, Stone RB, McAdams DA (2014) Function-based biologically inspired design. In: Goel A, Stone RB, McAdams DA (eds) Biologically inspired design. Springer, London

Nagel JKS, Pittman P, Pidaparti R, Rose C, Beverly CL (2016) Teaching bio-inspired design using C-K theory. Bioinspired, Biomimetic and Nanobio-materials 6(2):77–86. https://doi.org/10.1680/jbibn.16.00013

National Academy of Engineering (NAE) (2005) Educating the engineer of 2020: adapting engineering education to the new century. The National Academies Press, Washington, DC

Nelson B (2008) Biologically-inspired design: a unique multidisciplinary design model. In: American Society for Engineering Education annual congress and exposition, Pittsburgh, PA, 22–25 June 2008

Nelson B, Wilson J, Yen J (2009) A study of biologically-inspired design as a context for enhancing student innovation. In: ASEE/IEEE frontiers in education conference, San Antonia, TX, 2009

Otto K, Wood K (2001) Product design: techniques in reverse engineering, systematic design, and new product development. Prentice-Hall, New York

Pahl G, Beitz W (1996) Engineering design: a systematic approach, 2nd edn. Springer-Verlag, Berlin, Heidelberg, New York

Pahl G, Beitz W, Feldhusen J, Grote KH (2007) Engineering design: a systematic approach, 3rd edn. Springer Verlag

Panchal JH, Allen JK, Mistree F (2012) Managing dilemmas embodied in 21st century engineering. In: ASME international conference on design education, DETC2012-71168, Chicago, IL, 2012

Pappas E, Pierrakos O, Nagel R (2013) Using bloom's taxonomy to teach sustainability in multiple contexts. J Clean Prod 48:54–64. https://doi.org/10.1016/j.jclepro.2012.09.039

Partnership for 21st Century Skills (2005) The road to 21st century learning: a policymaker's guide to 21st century skills. Washington, DC

Patton MQ (2002) Qualitative research & evaluation methods, 3rd edn. Sage, Thousand Oaks, CA

Pidaparti RM, Nagel JKS (2018) C-K theory based bio-inspired projects in a sophomore design course. Paper presented at the ASEE southeastern section conference, Daytona Beach, FL

Radcliffe DF (2005) Innovation as a meta graduate attribute for engineers. Int J Eng Educ 21(2):194–199

Salgueiredo CF (2013) Modeling biological inspiration for innovative design. In: i3 conference, Paris, France, 2013

Sarkar P, Phaneendra S, Chakrabarti A (2008) Developing engineering products using inspiration from nature. J Comput Inf Sci Eng 8(3):1–9

Schmitt OH (1969) Some interesting and useful biomimetic transforms. In: International biophysics congress, Boston, Massachusetts, 1969, p 297

Seipel J (2011) Emphasizing mechanical feedback in bio-inspired design and education. In: ASME 2011 International mechanical engineering congress and exposition, Denver, CO, 2011

Shai O, Reich Y, Hatchuel A, Subrahmanian E (2009) Creativity theories and scientific discovery: a study of C-K theory and infused design. In: International conference on engineering design (ICED), Stanford, CA, 2009

Siddique Z, Panchal JH, Schaefer D, Haroon S, Allen JK, Mistree F (2012) Competencies for innovating in the 21st century. In: ASME international conference on design education, DETC2012-71170, Chicago, IL, 2012

Srinivasan V, Chakrabarti A (2009a) An empirical evaluation of novelty-sapphire relationship. Paper presented at the ASME 2009 design engineering technical conferences and computers and information in engineering conference, San Diego, CA

Srinivasan V, Chakrabarti A (2009b) SAPPhIRE—An approach to analysis and synthesis. In: International conference on engineering design, Stanford, USA, 2009

Steele JE (1960) How do we get there?, Bionics symposium: living prototypes–the key to new technology, 13–15 September 1960, WADD technical report 60-600, Wright Air Development Division, Wright-Patterson Air Force Base, OH, pp 488–489. In: Reprinted in The Cyborg handbook, ed. by CH Gray. Routledge, New York, 1995, pp 55–60

Stone DRB (2013) ME 513: bio-inspired design. Oregon State University, Mechanical, Industrial, and Manufacturing Engineering. http://mime.oregonstate.edu/cd/me-513-bio-inspired-design. Accessed 28 June 2018

Stone R, Wood K (2000) Development of a functional basis for design. J Mech Des 122(4):359–370

Thomas A, Breitenberg M (2007) Engineering for non-engineers: learning from "Nature Designs". In: American Society for Engineering Education annual congress and exposition, Honolulu, HI

Ullman DG (2002) The mechanical design process, 3rd edn. McGraw-Hill Inc, New York

Ulrich KT, Eppinger SD (2004) Product design and development. McGraw-Hill/Irwin, Boston, MA

Vattam S, Wiltgen B, Helms M, Goel A, Yen J (2010) DANE: fostering creativity in and through biologically inspired design. In: First international conference on design creativity, Kobe, Japan, 2010, pp 127–132

Vattam SS, Goel AK, Rugaber S, Hmelo-Silver CE, Jordan R, Gray S, Sinha S (2011) Understanding complex natural systems by articulating structure-behavior-function models. Educ Technol Soc 14(1):66–81

Vincent JFV, Mann DL (2002) Systematic technology transfer from biology to engineering. Philos Trans R Soc Lond A 360:159–173

Wanieck K, Fayemi P, Maranzana N, Zollfrank C, Jacobs S (2017) Biomimetics and its tools. Bioinspired, Biomim Nanobiomaterials 6(2):53–66. https://doi.org/10.1680/jbibn.16.00010

Weaver J, Kleinke D, Lynch-Caris T (2012) Extending the TRIZ methodology to connect engineering design problems to biological solutions. In: Proceedings of the NCIIA 16th annual meeting, San Francisco, CA, 2012

Weber H-J (2018) Konstruktionsbionik. htw saar. https://www.htwsaar.de/cecsaar/angebot/master/konstruktionsbionik. Accessed 28 June 2018

Weissburg M, Tovey C, Yen J (2010) Enhancing innovation through biologically inspired design. Adv Nat Sci 3(2):1–16. http://dx.doi.org/10.3968/j.ans.1715787020100302.001

Yen J, Helms M, Vattam S, Goel (2010) A evaluating biological systems for their potential in engineering design. In: 3rd International conference on bionics engineering, Zhuhai, China, 2010

Yen J, Weissburg MJ, Helms M, Goel AK (2011) Biologically inspired design: a tool for interdisciplinary education. In: Bar-Cohen Y (ed) Biomimetics: nature based innovation. CRC, Boca Raton, FL

Yen J, Helms M, Goel A, Tovey C, Weissburg M (2014) Adaptive evolution of teaching practices in biologically inspired design. In: Goel AK, McAdams DA, Stone RB (eds) Biologically inspired design: computational methods and tools. Springer, New York

Designing Knowledge Sharing Interfaces with Improved Interaction: Haptics and Web3D

Felix G. Hamza-Lup

Abstract As human beings, from an early age, learning and discovery are enabled by senses: visually exploring the 3D world around, holding and touching objects or listening to others and asking questions. In a traditional education environment, training setup or collaborative work, the majority of the interaction modalities and written materials are still designed for a 2D visual space (e.g., books, course web pages, articles, videos, and images). With the rapid advances in technology, new interaction modalities become widely available and have the potential to improve, and often revolutionize the way we interact, learn and cooperate with each other. The chapter explores new interaction paradigms and systems enabling the shift from a 2D visual space to 3D user interfaces, as well as the addition of a new interaction modality, the tactile or touch modality, through haptic user interfaces. The International Standards Organization's extended 3D (X3D) open standard provides a considerable opportunity to design and implement innovative and revolutionary Web-based interfaces to support and augment engineering courses and design activities. Haptics devices and associated APIs are enabling a revolution in engineering interaction design and engineering education. Several use-cases and example Web3D applications are presented to provide an overview of the existing trends. A concise presentation of the human tactile system, followed by a survey on current haptics hardware accompanied by several successful implementations for engineering education, highlights the near future trends.

Keywords Web3D · Haptics · Engineering education · Engineering design · Simulation · Intelligent dialogue

1 Introduction

Haptic technology has evolved at a rapid pace over the last decade and various consumer-based haptic devices have emerged on the market. Affordability of haptic hardware was enabled mainly due to the proliferation of computer games and the

F. G. Hamza-Lup (✉)
Computer Science, Georgia Southern University, Savannah, GA 31419, USA
e-mail: fhamzalup@georgiasouthern.edu

© Springer Nature Switzerland AG 2019   179
D. Schaefer et al. (eds.), *Design Education Today*,
https://doi.org/10.1007/978-3-030-17134-6_8

reduced cost of hardware components. As the surrounding world is highly tactile and three-dimensional (3D), many tasks routinely accomplished are tactile in nature and familiar from a 3D perspective. With the arrival of XR (Cross Reality—encompassing Augmented Reality, Mixed Reality, and Virtual Reality), the time has come for the tactile and the 3D visual modalities to naturally become part of the current and near future user interfaces. From applications for complex device manipulation (e.g., nano-manipulation, robotics), to training modules in a distance learning or technology-enhanced course, these technologies are redefining how users of all ages engage with others, collaborate and learn.

The chapter is organized as follows: in Sect. 2 an overview of the X3D ISO standard and architecture is provided, followed by several examples of Web-based applications designed to support learning and collaborative activities in various engineering and engineering design fields: from mechanical engineering to urban design. The same section explores the potential of Web3D in energy-efficiency assessment vis-à-vis building architecture and design.

Section 3 explains the haptic paradigm and provides a succinct explanation of the human tactile system. Next, the section explores state-of-art developments in haptic hardware systems and haptic rendering software. We explore applications of the visual-haptic paradigm designed to improve education and understanding of various abstract concepts fundamental to engineering disciplines. In conclusion, near future trends in 3D visual-haptic interfaces application in engineering design activities and intelligent dialogue, are explored as a trend that will accelerate in the foreseeable near future.

2 Web3D in Engineering Education

The web has become an enormous repository of data and the number of users has grown significantly over the past few decades, yet most of the web visual content mainly consists of 2D representations, static images, and dynamic multimedia. Presently, 3D display technology is in the beginning phase in numerous sectors (e.g., military simulation, medical, photography, gaming, education, and entertainment) and is expected to grow to over $112 Billion by 2020 with a projected compound annual growth rate of over 26% (Allied 2018). As 3D display devices are becoming available at a rapidly increasing pace, 3D content and development systems are lagging behind since most user interfaces have been designed with the flat screen paradigm in mind. Specifically, Web-based 3D content development and management have the potential to significantly improve not only the design of various applications but also the way users interact with each other online. Novel 3D user interfaces that enhance social interaction and promote learning and cooperation in a variety of engineering and design tasks can revolutionize knowledge sharing.

2.1 Web 3D Standards—Architecture Overview

Extended 3D (X3D) is an International Standards Organization (ISO) standard that provides a considerable opportunity to design and implement innovative user interfaces on the Web in order to support and augment engineering knowledge sharing and collaborative design activities. X3D is being developed by the Web3D Consortium (originally the VRML Consortium) as a mature and refined standard, improving on numerous issues from the past (e.g., XML compliance). It uses a scene graph construct to define the current scene, and allows interconnections with the Scene Access Interface (SAI) and other language bindings for advanced system functionality, as illustrated in Fig. 1.

The X3D standard uses XML encoding and is backward compatible with VRML. It provides a means to encrypt and compress the entire scene graph and all associated data. The scene graph contains all the components of the scene (i.e., nodes and their associated fields). The scene graph is parsed by a software component that usually runs in the browser (as a plug-in). The Scene Authoring Interface (SAI) provides extended functionality through a variety of scripting and programming languages (e.g., ECMAScript, Java, C/C++) as illustrated in Fig. 2.

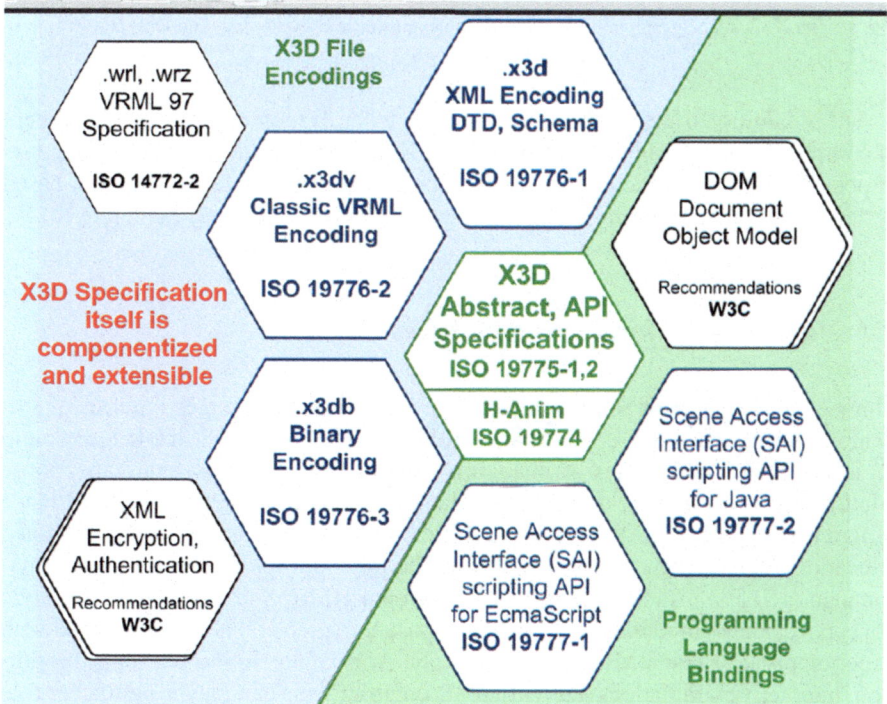

Fig. 1 X3D file encodings and programming language bindings (Brutzman and Daly 2007)

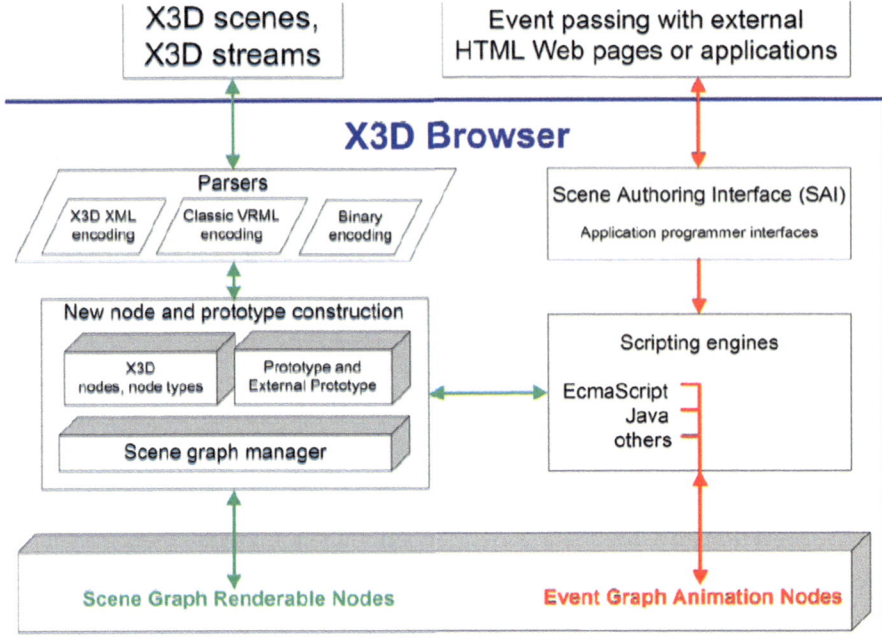

Fig. 2 The X3D scene graph and the scene authoring interface (Brutzman and Daly 2007)

The possibility to add extended functionality to the 3D content allows development of complex behavior and functionality in the X3D web interface, opening a large set of possibilities for interaction design and development. Complex n-Tier Web-based systems with extended middle-tier and database functionality may be deployed.

2.2 Interactive Engineering and Web3D

Many engineering programs throughout the nation face challenges in terms of student retention and success (Alter and Walser 2007; Brus et al. 2004; Steenkamp et al. 2017). As technology is changing the industry and jobs' landscapes more adult learners are seeking engineering education, many through an online/distance program (Rosen 2009). With the proliferation of computer-aided and web-based educational technologies in the classroom, the potential for Virtual Learning Environments (VLEs) (Annetta et al. 2010) is growing fast. Such tools may reinforce concepts from lectured material, provide exposure to practical issues associated with experiments and present visual, realistic applications of theoretical concepts. In addition, interactive simulations and virtual 3D environments can play a significant role in facilitating learning through engagement, immediate feedback and in creating real-world scenarios.

An example, Virtual Interactive Engineering on the Web (VIEW) (Goeser et al. 2011), provides a set of 3D web-based laboratories and modules designed for several engineering domains: materials engineering—Tensile Testing Laboratory and mechanical dissection—Small motors examination.

Tensile Testing Laboratory (Hamza-Lup et al. 2009), is an X3D module designed to introduce engineering students to the testing techniques required to evaluate certain mechanical properties of materials such as the elastic modulus, yield strength, ductility, and toughness. Tensile testing also emphasizes the importance of the load-strain and stress-strain curves in the evaluation of important mechanical properties of materials. New cutting-edge technologies in metals and metal alloys, composites, smart materials, and biomaterials, require a keen understanding and knowledge of material properties, and their testing is essential to engineering education. The simulator allows the user to choose a sample material slice and run it through the tensile test. During the stretching procedure, the simulator shows the load versus strain values and allows the user to inspect the fracture line in 3D by zooming on the slice as illustrated in Fig. 3—bottom image.

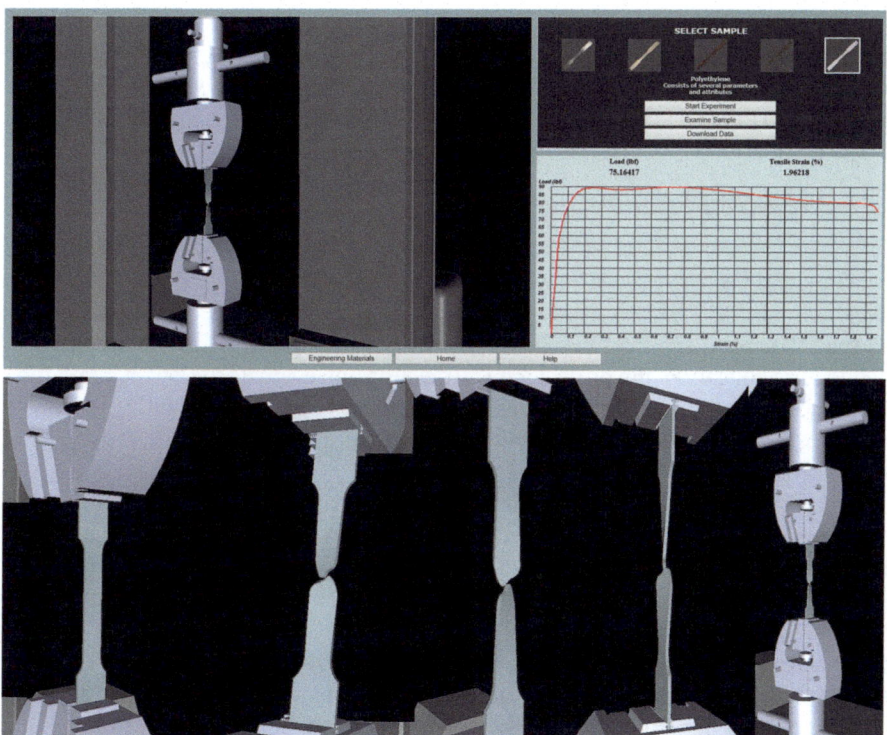

Fig. 3 Tensile testing simulation (device and load vs. strain plot—top image, 3D review of material fracture line—bottom image)

Mechanical dissection (Goeser et al. 2009) is an engineering activity that can satisfy a student's curiosity of how and why the components of given devices can convey specific motions to achieve the desired result. Mechanical dissection has the potential to increase student retention in engineering (Smith et al. 2017). Hence, several university engineering programs have developed mechanical dissection laboratories. However, such laboratories are not always feasible due to the lack of space, personnel, associated time and costs requirements. These issues can be addressed using a Web3D simulation, for example, the decomposition of an electric toothbrush system (as illustrated in Fig. 4), which only requires the use of an existing computer. Such 3D simulations may be used as complementary materials in *"Introduction to Engineering"* courses and may be accessed from any location, including the student home computer or handheld device.

Manufacturing engineering is another aspect worth exploring in conjunction with 3D, from CAD X3D viewers (Fraunhofer 2018) enabling access to the part structure in a tree or accordion view as illustrated in Fig. 5, to 3D complex object decomposition.

Additionally, metadata and annotations associated with each part can be explored for a better understanding of the parts before 3D Printing, CNC Machining, Injection Molding or Stereolithography. Furthermore, X3D projects extend to biomedical engineering through 3D printing and rapid prototyping of anatomically correct 3D models to help surgeons fully understand the state of an organ in a preoperative state (Bramlet 2018).

2.3 Architecture and Engineering—Urban Design

Structure and Form Analysis System (SAFAS) (Polys et al. 2013, 2015; Jones et al. 2014; Setareh et al. 2015) is an interactive 3D design tool and curriculum for architects learning structural engineering. Developed as a pedagogical tool for undergraduate architecture education, the project aims to develop and evaluate the use of simulation tools in architecture design education. The systems provide the ability to view a variety of design possibilities, allowing users virtual work-through different types of design problems and solutions.

Users can observe and compare the similarities and differences of the deformations and stresses of the structural members across diverse design conditions. The system improves the user's understanding of the relationships between structure and form and provides the ability to transfer these skills to future design problems (Polys et al. 2018).

3D sensor data representation and Web3D visualization can facilitate designing the cities of the future, and engineering a sustainable, resilient world. Aerial photogrammetry, in particular, is an efficient method to collect high-resolution, texture information that is appropriate for 3D modeling and visualization of large city landscapes. Unmanned Aerial Vehicles (UAV) photogrammetry systems consist of airborne drones, airborne Global Navigation Satellite Systems and Inertial Navigation

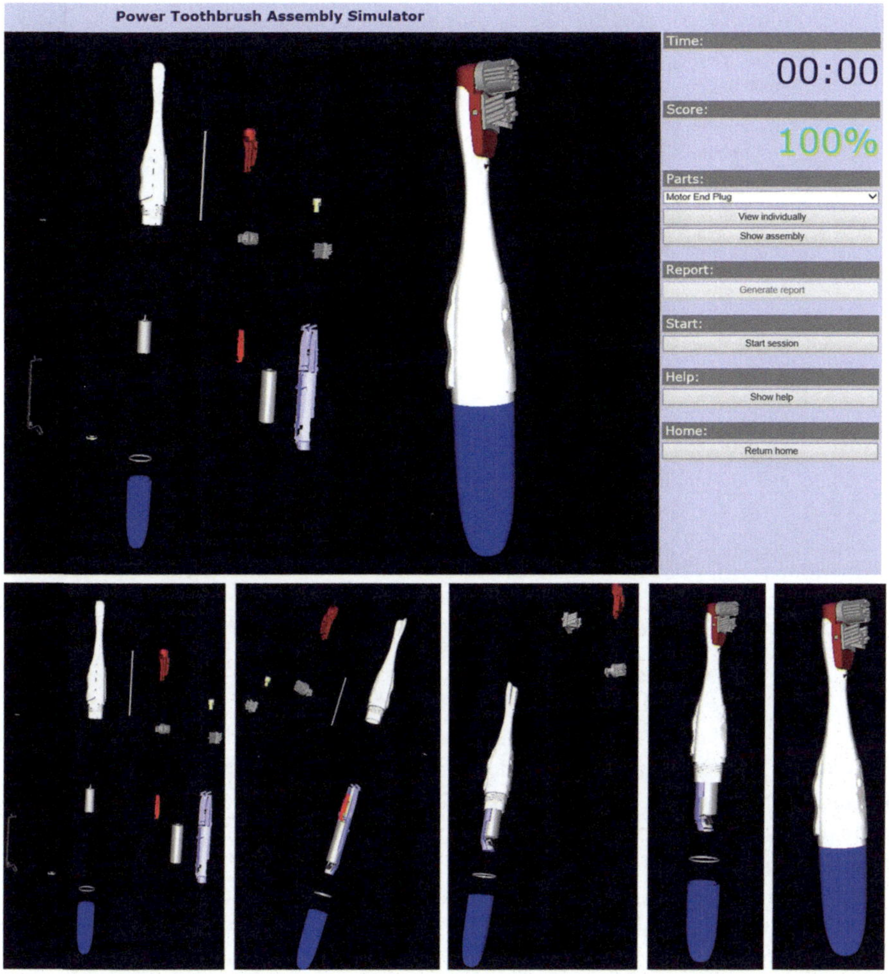

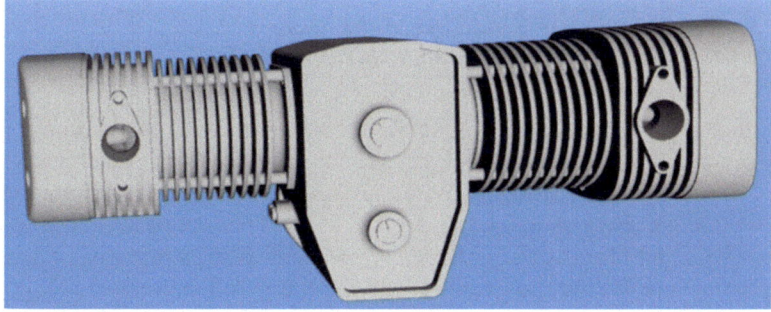

Fig. 4 Mechanical dissection of the electric toothbrush assembly

Fig. 5 CAD viewer (Fraunhofer 2018)

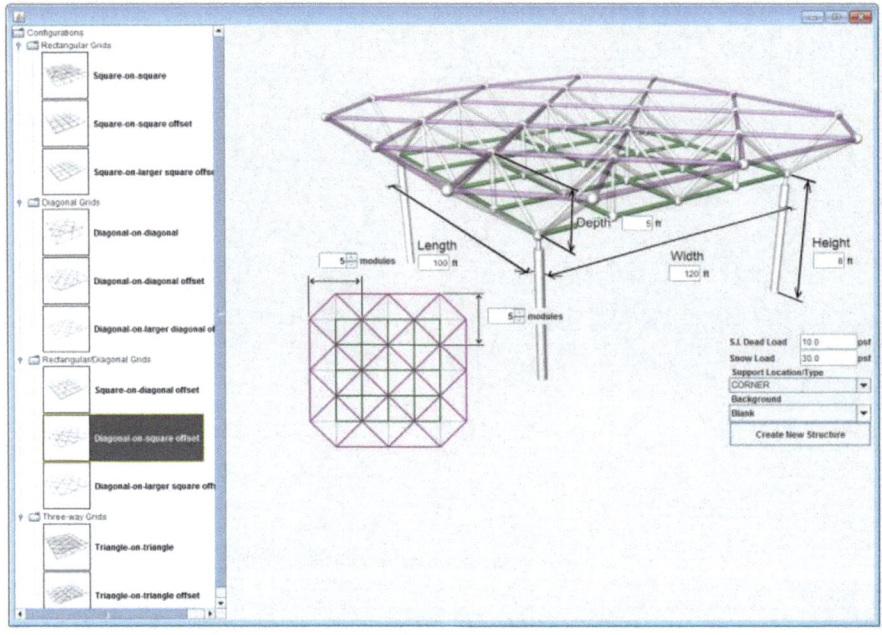

Fig. 6 Space frame structure in SAFAS (Polys et al. 2015)

Systems, and other components, which can provide aerial images as well as, position and orientation data (Colomina and Molina 2014). UAVs offer advantages such as flexibility, ease of operation, convenience, reliability, and low costs (Eisenbeiss 2009; Udin and Ahmad 2012) (Fig. 6).

X3D enables development tools that process and represent the collected data in 3D, enhancing the urban design and engineering projects ranging from automatic 3D city modeling (Bitmanagement 2017) to urban planning (Kocadag and Hamza-Lup 2013). Designing advanced 3D visual tours for travelers is also possible (Bitmanagement 2018), allowing route planning as illustrated in Fig. 7.

2.4 Assessing Energy Efficiency in Design

Designing energy-efficient buildings as well as retrofitting existing buildings to satisfy energy-efficiency standards requires the cooperative effort of architects, engineers, policy makers, and investors. While sustainable methods of construction have made tremendous progress in the last decade, using 0-carbon passive heating technologies (Voss and Musall 2012), thermally deficient western methods of construction hardly tap into the huge potential of applying energy-efficiency technology. Civil and Environmental engineers can nowadays take advantage of a wide range of big

Fig. 7 3D urban planning—3D city model of Savannah (Kocadag and Hamza-Lup 2013)

data processing tools and simulations that represents data collected from a myriad of smart sensors.

3D sensor data representation can be facilitated by Web3D visualization to enable better designs, by providing means to assess thermal comfort. Thermal comfort is a feeling of personal satisfaction within the thermal environment and depends on many factors (like age, gender etc.). It is an important factor for health and well-being as well as productivity since, a lack of thermal comfort, causes stress among building occupants. While a warmer environment can make people feel tired, a too cold environment may bring distraction and restlessness.

Conventional models of thermal representations are in use by construction professionals and HVAC (Heating, Ventilation, and Air Conditioning) engineers for many years and they rely on proprietary thermal analysis software products like Solid-Works, Ansys Advantage, Ansys CFX and many others that support CAD-based models integration. However, most of the simulation systems are proprietary and too complex to be applied in conjunction with real-time data.

A web-based thermal maps simulation (Hamza-Lup and Maghiar 2016) may be employed during the early planning phase to determine the order of magnitude of impacts related to design alternatives (as illustrated in Fig. 8).

The type of early design impact analysis may represent one of the key saving aspects during pre-construction, construction and operation phases while promot-

Fig. 8 Thermal X3D modeling of a commercial building (Hamza-Lup and Maghiar 2016)

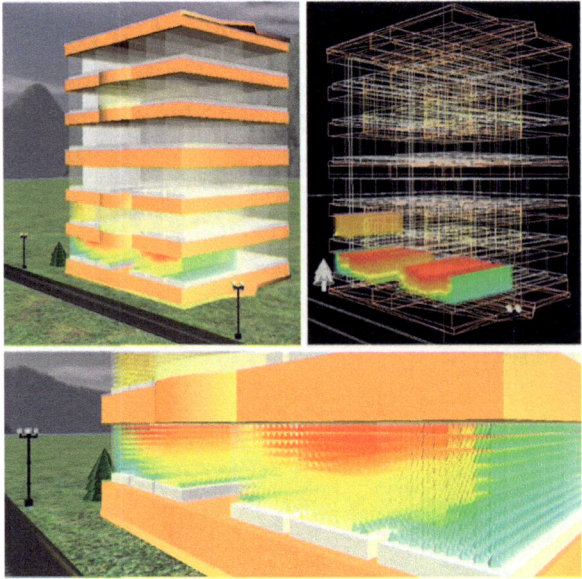

ing the sustainable design of the respective residential or commercial building. The proposed representation and modeling technique can also be distributed to the contractors and owners in order to improve the communication intent of the design as the interface is Web-based. 3D building data can be extracted from CAD architectural plans in recent buildings, and employing existing 3D scanning methods for older buildings (Hullo et al. 2015).

2.5 *Near Future Developments*

Spatial visualization skills are fundamental to a variety of disciplines but are particularly important for STEM disciplines (Uttal and Cohen 2012). From studying isomers (compounds with identical molecular compositions, but different spatial configurations) in chemistry, to engineering and architecture, enabling complex concepts understanding (Duffy 2017), these skills rely on the capacity to mentally manipulate 3D spatial forms. Research proved that spatial skills respond positively to training and educational intervention (Wright et al. 2008), hence providing users with advanced 3D Web-based interfaces and simulations can improve not only their mental models but also their efficiency through online/remote collaboration.

As 3D content has the potential of becoming widespread, and display technology allows the implementation and visualization of complex Cross Reality (XR) projects, the near future will blur the boundaries of reality and virtuality allowing the design of real spaces with virtual objects embedded using Augmented Reality (AR) techniques, as well as the design of virtual spaces with real objects scanned in place using mobile 3D scanning systems (Lansard 2018) or advanced automated photogrammetry applications (Bourke 2012).

Mobile 3D scanning technology (Sense 2018) and X3D will facilitate reverse-engineering and 3D data generation improving many aspects of engineering, from 3D post-manufacturing inspection processes and manufacturing speed-up to the deployment of advanced learning tools for understanding engineering processes and systems.

Current efforts to provide X3D with native support in HTML5 browsers are underway through the X3DOM project (X3DOM 2018) and other initiatives like the Cobweb browser and Titania X3D editor (Titania 2018). The very near future will allow a direct design with 3D components from designated libraries of 3D objects. Projects like Blend4Web, provide a free and open-source framework for creating and displaying interactive 3D computer graphics in web browsers (Blend 2018) while new initiatives exist and connect 3D with HTML5. A notable effort, WebGL (2018) provides a cross-platform for a low-level 3D graphics API based on OpenGL ES and is open to ECMAScript via the HTML5 Canvas element.

3 Haptic Interfaces in Engineering Education

Haptic systems connect the human tactile system with computers and enable the introduction of revolutionary user interfaces in a variety of fields that involve interaction with forces. Together with the 3D visual modality, the tactile communication channel allows development of revolutionary user interfaces with the potential to transform the way we interact with computer-based systems.

3.1 What is Haptics?

Haptic is rooted in the Greek word "*haptesthai*", meaning "*touch*". The word "haptic" was introduced in the English language in the late nineteenth century as a medical synonym for "tactile" and later it had developed a psychological sense, describing visually impaired individuals whose perception depended primarily on touch rather than sight. Recently the word "*haptic*" has been used to denote computer-based applications and systems that allow tactile interaction with virtual objects both rigid and deformable (i.e., plastic and elastically deformable objects).

Touch is one of our five senses that has been little explored by technology in the past 50 years, compared with the visual and auditory sense—if we consider the evolution of technology for the visual system (e.g., television and visual media from monochrome to color, from low resolution to HDTV) and the auditory system (e.g., radio-based systems, and the myriad of sound based systems e.g., sound synthesizers, amplifiers etc.). Nevertheless, the past decade has seen the considerable interest and a significant evolution from the technology standpoint vis-à-vis tactile systems.

Exploring the tactile sense, one can observe that it is significantly different than all the other senses; it is the only sense that allows full body reception through the skin—input (all the other senses and associated sensors are clustered on the head: mouth, nose, ears, eyes) and several output channels through the movement of the body/limbs. A basic categorization of the human tactile systems is:

- *Kinesthetic (Force) feedback.* Represents feedback perceived by sensory organs in the muscles and ligaments that are sensitive to forces acting on the musculoskeletal system. Sensory organs called *proprioceptors* are stretch sensors, embedded in the ligaments and muscles that allow perception of the position of different parts of the body. The proprioceptors are essential sensors for maintaining balance and coordination and are necessary for effective interaction with the environment. Loss or impairment of proprioception has a negative impact on human motor coordination and balance (Proske and Gandevia 2012).
- *Cutaneous* (tactile, vibrotactile and thermal) feedback. Such feedback is perceived by sensory organs in the skin and depends on specialized sensors in the skin: *thermoreceptors* for heat and cold, *nociceptors* for pain, and *mechanoreceptors* for pressure. The mechanoreceptors are Meissner corpuscles, Pacinian corpuscles, Merkel's disks, and Ruffini endings, with their main features succinctly described in Table 1 (McGlone and Reilly 2010; Johnson 2001).

The high density of cutaneous sensors i.e., as many as $2500/cm^2$ on the palms of the hand (Afifi and Bergman 2005) allows perception of small forces (below 0.1 N) as light touch, while the musculoskeletal system typically does not get excited by these small forces, enabling perception of much higher forces (e.g., the human handgrip force ranges from 300–400 N depending on gender, age and other factors (Massy-Westropp et al. 2011)).

Kinesthetic (Force) feedback is important in the perception of gross shape, weight, impact forces and it is critical for locomotion—walking, running, etc. While tactile feedback is essential for all touch-based activities (i.e., assessing object surface properties like roughness and temperature), the force feedback is usually dominant for the majority of the human activities.

3.2 Haptic Interfaces: The Hardware

Haptics interfaces provide users with cutaneous feedback and/or kinesthetic/force feedback during interaction with computer-generated virtual elements or remote objects manipulation (robotic telemanipulation).

Haptic interfaces come in a wide variety of types and shapes from vibrotactile systems to complex robotic arms that track the position and orientation of the user's arms. The tracking information is used to compute collisions between the user's

Table 1 Mechanoreceptors and associated features

Feature	Meissner corpuscles	Pacinian corpuscles	Merkel's disks	Ruffini endings
Location	Superficial dermis	Dermis and subcutaneous	Basal epidermis	Dermis and subcutaneous
Mean receptive area (mm^2)	13	101	11	59
Spatial resolution	Poor	Very poor	Good	Fair
Sensory units (%)	43	13	25	19
Response frequency range (Hz)	10–200	70–1,000	0.4–100	04–100
Min. threshold frequency (Hz)	40	200–250	50	50
Sensitive to temperature	No	Yes	Yes	At >100 Hz
Spatial summation	Yes	No	No	Unknown
Temporal summation	Yes	No	No	Yes
Physical parameter sensed	Skin curvature, velocity, local shape, flutter, slip	Vibration, slip, acceleration	Skin curvature, local shape, pressure	Skin stretch, local force

controlled pointer(s) representation in the virtual space and other virtual objects, allowing the user to "virtually" touch and feel such objects (as illustrated in Fig. 9).

Due to their application and complexity, haptic interfaces can be categorized on several dimensions. *Vibrotactile* haptic interfaces are usually simulating cutaneous force actions, while *force feedback* systems use servomotors to restrict the user's hand and/or fingers. *Unipoint* interaction, as well as *Multipoint* interaction, is possible, albeit multipoint interaction haptic interfaces are significantly more costly and complex than unipoint systems. Another categorization can be based on the mechanics of interactions:

- *Linkage-based* consist mainly of mechanical linkages e.g., 3D Systems: Phantom Omni and Desktop products; Force Dimension: Delta, Omega and Sigma7. An armature-based device like Mirage F3D-35 offers high powered 6 DOF and 3 force DOF capabilities (Dimension 2018; 3DSystems 2018).
- *Tension-based* systems consist of wires and pulleys, having the advantage of less weight e.g., SPIDAR-8, Mantis (Akahane et al. 2013).
- *Magnetic*, a more recent interface is based on magnetic forces and has the advantage of no weight hence no inertia e.g., Maglev200 (Maglev 2018).

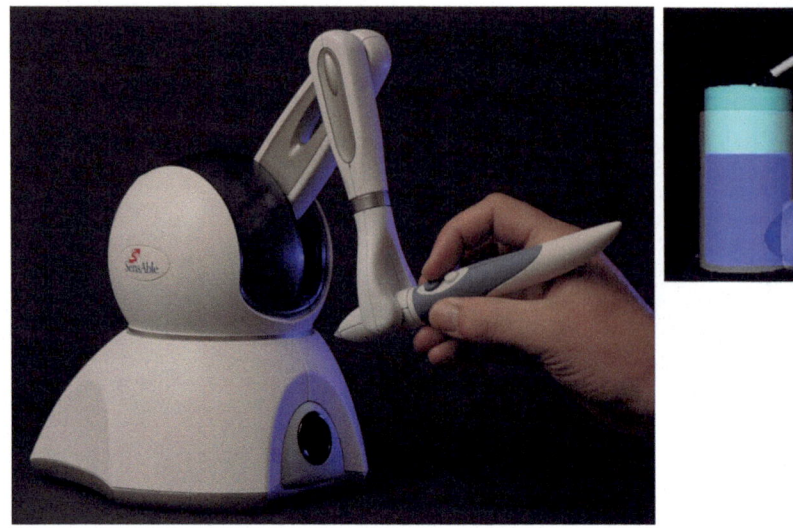

Fig. 9 Haptic interface-tactile feedback on pointer collisions with virtual objects

- *Acoustic* or *vibration based*, conveying forces through vibrations or acoustic waves (Takasaki et al. 2000).

Haptic interfaces have specific characteristics, the most important being enumerated in Table 2.

Resistive devices use brakes to restrict a user's motion, while *Active* devices use motors to restrict a user's motion and actively move their body around. Active devices

Table 2 Haptic interfaces characteristics (Hamza-Lup et al. 2019)

Haptic characteristic	Ideal quantity
Degrees of freedom	High
Workspace	High
Input position resolution	High
Output position resolution	High
Continuous force	High
Maximum force	High
Maximum stiffness	High
System latency	Low
Haptic update rate	High
Inertia	Low
Dexterity	High
Isotropy	High

can simulate a wider range of interactions, but they generally need to be more powerful than passive devices and are more difficult to control.

Bimanual haptic interaction is also possible and opens up the possibility of a wide range of natural actions employing coordination between the hands. Bimanual haptics refers to haptic interaction through both hands of the same person. In many tasks, we find in common to use both hands one being the dominant hand while the other is the non-dominant one. Guiard's Kinematic Chain (Guiard 1987) approximates the dominant hand and non-dominant hand with two motors assembled in series. The bimanual operation is vital in many engineering tasks that require precision manipulation (Hinckley et al. 1997), as well as for designing two-handed interaction interfaces (Cutler et al. 1997). Several force feedback systems can be adapted for bimanual haptic interaction and user's hands asymmetry must be considered in the design (Talvas et al. 2014).

3.3 Haptic Rendering—The Software

Haptic rendering is a software enabled process that allows the display of computer-controlled forces on the user to make him/her experience the actual feel of virtual objects and/or forces. If a parallel is drawn between graphics rendering and haptic rendering, the haptic rendering would be the equivalent of graphics rendering but from a tactile/force feedback perspective.

The APIs for haptic rendering follow an architecture similar to the one presented in Fig. 10. It is easy to observe the central role that haptic and visual devices play in the multisensorial application development. Moreover, there is a need for other devices as well, that assures audio rendering for example, and may respond to other specific application needs. Each API is responsible for the implementation of the interfaces with those devices and for their synchronization with the visual component. Meta-languages, such as XML or X3D are used to describe the scene graph structure. Frequently, the C++ and the Python language is used for enriching the scene graph structure specified by a meta-language, adding functionality by scripting modules.

To exemplify, Open Haptics (OpenHaptics 2018) is a commercial set of libraries that enables software developers to add haptics and 3D navigation to a broad range of applications. Other platforms like Simulation Open-Framework Architecture (SOFA 2018), an open-source simulation framework, are dedicated to the development of algorithms for deformation. Computer Haptics and Active Interfaces (CHAI3D 2018) is an open-source designed to facilitate the development of 3D modeling applications augmented with haptic rendering. It supports several commercial haptic interfaces such as Servo2Go and Sensoray 626 I/O board, IEEE1394 interface. CHAI3D provides an easy solution to interface any haptic device with a specific computer-based application.

A comparative study of APIs and frameworks for haptic application development (Popovici et al. 2012) selects the H3D API (H3D 2018) as the winner on several dimensions. H3D is dedicated to haptic modeling that combines the OpenGL and

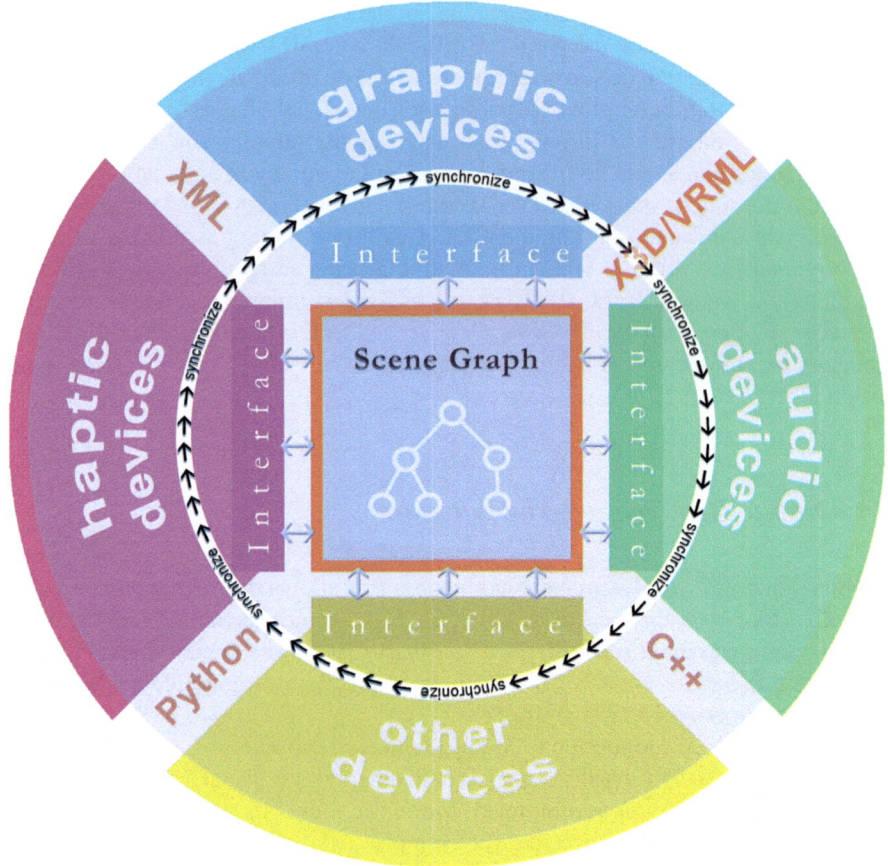

Fig. 10 Conceptual architecture of a multisensorial haptic platform

X3D standards together with haptic rendering in a single scene graph that mixes haptic and graphics components. H3D is independent of the haptic device multi-platform that allows audio and 3D stereoscopic device integration. H3D is conceived to support rapid prototyping. Combining X3D, C++ and the Python scripting language, H3D improves the speed of execution, when performance is critical, as well as the high speed of development, when rapid prototyping is required.

3.4 Force Simulation in Engineering Design and Education

In the framework of Engineering, force feedback can be exceptionally useful in a variety of simulation and training systems, as well as in practical systems implementation. Engineering educators are constantly challenged with developing the student's ability to connect theory with reality. Students typically enter and leave college level physics courses with faulty mental models (Halloun 1985) and such mental models are later used as building blocks for more advanced Engineering concepts. Mental models are employed by engineering students to generate explanations (Fazio et al. 2013) and they play a significant role in cooperation and coordination of team activities in complex environments and contexts (Casakin and Badke-Schaub 2015). The addition of multiple senses can reduce the distance between the concept that a student has formed about a certain phenomenon of the physical world, and the reality, enabling the correct understanding of fundamental concepts.

3.4.1 Hydraulics and Engineering

An important principle of hydraulics taught in high schools is the Pascal's law, or Pascal's principle, which states that for all points at the same absolute height in a connected body of an incompressible fluid at rest, the fluid pressure is the same, even if additional pressure is applied on the fluid at some place. Moreover, the concept of pressure and its dependency on force magnitude and surface area is fundamental for any engineer that works with hydraulic systems. A project deployed in this sense, Haptek16 (Hamza-Lup and Adams 2008), uses force feedback systems to enable students to experiment and gain a deeper understanding of the laws of hydraulics. As depicted in Fig. 11 the user employs a haptic device to apply a force on the left cylinder and consequently feels a force proportional with the pressure in the system.

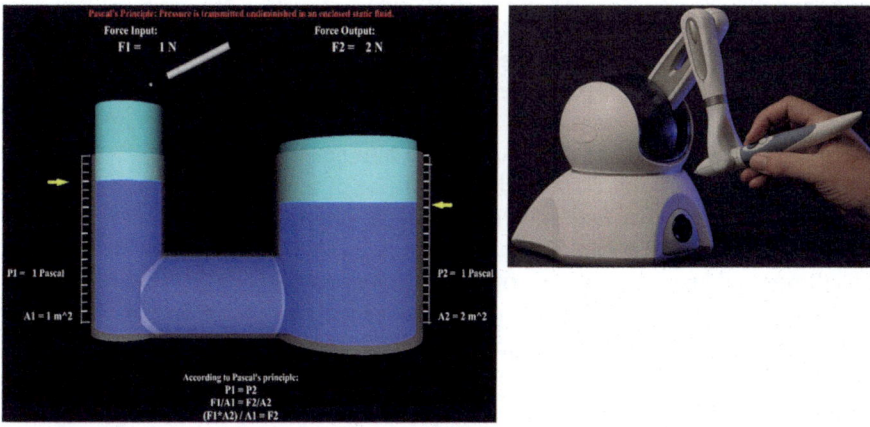

Fig. 11 The user applies pressure on the left cylinder and feels the force feedback generated by that pressure

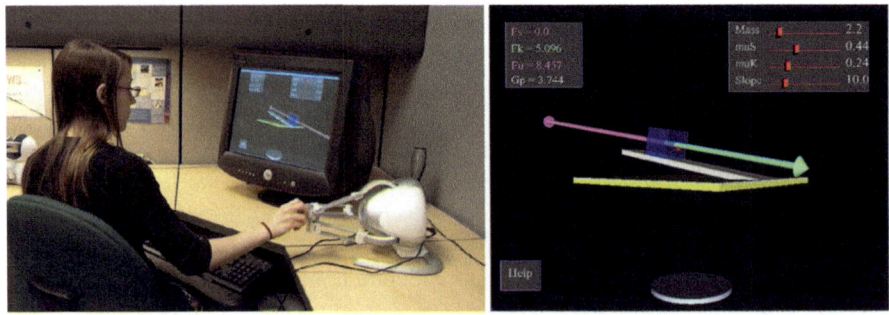

Fig. 12 Student using the haptic friction simulator: room view (left) and screen snapshot (right)

3.4.2 Static and Kinetic Friction in Mechanical Engineering

Static friction is defined by the inequality $Fs \leq \mu_s * N$, where Fs is the force of static friction, μ_s is the coefficient of static friction, and N is the normal force. The maximum value of the static friction Fs_{max} is equal to $\mu_s * N$. The fact that the static frictional force is described by inequality is the ultimate source of difficulty for many students. Since all forces they have seen before are described by equalities, they tend to set $Fs = Fs_{max}$. Depending on the problem, students may not be able to realize their mistake (e.g., when an object is pushed with a force greater than Fs_{max}). If the force applied to an object is less than Fs_{max}, the use of the incorrect equality $Fs = Fs_{max}$ yields the nonphysical result that an object will move in the opposite direction of the force being applied. Such dynamic components, required for the understanding of kinetic and static friction coefficients, cannot be illustrated in a textbook. Visual-Haptic simulators have been proposed to alleviate this problem (Hamza-Lup and Baird 2012), employing simple force feedback systems as illustrated in Fig. 12.

The user can control parameters such as the block's mass, the coefficients of friction, and the slope of the plane, allowing him/her to "see and feel" the effects each parameter has on the forces. The magnitude of the force vectors is displayed dynamically in 3D illustrating how these forces vary in response to configuration changes. The interaction can be recorded in a sequence of screenshots or small movies and used later to complement course material or laboratory sessions.

Simulators for understanding concepts like torque and a lever arm and their role in rotational equilibrium have been proposed (Neri et al. 2018). The simulator also explores the dynamics of pulley systems using a haptic Double Atwood machine and rolling reels to show how linear acceleration of a reel center of mass will increase when the radius value of the inner pulley increases for a given downward force.

The complexity of aerodynamic forces in aerospace engineering has also spawned haptic simulators to facilitate understanding abstract concepts by allowing users to change the speed and angle of attack on an airplane wing and feeling the effects of the forces generated in the wings (Lopes and Vaz de Carvalho 2010).

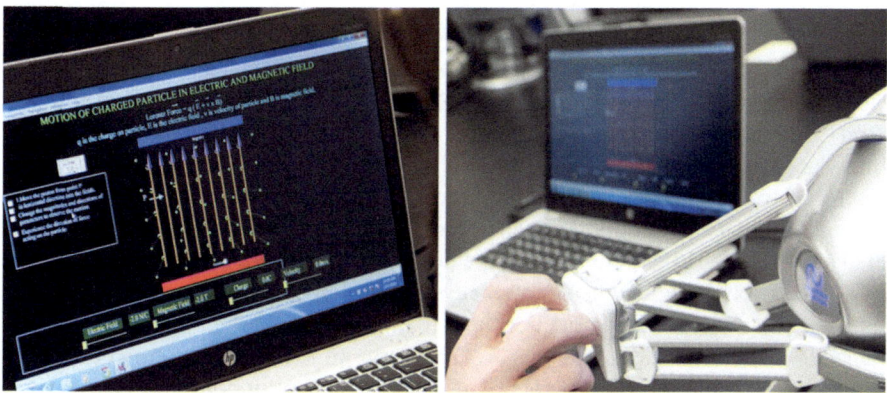

Fig. 13 A game-like scenario—user "grabs" a charged particle and moves it through an electro-magnetic field, feeling the electromagnetic forces acting on it

3.4.3 Electromagnetism and Electrical Engineering

An accurate understanding of the electromagnetism paradigms is fundamental and establishes important knowledge building blocks for understanding a variety of systems and phenomena (electric engines, current propagation, etc.). Electromagnetic force usually exhibits electromagnetic fields, such as electric fields and magnetic fields, and is a fundamental interaction in nature. An example is a Lorentz force affecting a charged particle as it passes through electric and magnetic fields. In addition to their geometric three-dimensionality, one of the confusing parts of electricity and magnetism involves the implementation of the numerous "right-hand rules" that determine the direction of forces or fields. A visual-haptic interface to practically experience and feel the Lorentz force on a charged particle moving in electric and magnetic fields while observing the directions of the field vectors is desired. A game-like scenario can give the user the impression of physically grabbing a small charged particle and moving it across an electromagnetic field of varying polarity at different speeds. The graphical user interfaces accompanied by the haptic feedback provide cues to the direction, orientation and magnitude of different force vectors, conveying the participant a better understanding of the forces involved in the process.

Such interfaces convey a better understanding of the magnitude and direction of the forces involved. As a particle is moved through regions of space containing electric and magnetic fields of varying strengths and directions, the tactile feedback from the device helps strengthen the ideas of velocity-dependent forces and demonstrate the fact that a uniform magnetic field does not change the speed of a charged particle, but only its direction (as illustrated in Fig. 13).

Experimental results with such interfaces show improved scores for the participants and provide an optimistic base for the development of haptic user interfaces for abstract concept exploration and understanding.

4 Conclusion

There are many 3D design and modeling software platforms on the market today. From Computer Aided Design (CAD) software (e.g., SolidWorks, Catia, AutoCAD) widely used in mechanical engineering, automotive/aerospace engineering and manufacturing, to 3D Modeling tools (e.g., Maya 3D, Blender, Autodesk 3D Studio Max, Cinema 4D) that explore the artistic side of the design. Web3D and the associated standards (e.g., X3D, WebGL) facilitate deployment of 3D content in a web-based environment, enabling extensive interaction and collaboration among the engineers and designers involved in product design and development.

Generation Z the "digital natives" will be very influential in dictating changes in systems' design in the years to come and intelligent, multimodal interfaces are the next demand. Such user interfaces will allow knowledge sharing through intelligent dialogue (Hamza-Lup and Goldbach 2019) on multiple communication channels and several dimensions, some augmented by intelligent software agents.

Designing an exceptional user interface in the context of today technology requires the understanding of novel technology capabilities. We experience the world around us in 3D, we feel the objects that we touch and more importantly, we interact with each other exchanging ideas and knowledge. XR and haptic technologies are blurring the boundaries of reality and virtuality, allowing virtual objects to blend into reality and real objects to merge with virtuality. Such capabilities are calling on new design challenges and opportunities for everything that surrounds us. From simple user interfaces designed for education and training to complex user interfaces designed for aircraft control, the challenge resides in the understanding of the human sensory system in conjunction with the capabilities of the near future multisensory computer-based systems. Future designers must acquire a comprehensive understanding of human and device capabilities, and they have to keep an open mind and boldly explore revolutionary interaction techniques that have never been explored before.

Acknowledgements We would like to acknowledge Don Butzman and Nicholas Polys for discussions about the X3D standard and novel interaction techniques in the 3D space. This work was also facilitate by the Erasmus+ program for international cooperation between Georgia Southern University, US and Transilvania University of Brasov, Romania.

References

3DSystems (2018) "The Touch haptic device" former Phantom Omni. https://www.3dsystems.com/haptics-devices/touch. Accessed 6 Sept 2018

Afifi K, Bergman RA (2005) Functional neuroanatomy: text and Atlas, 2nd edn. McGraw Hill Professional, 18 Feb 2005

Akahane K, Hyun J, Kumazawa I, Sato M (2013) Two-handed multi-finger string-based haptic interface SPIDAR-8. In: Galiana I, Ferre M (eds) Multi-finger haptic interaction. Springer series on touch and haptic systems. Springer, London

Allied A (2018) Global 3D display market (Type, Technologies, Access Methods, Application, and Geography)—size, share, global trends, company profiles, demand, insights, analysis, research, report, opportunities, segmentation and forecast, 2013–2020, Oct 2018

Alting A, Walser A (2007) Retention and persistence of undergraduate engineering students: what happens after the first year? In: Proceedings of the American society for engineering education annual conference and exposition, Honolulu, HI. https://peer.asee.org/2344

Annetta LA, Folta E, Klesath M (2010) V-learning: distance education in the 21st century through 3D virtual learning environments, Springer. ISBN 9048136202

BitManagement (2017) Automatic construction of 3D cities. http://www.web3d.org/case-studies/automatic-construction-3d-cities/automatic-construction-3d-cities. Accessed 15 Sept 2018

BitManagement (2018) City of Vienna. http://www.web3d.org/example/historic-vienna. Accessed 15 Sept 2018

Blend (2018) Blend for web—unleashing the power of 3D internet. https://www.blend4web.com/en/. Accessed 15 Sept 2018

Bramlet M (2018) Pulmonary atresia, VSD with confluent branch pulmonary arteries and PDA. https://3dprint.nih.gov/discover/3dpx-001608. Accessed 15 Sept 2018

Brutzman D, Daly L (2007) X3D: 3D graphics for web authors, Morgan Kaufmann Publishers

Brus C, Zhao L, Jessop J (2004) Visual-spatial ability in first-year engineering students: a useful retention variable? In: Proceedings of the American society for engineering education annual conference and exposition, Portland, OR

Bourke P (2012) Automatic 3D reconstruction: an exploration of the state of the art. GSTF J Comput (JoC) 2(3)

Casakin H, Badke-Schaub P (2015) Mental models and creativity in engineering and architectural design teams. In: Gero J, Hanna S (eds) Design computing and cognition '14. Springer, Cham

CHAI3D (2018) Computer haptics and active interfaces—CHAI3D. https://www.chai3d.org. Accessed 10 Oct 2018

Colomina I, Molina P (2014) Unmanned aerial systems for photogrammetry and remote sensing: a review. ISPRS J Photogramm Remote Sens 92:79–97

Cutler LD, Frohlich B, Hanrahan P (1997) Two-handed direct manipulation on the responsive workbench. In: Proceedings of ACM symposium on interactive 3D graphics, pp 107–114

Dimension F (2018) Force dimensions haptic devices. http://www.forcedimension.com/. Accessed 3 Oct 2018

Duffy G (2017) Spatial thinking in the engineering curriculum: an investigation of the relationship between problem solving and spatial skills among engineering students. Doctoral thesis, DIT

Eisenbeiss H (2009) UAV photogrammetry. PhD Thesis, Swiss Federal Institute of Technology Zurich, Zurich, Switzerland

Fazio C, Battaglia OR, Paola B (2013) Investigating the quality of mental models deployed by undergraduate engineering students in creating explanations: the case of thermally activated phenomena. Phys Rev ST Phys Educ Res 9(2), American Physical Society

Fraunhofer (2018) CAD viewer. http://www.web3d.org/example/cad-scene-graph-viewer. Accessed 10 Sept 2018

Goeser PT, Johnson WM, Hamza-Lup FG, Schaefer D (2011) VIEW—a virtual interactive web-based learning environment for engineering. IEEE Adv Eng Educ J Spec Issue Res e-Learn Eng Educ 2(3). ISSN: 1941-1766.

Goeser PT, Johnson WM, Hamza-Lup FG, Sopin I, Sanchez C, Hager P (2009) A different view: virtual interactive engineering on the web. In: Proceedings of the American society for engineering education annual conference and exposition, June 14–17, Austin, Texas

Guiard Y (1987) Asymmetric division of labor in human skilled bimanual action: the kinematic chain as a model. J Motor Behav 19(4):486–517

Halloun A, Hestenes D (1985) The initial knowledge state of college physics students. Am J Phys 53:1043–1055

Hamza-Lup FG, Adams M (2008) Feel the pressure: e-learning system with haptic feedback. In: The 16th symposium on haptic interfaces for virtual environments and teleoperator systems (Haptics 2008), March 13–14. Reno, Nevada, pp 445–450

Hamza-Lup FG, Baird WH (2012) Feel the static and kinetic friction. EuroHaptics 2012, Part I, LNCS 7282, 12–15 June. Tampere, Finland, pp 181–192

Hamza-Lup FG, Goldbach IR (2019) Survey on intelligent dialogue in e-learning systems. In: Proceedings of the international conference on mobile, hybrid, and on-line learning, Feb 24–28. Athens, Greece, pp 49–52. ISBN: 8-1-61208-689-7

Hamza-Lup FG, Maghiar M (2016) Web3D graphics enabled through sensor networks for cost-effective assessment and management of energy efficiency in buildings. Graph Models J Elsevier 88:66–74. www.sciencedirect.com/science/article/pii/S1524070316300042

Hamza-Lup FG, Goeser PT, Johnson WM, Thompson T, Railean E, Popovici DM, Hamza-Lup G (2009) Interactive 3D web-based environments for online learning: case studies, technologies and challenges. In: International conference on mobile, hybrid, and online learning, Feb 1–7. Cancun, Mexico, pp 13–18

Hamza-Lup FG, Bergeron K, Newton D (2019) haptic systems in user interfaces—state of the art survey. In: Proceedings of ACMSE 2019, 18–20 April 2019. Kennesaw, GA, USA

Hinckley K, Pausch R, Proffitt D, Patten J, Kassell N (1997) Cooperative bimanual action. In: Proceedings of ACM SIGCHI conference on human factors in computing systems, 1997, pp 27–34

Hullo JF, Thibault G, Boucheny C, Dory F, Mas A (2015) Multi-sensor as-built models of complex industrial architectures. Remote Sens 7:16339–16362

H3D (2018) Haptics 3D by Sense graphics "H3D application programming interface manual". http://www.h3dapi.org. Accessed 2 Oct 2018

Johnson KO (2001) The roles and functions of cutaneous mechanoreceptors. Current Opin Neurobiol 11(4):455–61. Review. PubMed PMID: 11502392

Jones BD, Setareh M, Polys NF, Bacim F (2014) Application of an online interactive simulation tool to teach engineering concepts using 3D spatial structures. Int J Web-Based Learn Teach Technol (IJWLTT) 9(3)

Kocadag F-AL, Hamza-Lup FG (2013) X3D in Urban Planning – Savannah in 3D. In: The Proceedings of the 51st association for computing machinery southeast conference, April 4–6, Savannah, Georgia, US. ISBN: 978–1–4503–1901–0

Lansard M (2018) The 8 best 3D scanning apps for smartphones and iPads. https://www.aniwaa.com/best-3d-scanning-apps-smartphones/. Accessed 18 Oct 2108

Lopes D, Vaz de Carvalho C (2010) Simulation and haptic devices in engineering education. Elektronika ir Elektrotechnika 102(6). http://eejournal.ktu.lt/index.php/elt/article/view/9383. Accessed 17 Oct 2018

Maglev M (2018) Maglev 200™ system—high-fidelity interaction with virtual or remote environments. http://butterflyhaptics.com/products/system/. Accessed 2 Oct 2018

Massy-Westropp NM, Gill TK, Taylor AW, Bohannon RW, Hill CW (2011) Hand grip strength: age and gender stratified normative data in a population-based study. BMC Res Notes 4:127. https://doi.org/10.1186/1756-0500-4-127

McGlone F, Reilly D (2010) The cutaneous sensory system. Neurosci Biobehav Rev 34(2010):148–159

Neri L, Noguez J, Robledo-Rella V, Escobar-Castillejos D, Gonzalez-Nucamendi A (2018) Teaching of classical mechanics concepts using visuo-haptic simulators. Educ Technol Soc 21(2):85–97

OpenHaptics (2018) OpenHaptic by 3DSystems "OpenHaptics developer edition". https://www.3dsystems.com/haptics-devices/openhaptics. Accessed 10 Oct 2018

Polys NF, Bacim F, Setareh M, Jones B (2013) Building novel Web3D user interfaces: a case study from architecture and structural engineering. In: Proceedings of the 18th international conference on 3D web technology (Web3D '13). ACM, San Sebastian ES, pp 135–138

Polys NF, Bacim F, Setareh M, Jones B (2015) SAFAS: unifying form and structure through interactive 3D simulation. Eng Design Graph J 79:1–23

Polys NF, Newcomb C, Schenk T, Skuzinski T, Dunay D (2018) The value of 3D models and immersive technology in planning urban density. In: Proceedings of the 23rd international ACM conference on 3D web technology (Web3D '18). ACM, New York, NY, USA, Article 13, 4 pp. https://doi.org/10.1145/3208806.3208824

Popovici DM, Hamza-Lup FG, Seitan A, Bogdan CM (2012) Comparative study of APIs and frameworks for haptic application development. In: 12th international conference on cyberworlds, Darmstadt, Germania, 25–27 Sept 2012, pp 37–44

Proske U, Gandevia SC (2012) The proprioceptive senses: their roles in signaling body shape, body position and movement, and muscle force. Physiol Rev 92(4):1651–1697

Rosen A (2009) e-Learning 2.0: proven practices and emerging technologies to achieve real results, AMACOM, 2009. ISBN 0814410731

Sense S (2018) Take your world from physical to digital with 3D scanning. https://www.3dsystems.com/3d-scanners/sense-scanner. Accessed 2 Sept 2018

Setareh M, Jones B, Ma L, Bacim F, Polys N (2015) Application and evaluation of double-layer grid spatial structures for the engineering education of architects. J Architect Eng 21(3)

Smith A, Knizley A, Luck R (2017) A product dissection project designed for student motivation and retention in an introduction to engineering course. American Society for Engineering Education, San Juan, Puerto Rico, 2–5 Mar 2017

SOFA (2018) Simulation open framework architecture—SOFA. http://www.sofa-framework.org. Accessed 2 Oct 2018

Steenkamp H, Nel AL, Carroll J (2017) Retention of engineering students. In: 2017 IEEE global engineering education conference (EDUCON), Athens, 2017, pp 693–698. https://doi.org/10.1109/educon.2017.7942922.

Titania (2018) Titania X3D editor—bring colours to your life. http://create3000.de/users-guide/. Accessed 5 Oct 2018

Takasaki M, Nara T, Tachi S, Higuchi T (2000) A tactile display using surface acoustic wave. In: Proceedings 9th IEEE international workshop on robot and human interactive communication, Osaka, Japan, 2000, pp 364–367

Talvas A, Marchal M, Lecuyer A (2014) A survey on bimanual haptic interaction. IEEE Trans Haptics 7(3):285–300

Udin WS, Ahmad A (2012) Large scale mapping using digital aerial imagery of unmanned aerial vehicle. Int J Sci Eng Res 3:1–6

Uttal DH, Cohen CA (2012) Spatial thinking and STEM education. In: The psychology of learning and motivation, vol 57, pp 147–181. ISSN: 0079–7421

Voss K, Musall E (2012) Net zero energy buildings—international projects of carbon neutrality in buildings, 2nd edn, Nov 2012. Institut für Internationale Architektur-Dokumentation GmbH & Co. KG, München. ISBN 978-3-920034-80-5

WebGL (2018) The Khronos project—WebGL. https://www.khronos.org/webgl/. Accessed 5 Oct 2018

Wright R, Thompson WL, Ganis G, Newcombe NS, Kosslyn SM (2008) Training generalized spatial skills. Psychon Bull Rev 15:763–771

X3DOM (2018) What is X3DOM, and what can it do for me? https://doc.x3dom.org/gettingStarted/background/index.html. Accessed 5 Oct 2018

Development of Complex System Design Oriented Curricula: The Example of the Grande Ecole CentraleSupélec

Marija Jankovic, Didier Dumur, John Cagnol and Valérie Ferreboeuf

Abstract Demand for complex system and Systems Engineering curricula is on the rise. This trend is in particular supported by current technology and economic developments such as the Internet of Things, Artificial intelligence, Servitization, the need to develop Product Service Systems (PSS) as well as System of Systems (SoS). In this context, one of France's top engineering schools has been developing a complex-system-focused curriculum. One of the majors is focusing on complex system design: Large-scale interacting systems. We give details of the curriculum orientations, we discuss the specificity of the education system of engineering schools (Grandes Ecoles) and the process that has been proposed to develop this new educational program.

Keywords Complex engineered system design · Curriculum design · Grandes Ecoles · French higher education

1 Introduction

"Demand for Systems Engineering (SE) at different levels of policy, analysis, architecture, and phases of implementation is rapidly growing, while well trained SE resources are scarce needing intense development" (Airbus ADS R&T

M. Jankovic (✉) · V. Ferreboeuf
Laboratoire de Génie Industriel, CentraleSupélec, Paris, France
e-mail: marija.jankovic@centralesupelec.fr

V. Ferreboeuf
e-mail: valerie.ferreboeuf@centralesuepelc.fr

D. Dumur
Laboratoire des Signaux et Systèmes, CentraleSupélec, Paris, France
e-mail: didier.dumur@centralesupelec.fr

J. Cagnol
Laboratoire Mathématiques et Informatique pour la Complexité et les Systèmes, CentraleSupélec, Paris, France
e-mail: john.cagnol@centralesupelec.fr

© Springer Nature Switzerland AG 2019
D. Schaefer et al. (eds.), *Design Education Today*,
https://doi.org/10.1007/978-3-030-17134-6_9

Days—Keynote by Prof. Dipl.-Ing. Heinz Stoewer, M.Sc., SAC GmbH, Paris, 18. November, 2016). The need for complex systems engineering curricula has been increasing considerably. For the past three years, one of the major French engineering schools, the Grande École, CentraleSupélec, has been working on the development of a completely new program entirely focused on complex systems. Moreover, one of the majors that is proposed within this curriculum is entirely concentrating on complex systems engineering: Large-scale interacting systems.

The system of French engineering schools is specific to the French national education system and not necessarily known in other countries. Hence, we propose the following organization of the paper: Sect. 2 addresses the system of Grandes Écoles, in order to present its positioning with regard to the classical university system. Section 3 details the process that was discussed and chosen to undergo to identify and design future complex systems engineering curriculum. Current scientific and industrial challenges as well as proposed content for the Large-scale interacting systems curriculum is discussed in Sect. 4. Finally, we present our conclusions in Sect. 5.

2 Grandes Écoles: The French Education System

In France, higher education is made of two systems that coexist: the classical University system and the "Grandes Écoles" system. Details and equivalence of these two structures can be seen in Fig. 1.

In the University system (on the right), high school graduates take the government-run national examination (*Baccalaureate*) and can enroll into a University if they pass that institution's exam. This system is consistent with the European structure, called LMD (License–Master–Doctorate). Essentially, there are three years of Bachelor of Sciences studies (so-called License), after which students can enroll in a Master of Science program followed by the possibility of entering into Ph.D. studies.

In the Grandes Écoles structure, only students who passed the *Baccalaureate* and have excellent academic records can enroll in what is called "Classes préparatoires" (preparatory classes). This is a specific intensive mathematics and physics program preparing students who wish to attend one of the engineering schools (*Ecoles d'ingénieur*). After two or three years, there are different nation-wide exams that are organized by the various schools. The students who pass these exams are selected for the engineering schools' program based on their ranking and the number of places available in each institution.

If we compare the number of students in the University system, there are 340,000 students in total that enrol in the first year in France, and 110,000 of these students have chosen scientific programs (see Table 1) (Ministère de l'Enseignement Supérieur 2018). However, in the Grandes Écoles system, only 43,000 of selected students enter Classes préparatoires aux Grandes Écoles (CPGE), out of them 25,000

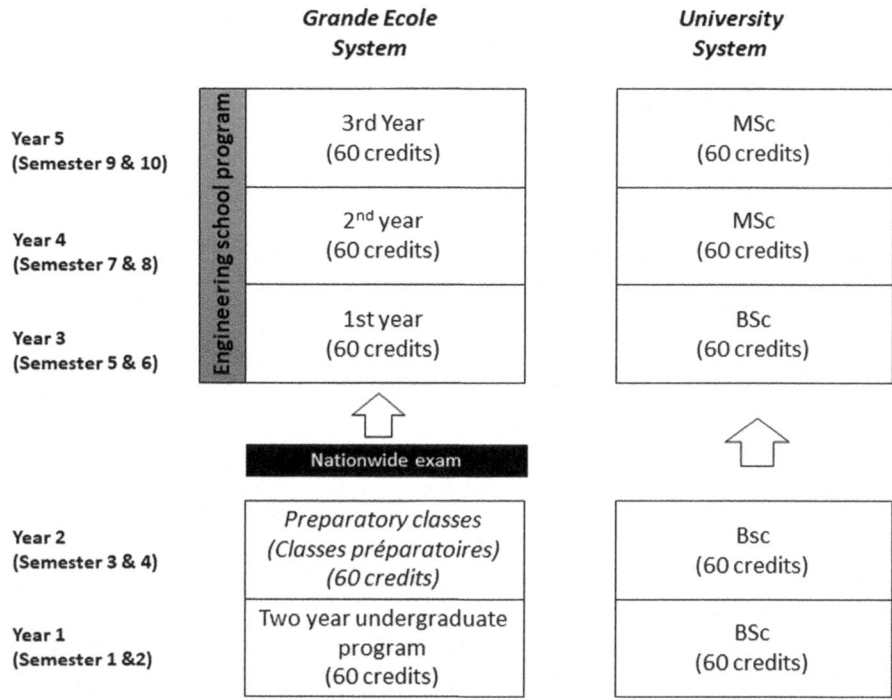

Fig. 1 Two educational programs in France: Grande Écoles (left) and University (right)

Table 1 Number of students enrolling into University and CPGE (Classes préparatoires)

	1st year	Science
University	340,000	110,000
CPGE (Classes préparatoires)	43,000	25,000

students have chosen scientific programs (in the majority of cases they aim for entering the engineering schools—Écoles d'ingénieurs) (Ministère de l'Enseignement Supérieur 2017).

One can see that for the overall population that aims at enrolling into scientific and engineering programs there are around 135,000 students in total (both at university and in *classes préparatoires*) (see Table 2). The students that aim at entering CentraleSupélec must take the exam after preparatory classes entitled "Concours CentraleSupélec". Around 10,000 students take this exam each year. The number of students who are eventually selected and enrol into CentraleSupélec is around 1,000.

Table 2 Selection ratio for CentraleSupélec in particular

Overall students in the Grandes Eoles system	Students that prepare the examination for CentraleSupélec	CentraleSupélec
135,000	10,000	1000
		1.3% of students

3 Design of a New Curriculum: Complex System Orientation

3.1 Historical Context for the Overall Design Methodology

Ecole Centrale Paris was founded in 1829 by Alphonse Lavallée and for over a century has offered cutting-edge training for high-level industry engineers. In recent years, the school has consistently placed second in national rankings. Students completing the intense French preparatory schools frequently choose Centrale Paris over almost any other French engineering school, which results in a student recruitment that is highly competitive and selective.

Supélec was founded in 1894 by Eleuthère Mascart. Its original focus was on electrical engineering, but over the years, the school has established a solid reputation for training in the fields of energy and systems. It is also ranked among the most competitive engineering schools in France.

On January 1st, 2015, Ecole Centrale Paris and Supélec merged to create CentraleSupélec, a new state-of-the-art engineering school within the University of Paris-Saclay. The curricula of the two former schools are currently being phased out and a brand-new engineering curriculum will be offered in the fall of 2018. The design of this new curriculum has been three years in the making. It was launched in the spring of 2015 with an extended assessment of societal needs as well as corporate-world challenges. We examined how high-level engineers with aptitudes for abstraction could fulfill these needs and how they should be trained to do so. The goal of this comprehensive assessment was to obtain:

- The new skills the students should acquire during their future studies;
- Guiding principles in the pedagogy and curriculum implementation;
- The majors (disciplinary fields or industrial sectors) as well as next-generation jobs and careers.

To achieve this, the school put together several workgroups to conduct interviews with top executives from various international corporations. In 2015, interviews were conducted with companies of different sizes and from diverse fields. The interviews focused on three main issues: (i) the environment of the company, how it will evolve, the future disruptions, the new challenges; (ii) the jobs and careers within the company and its sector; (iii) the skills required and how high-level engineers such as Centrale and Supélec students can create and enhance their value. The results of these interviews were highly comprehensive and clearly defined the skills necessary for the CentraleSupélec graduates. The main findings were summarized in an eight-page internal report (Ferreboeuf et al. 2015).

In addition, an extensive analysis of existing databases and reports was carried out (Manyika et al. 2013; Direction Générale des Entreprises 2014; Alanou et al. 2015; EY and LinkedIn 2015; France Stratégie 2015; La Stratégie Nationale de Recherche (SNR) 2015; The Boston Consulting Group 2015; Blanchet 2016). The school also worked with the Boston Consulting Group (BCG) on the mega-trends of society and business (The Boston Consulting Group 2015). This collaboration highlighted the key technologies that will play a significant role in the transformation of companies and society.

Following this analysis, 700+ alumni were polled in different industries. This quantitative study provided a ranked list of key technologies, industrial sectors, and jobs.

Although other higher education institutions cannot be considered as competitors to CentraleSupélec, the methodology used in business to carry out a competitor analysis was put into place. The goal was to find positive differentiation criteria and emphasize them in the new curriculum when possible and relevant. The main findings were summarized in a report by Da Costa et al. (2015). Committees were formed to review the strong points of the two former schools, which is summarized in a report by Aubin et al. (2015). Creativity focus groups were conducted to allow faculty, students and administrative staff to describe what their dream school would be. Finally, a study was conducted to understand the desires of prospective students, current students, and young graduates. Alanou et al. (2015) summarized these results.

3.2 Skills

The interviews, the report examination, and the alumni surveys all emphasize that engineers must work in a completely internationalized environment on complex products where digital and data volumes are essential. They all underlined the absolute necessity to master complex systems. These systems are composed of several entities whose piece-by-piece study is not sufficient to grasp the understanding of their overall behavior and to act upon it. Three "personal skills" were considered essential:

- adaptability in terms of (i) personal attitude and (ii) the mastery of agile tools;
- the ability to work collaboratively with many participants who have diverse skills;
- the ability to communicate and convince.

To fulfill these prerequisites, CentraleSupélec will train engineers who can master science and technology with extreme conceptualization and abstraction capacities as well as key skills in the field of complex systems. Graduates should be innovators and leaders who undertake initiative and action; they should be able to create economic and social value at an international level. They should be comfortable and innovative with major technological and societal changes, especially in the digital world. They should be humanists and sensitive to social issues. They should have a sense of responsibility and respect for others.

Different surveys, interviews, and analysis have been used as a support to define nine major competences for future engineers of CentraleSupélec:

- analyze, design and implement complex systems with scientific, technological, human and economic components (henceforth denoted C1);
- acquire and develop in-depth expertise in a scientific or sectorial field and/or job;
- act, initiate, and innovate in the science and technology environment;
- create value for a company and its customers;
- thrive in a multicultural and international environment;
- thrive and innovate in the digital world;
- convince others;
- be a leader: lead a team, carry out a project;
- think and act as an accountable, ethical professional.

The acquisition of these skills is tiered and fundamental for the curriculum; it is required of all students and every course will contribute to its development through specific modules. The statement of objectives of the curriculum is detailed in (Cagnol et al. 2016). All nine skills are carefully described and partitioned in learning outcomes.

The importance and the challenge to acquire skill C1 specifically made it necessary to come up with new teaching strategies that will be the aim of future publications.

3.3 Majors

The school put together a workgroup to answer the following questions:

- what are the industrial sectors where the alumni of Centrale and Supélec work today and what are the promising sectors where the CentraleSupélec graduates will be working in the next generation?
- what are the challenges and scientific/technological transformations to come in these sectors?
- what will be the favorable jobs and careers for our students?

Several official reports as well as interviews and surveys were organized in order to answer these questions (see Table 3).

We used the European national classification of activities NACE to segment the industrial sectors. For each sector, we defined two indices:

- a *presence* index based on the current existence of alumni from Centrale or Supélec in that sector using (scale: 0–6);
- a *growth* index (scale 0–8) based on

Table 3 Different sources used in the analysis in the design of the overall curriculum

	Source					
	IESF Report	France-Strategie Report	Other	Interviews	Alumni survey	Workshops with faculty
Question 1	X	X			X	
Question 2			X	X	X	X
Question 3	X		X	X	X	

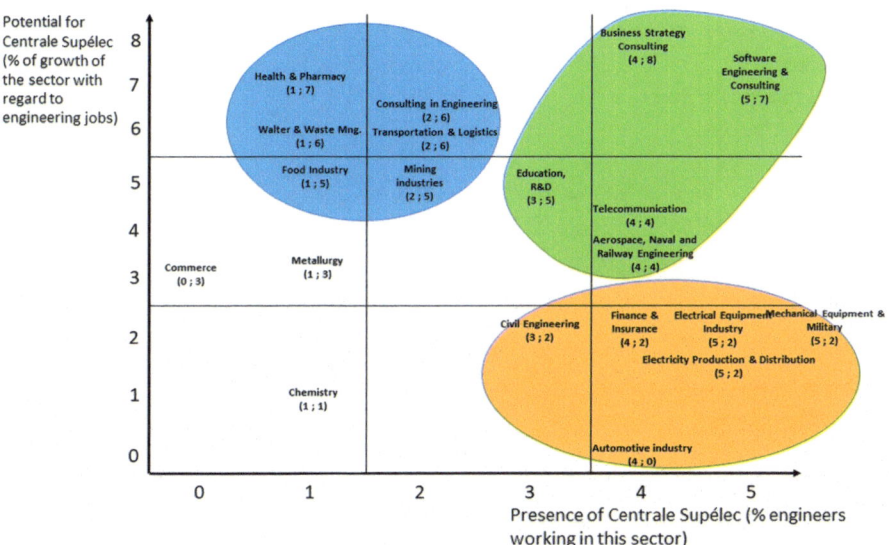

Fig. 2 Industry sector analysis

– the sectorial job growth estimations by France Stratégie (2015) which were established thanks to the data gathered by the National Institute of Statistics and Economic Studies INSEE,

– the report from the alumni on the needs for high-level engineers in the sector.

This evaluation has allowed for the following industry sector positioning and analysis (Fig. 2).

This matrix provides a fundamental vision of three categories of sectors:

• the yellow category which CentraleSupélec currently addresses but which may not be the main driver in the future,

• the green category which must absolutely be preserved,

• and the blue category which is currently not extensively addressed but is very promising in the future.

In addition, based on Manyika et al. (2013) and Direction Générale des Entreprises (2014), we found 15 key fields and key technologies to be relevant to the transformation and issues for different business sectors:

- Complex Systems
- Virtualization
- 3D-printing
- Data Sciences/big data
- Knowledge automation and Artificial Intelligence
- Digital Business
- Internet of things, Connected objects
- Cybersecurity
- Smart networks/grids
- Robotics
- Biotechnologies
- Autonomous vehicle/flux automation, planning
- Low techs
- Energy transition
- Advanced materials

Based on the Industry sector analysis (see Fig. 2), the key fields and technologies and (La Stratégie Nationale de Recherche (SNR) 2015), eight major sectors were defined:

- Energy
- Large-Scale Interacting Systems
- Computer Science and Digital Transformation
- Mathematics and Data Science
- Physics and Nanotechnology
- Biotechnology and Environmental Engineering
- Internet of Things and Communicating Systems
- Civil Engineering and Transportation

In addition, to identify the future favorable jobs for our student, we used the same type of methodology with two major steps:

1. a qualitative analysis based on (EY and LinkedIn 2015; Blanchet 2016; LinkedIn Talent Solutions 2016a, b) and interviews led us to define global trends for jobs and careers and to outline eight jobs types for our future graduates:

   - Operations Management
   - Project Management
   - Business analytics
   - Entrepreneur
   - Complex Systems
   - Innovation/Development/Service
   - Research

– Sales

2. a quantitative analysis based on an Alumni survey confirmed the relevance of these eight job types and allowed us to assess the need for each job type now and in the future.

The details of the methodology and thorough descriptions are given in the report of Dumur and Ferreboeuf (2015).

All eight majors make full use of the aforementioned skills, including the skill set regarding complex systems. Due to the necessity to integrate these competences, the decision was made that all students will be trained in system thinking and in complex system design. Moreover, all students have specific training with regard to soft skills needed for system engineers. In the first and second year, students attend specific workshops that aim at developing student soft skills such as Problem solving, Team work and Project management, Communication skills and Leadership skills. These workshops are organized and managed by both academics and professionals. In addition to these workshops, students have several mandatory projects to accomplish in order to further foster soft skill development. In the third year, soft skills are organized around a 6 week period representing Professional focus (or minor) that are based upon the list of the jobs previously mentioned. Each Professional focus targets a specific soft skill set with regard to job types that students are interested in.

The industry interviews have underlined the challenges intrinsic to complex system design such as modelling inherent couplings, incorporating uncertainty modelling, large-scale optimization, multi-disciplinary design optimization and emergent behavior (Minai et al. 2006; Bloebaum and McGowan 2010). Moreover, in order to further refine these competences, a major *Large-Scale Interacting Systems* was proposed. We will describe the positioning and the context of the curriculum of this major in more detail in Sect. 4.

4 The "Large-Scale Interacting Systems" Major

4.1 Underlying Challenges

Current technological developments, e.g. autonomous vehicles as well as the Internet of Things (IoT), are increasing the system connectivity, thus highlighting the difficulty of system mission design and growing complexity in company partnership building. This connectivity has given rise to the System of Systems (SoS). Hence, new methodologies and approaches are needed for predicting and monitoring system health, creating resilient designs, and guaranteeing cyber security. SoS design implies that future businesses are multi-actor and shared, indicating the need to develop particular value-driven approaches; those that allow for simultaneous exploration of business and technical feasibility are crucial for managing time to market and gaining flexibility in market positioning. Moreover, virtual engineering and digital revolution supports are essential for managing Time to Market, thus highlighting the need for Model-Based approaches and increase in Simulation and Visualization in complex system design.

Consequently, these changes and Industry 4.0 trends are currently broadening the role of system engineers in companies requiring both system synthesis but also, due to their complexity, the blending of different domains such as health management, operational system management, maintenance, market, and serving as a link between these different domains even on a higher level of one system. Some of the technological and economic trends considerably impacting the design of complex systems and their operations are: Internet of Things (IoT), Autonomous vehicles, Big data, Machine learning and artificial intelligence, "servitization", etc.

One of the increasingly present challenges is **Internet of Things (IoT)**. IoT aims at connecting all our devices and managing data flows from these devices. Giusto et al. (2010) define the Internet of Things as a "a new paradigm in which things or objects provided with Radio Frequency Identification (RFID), tags, sensors, actuators, etc.—through unique addressing schemes—being able to interact with each other and cooperate with their neighbors to reach common goals". In general, IoT is presented as organized around four major layers: (1) Physical objects, (2) the Network layer, (3) IoT operating services and (4) the Service layer. In the case of complex systems, this means that the mission and system perimeter are not necessarily predefined and need to be explored during the systems engineering projects. Thus, system engineering that allows for a rigorous exploration of the system definition perimeter is needed. Moreover, the connectivity of systems imposes that business is developed conjointly, implying the need for new business models and feasibility methods and analysis allowing for conjoint exploration of business and system architecture. One can see that IoT challenges are focusing also on providing IoT services on the latest level i.e. increasing "**servitization**".

International Data Corporation estimates that in 2011 the data created and copied will reach the volume of 1.8 ZB ($\sim 10^{21}$) increasing by nine times every five years (Gantz and Reinsel 2011). Collecting data stemming from numerous devices, and often having multiple devices at the same time, to propose related services, novel data analyzing techniques are needed. In the current language, we use the term "**Big data**". IDC underlines that "big data technologies describe a new generation of technologies and architectures, designed to economically extract value from very large volumes of a wide variety of data, by enabling the high-velocity capture, discovery, and/or analysis" (Gantz and Reinsel 2011).

Big data analysis or analytics is concerned not only in identifying the hidden relationships in data but also in efficiently processing large volumes of data. Moreover, **machine learning** becomes essential in extracting this hidden meaning from large data volumes. O'Leary (2013) states that the major function of the artificial intelligence in the context of big data is data categorization, i.e. structuring data into meaningful chunks. In complex systems, collecting the data in different life cycles is critical. Current methods that support and manage the system efficiency, resiliency, security, cybersecurity, etc. will need to integrate these new methods in the systems engineering process.

Another technology trend aiming at shaping a completely different future is design of **autonomous vehicles**. This trend takes its roots in previously discussed challenges and promises considerable societal changes. Large-scale uptake of a shared and self-driving fleet of vehicles is changing the face of European cities. OECD International Transport Forum (OECD 2015) has stated that with future developments the same personal mobility can be obtained with 10% of the cars. These estimations and changes are of course global. In terms of policy recommendation with regard to these changes, it is considered that new vehicle types (integrating different types of technologies) and related business models will be required (OECD 2015). As previously discussed, the major variables in these developments are transportation system configurations, new types of vehicles that will be developed, new transportation services that will be proposed in relation with new vehicle types, and new business models that could considerably change the market competition.

These systems are not only Product Service systems (PSS) (Tan 2010), due to the fact that not only coupled service and product approaches need to be considered, but also their integration in the transportation system that is yet to be defined and is dependent on future rapid development of new technologies needed for a Systems of Systems (SoS) approach (Hein et al. 2018a). A **System of Systems (SoS)** is a system constituted of independent systems. These types of complex systems have been tackled mostly by the government or military where there is, in general, one contractor/client. Examples range from military ground forces to aircraft traffic management systems. In order to identify the difference between the Systems and System of Systems (SoS), Maier (1996) proposes three main characteristics: (1) operational independence: constituting systems have independent and autonomous behavior, (2) managerial independence: constituting systems are managed by different legal entities, and (3) evolutionary independence: the life cycle of one system does not depend upon the others. Moreover, the SoSs are often geographically distributed even if it is not always the case. Other characteristics have also been discussed such as integrating network as a way to connect systems and collaboration (DeLaurentis 2005). The difficulty of future SoS is that there is not one client or contractor underlying the need to be able to create and identify meaningful business partners and business models conjointly with underlying system feasibility (Hein et al. 2018a, b).

Previously discussed rapid technological progress related to autonomous vehicles, connected systems, but also human system integration challenges pose considerable design problems. Hence, the aim of the Large-Scale Interacting Systems major is to train students in engineering complex system design while addressing current and future challenges: the integration of artificial intelligence, cyber-physical systems, product and service system of systems, operational complex systems design, the need for concurrent development of business strategy and system architectures, etc. Moreover, the current knowledge development results in a need for life-long learning where future engineers must be prepared for high-level critical thinking rather than only having access to a body of knowledge (INCOSE 2007a, b).

4.2   *Objectives and Trends*

As detailed above, regardless of the major the students choose, the objectives of
CentraleSupélec's curriculum are all oriented towards providing key expertise in the
field of complex systems. However, among the eight sectors described in Sect. 3, we
will focus on the "Large-Scale Interacting Systems" major, which emphasizes the
focus on complex system skills and educate students who will be leaders in complex
engineered system design addressing current social challenges and technological
progress. It's important to point out that the "Large-Scale Interacting Systems" major
is at the heart of the complex systems engineering science, training future engineers
in appropriate methodologies and tools, namely, to design, operate and manage the
whole life cycle of systems of high complexity and size. To acquire abilities to master
the systems complexity in a strongly constrained environment requires developing
a global and exhaustive vision of products/processes/services along the whole life
cycle. The objective of the education program is to address systems at different levels:
(1) physical systems or sub-systems such as an engine, (2) technological systems
or cyber-physical systems (such as car or autonomous car), and (3) enterprise and
organization level (such as organizational system that will support the development
of systems or organization as a system such as air-traffic management control) (see
Fig. 3).

Focusing on these levels, the objective is to design systems which can be operated,
are resilient and can adapt themselves to the evolutions occurring during their use in
the short or long term. To characterize the system, it is thus necessary to take into
account its functional and behavioral nature as well as these various levels.

The engineer trained in the major will have a global vision of the system and will
be able to analyze the impact of modifications between the levels and within the
levels, considering not only physical components but also their interactions.

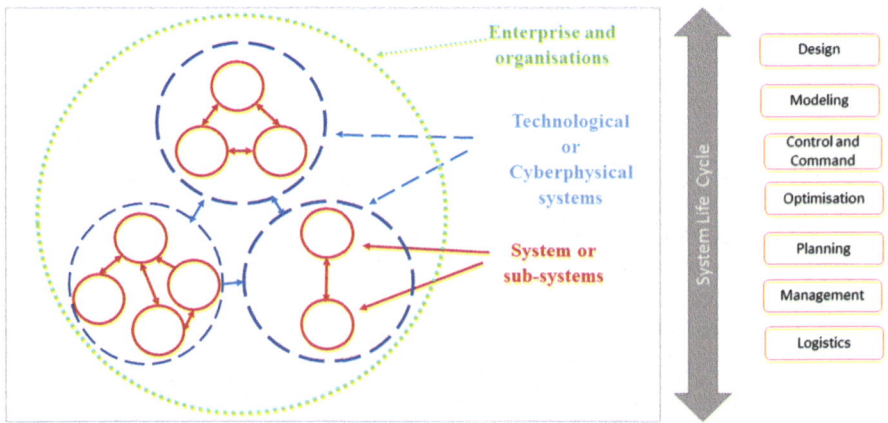

Fig. 3 Large scale interacting systems major as a three system levels perimeter

Several fundamental characteristics of complex systems must be addressed through the curriculum of this major:

- the heterogeneous nature of complex systems: they are constituted of heterogeneous elements, in terms of physical nature, temporal evolution (discrete and/or continuous) and time or dimension scale, leading to complex dynamics, and involving several disciplines such as control, mechanics, electronics, software design, industrial engineering, etc. This heterogeneous behavior can be seen at any of the three mentioned levels in Fig. 3.
- the life cycle of such systems: complex systems often have very long life cycles, e.g. some military boats designed to last 20 years are still in service far beyond the initially planned number of years. This illustrates the difficulty to predict the future and assess the implications of predictions related to the use of these systems.
- temporal evolution: long life cycles of these systems require a vision over a long period of time as well as an anticipation of all possible uses. Moreover, this variability depends on exogenous factors that need to be considered (for example the evolution of technologies, market, competitors, economy, etc.).
- interactions and interdependences of the elements within complex systems: complex systems are constituted of many interacting and interdependent components. These interactions and interdependences have an influence on the behavior of the whole system, leading to non-anticipated emergent behaviors. This complexity provokes new challenges because well-established tools used for the study and the optimization of low scale/dimension systems must be adapted from now on to phenomena/problems which appear only when complex interactions become established. In particular, the time-varying nature of networks topology, the causality of interconnections, the impact of the physical nature and of the constraints of interacting systems as well as large-scale optimization of variables of different nature are real challenges inherent to complex systems.

To address previously discussed points, several challenges have been identified as essential for this future education program:

- adequate modelling and representation of complex systems: the major limitation of system engineering is the so-called "Model Based System Engineering". The above-mentioned disciplines consider different types of modelling. Therefore, there is a need for integrating all possible systems modellings, as well as new approaches allowing to connect all the models and to organize them within coherent simulations and decision-making processes. Abstractions, models and tools which allow understanding and mastering the global system from its components and from their interactions must be defined. Techniques of model reductions and decomposition of complex dynamic systems are to be analyzed.
- prediction of the behavior of complex systems: nowadays computers capacities can address complex systems at a growing refinement level, allowing the integration of several parameters and interactions. These new technologies authorize the development of innovative approaches in terms of prediction of the complex systems future behavior, which the major will have to consider.

Complex system characteristics	Challenges			
	Complex system modelling	Behaviour prediction	Multi-level and multi-scale optimization and management	Complex system management and operation
Heterogeneous nature	++	+	+	
System life cycle		++	+	++
Temporal evolution		++		+
Interactions and interdependence	++	+	++	+

Fig. 4 Mapping between complex system characteristics and education challenges (the number of "+" shows the strength of relationship)

- multilevel and multi-scale optimization and management methods: one of the limitations is also the capacity to develop models linked to several interconnected levels. Hence, new modelling approaches allowing integration of many system parameters over several levels must be explored. These modelling techniques will further pave the way for new control strategies, taking interactions into account, such as decentralized, distributed, cooperative, and reconfigurable architectures.
- complex systems management and operation methods: complex systems require the implementation of management approaches that need to be adapted over time. Therefore, all the appropriate methods will have to be examined in this major, for example, multi-criteria optimization.

Challenges and complex system characteristics are tightly related. Figure 4 addresses these relationships. The strength of relationships is also expressed in this matrix showing that in order to apprehend and understand complex systems and their characteristics, several challenges in education have to be taught through the curriculum.

In this context, the "Large-scale interacting systems" major specifically aims at training engineers who will acquire strong skills in complex systems, in particular:

- system engineering,
- system design and control for industry and services, by means of advanced control techniques, for small and large-scale interconnected and interacting systems,
- design and management of "product and/or services" systems taking into account constraints over the whole life cycle (design, operation, use, decommissioning),
- design and management of large-scale industrial systems, systems of factories, eco-industrial estates, etc.,
- methods and tools supporting decision-making processes covering all the aspects of design, production, use and decommissioning of systems,
- systems performance optimization and management, implying the use of numerical tools for performance optimization and evaluation.

The objective is that these students start their professional life in one of the eight job types identified in Sect. 3, depending on the concentration they will choose within the perimeter of the "Large-Scale Interacting Systems" major. The previously discussed needs and trends have been a guide for developing a refinement to three potential concentrations for students within the perimeter of the "Large-Scale Interacting Systems" major: (i) Control Engineering, (ii) Design and System Sciences and (iii) Supply Chain and Operations Management. This refinement resulted from studies mentioned in Sect. 3, in particular, the industrial need to recruit engineers who should have a large body of knowledge in terms of complex systems, but who should have also developed deeper skills in one of the three domains of these concentrations. Although these concentrations may have elements in common, each one has its own specificity, being respectively control, design and management oriented.

The scopes of these concentrations are the following:

- The "Control Engineering" concentration trains engineers in having a global view of the problem inferred by complex dynamic systems with a specific focus on methodologies allowing their control at any of the three system levels mentioned in Fig. 3. It develops aspects connected to modelling, identification, estimation, optimization, design and integration of efficient and robust control laws, fulfilling technical, economic, human requirements, and adapted to the complexity and the heterogeneous nature of the systems. It also addresses aspects connected to the diagnostic, safety and reconfigurability of these systems,
- The "Design and System Sciences" concentration provides knowledge in the field of design and system sciences, in particular, approaches, methodologies and tools for the design of systems of products and/or services, starting from the definition of development strategies, system architecture design, product family and line design, up to the prototyping and/or industrialization phase, while ensuring the satisfaction of customer needs and/or potential usages,
- The "Supply Chain and Operations Management" concentration which aims to train engineers in the operations management sciences field, with appropriate approaches, methods, models and tools for planning, controlling and operating industrial systems (supply chains, production, distribution, maintenance, etc.) in order to improve and optimize their performances in an extended enterprise vision.

An overview of the "Large-scale Interacting Systems" major curriculum is given in the next section.

4.3 Curriculum Design

The CentraleSupélec curriculum is divided into academic periods of eight weeks, two academic periods per semester, with alternative periods between Academic terms and Engineering Challenge Terms. In accordance with this organization, the curriculum of the "Large-scale interacting systems" major is structured around three terms of eight weeks (one Engineering Challenge term and two Academic Terms), including two weeks in between dedicated to the training in relation to the eight job types mentioned in Sect. 3; and finally a five to six-month final internship.

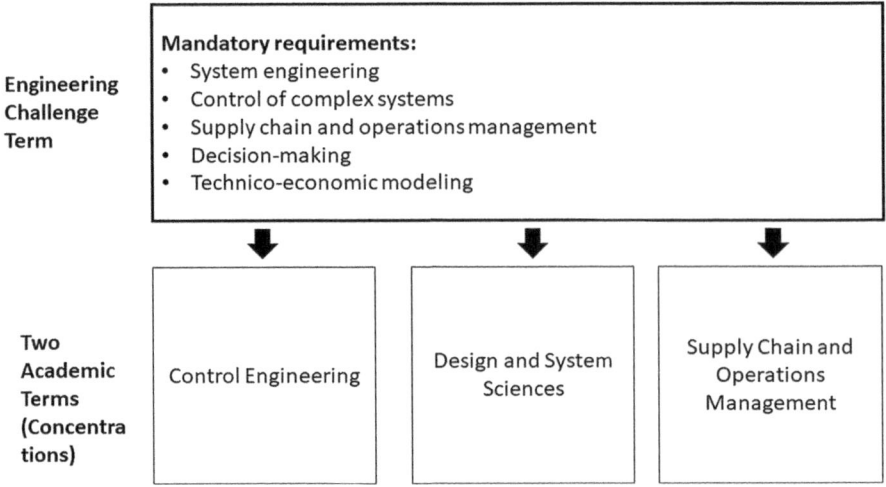

Fig. 5 Structure of the "Large-scale Interacting Systems" major

The overall structure is given in Fig. 5. The Engineering Challenge Term is focused on the core topics of the major. This term is mandatory for all students of the major and is constituted of modules which will provide the basic knowledge inside the perimeter defined above. The two following terms are focused on concentrations, even if some specific modules such as "optimization" or "machine learning" may be shared by the three concentrations.

The pedagogical activities are of different nature, which was also the case in the CentraleSupélec's former curricula. Along with scientific modules with lectures and labs foreign languages and some lab work will be required. Moreover, students will also have to complete a project (in parallel of the three terms mentioned above), in close relation with industrial partners who provide real case studies, needs analysis and precise specifications.

5 Conclusions

Complex systems curricula are more and more in demand due to the future challenges that need to be addressed in the areas of current technology and economic developments such as IoT, artificial intelligence, increase in development of product service systems and system of systems, and current Systems Engineering practices. Thus, new educational programs are needed. Over the last two years, one of the major engineering schools in France, the Grande École, CentraleSupélec, has been developing a program entirely focused on complex systems. Moreover, one of the majors, Large-Scale Interacting Systems, specifically focuses on complex systems with three concentrations: Control Engineering, Design and System Sciences and

Supply Chain and Operations Management. This state-of-the-art program is comprised of an innovative pedagogy, Academic Terms and accompanying Engineering Challenge Terms where students work closely with companies on real engineering problems in order that future graduates are prepared for the major technological and societal challenges of tomorrow.

References

Alanou V, Delle Vedove S, Font S (2015) Synthèse des enquêtes élèves et jeunes anciens. Centrale-Supélec

Aubin V, Dumur D, Gillet J-M (2015) Synthèse de l'enquête lignes de force. Centrale Supélec

Blanchet M (2016) Industrie 4.0: nouvelle donne industrielle, nouveau modèle économique. Paris

Bloebaum CL, McGowan A-MR (2010) Design of complex engineered systems. J Mech Des 132(12):120301–120301

Cagnol J, Aldebert P, Bourda Y, Aubin V, delle Vedove S, Dumur D, Herbin E, Ferreboeuf V, Fiorina J, Font S, Gillet J-M, Richecoeur F, Zeitoun A (2016) New curriculum. Statement of objectives, Centrale Supélec

Da Costa P, Fiorina J, Sciammana M (2015) Enquête concurrence. Centrale Supélec

DeLaurentis D (2005) Understanding transportation as a system-of-systems design problem. In: 43rd AIAA aerospace sciences meeting and exhibit

Direction Générale des Entreprises (2014) Etudes Technologies clés 2020

Dumur D, Ferreboeuf V (2015) Secteurs, défis technologiques et métiers porteurs. Synthèse du GT4, Centrale Supélec

EY and LinkedIn (2015) La révolution des métiers. Nouveaux métiers, nouvelles compétences: quels enjeux pour l'entreprise?

Ferreboeuf V, Morosini P, Richecoeur F (2015) Synthèse enquête employeurs. Centrale Supélec

France Stratégie D (2015) Les métiers en 2022

Gantz J, Reinsel D (2011) Extracting value from chaos. IDC iview 1142(2011):1–12

Giusto D, Iera A, Morabito G, Atzori L (2010) The internet of things: 20th Tyrrhenian workshop on digital communications. Springer Science & Business Media

Hein A, Jankovic M, Chazal Y (2018a) A methodology for architecting collaborative product service system of systems. In: IEEE 13th system of systems engineering conference, Paris, France

Hein A, Poulain B, Jankovic M, Fakhfakh S (2018b) Product service system design in a system of systems context: a literature survey. In: 15th International design conference—DESIGN 2018, Dubrovnik, Croatia

INCOSE (2007a) Systems engineering handbook: a guide for system life cycle processes and activities

INCOSE (2007b) Systems engineering vision 2020 (INCOSE-TP-2004-004-02). INCOSE

La Stratégie Nationale de Recherche (SNR) (2015) Stratégie nationale de recherche: Rapport de propositions et avis du conseil stratégique de la recherche

LinkedIn Talent Solutions (2016a) Global recruiting trends

LinkedIn Talent Solutions (2016b) Tendances du recrutement 2016 en France

Maier MW (1996) Architecting principles for systems-of-systems. In: INCOSE international symposium, vol 6, no 1, pp 565–573

Manyika J, Chui M, Bughin J, Dobbs R, Bisson P, Marrs A (2013) Disruptive technologies: advances that will transform life, business, and the global economy. McKinsey Global Institute

Minai AA, Braha D, Bar-Yam Y (2006) Complex engineered systems: a new paradigm. In: Complex engineered systems. Springer, pp 1–21

Ministère de l'Enseignement Supérieur, d. l. R. e. d. l. I. (2017) Les étudiants en classes préparatoires aux grandes écoles en 2017–2018. No 26

Ministère de l'Enseignement Supérieur, d. l. R. e. d. l. I. (2018) Les effectifs universitaires en 2017–2018. No 5

O'Leary DE (2013) Artificial intelligence and big data. IEEE Intell Syst 28(2):96–99

OECD (2015) Urban mobility system upgrade: how shared self-driving cars could change city traffic. Corporate Partnership Board Report

Tan AR (2010) Service-oriented product development strategies. DTU Management

The Boston Consulting Group (2015) Megatrends library: top fifty trends

A Strategic Design Approach for Anticipating the Future World

André Liem

Abstract In today's globalised economy, the search for competitive advantage in developing innovative products, services and systems, has been substituted by organisations' combined efforts to establish collaborative advantage. This means that the creation of 'stand-alone' design solutions would no longer meet the needs for solving complex problems within an environment in which technologies become more advanced and complex and where end-user needs and stakeholders' interests become more diverse. From an end-user perspective, effective and meaningful interactions are sought after, which are prospective and out of their frame of reference. From an organisational perspective, the growing complexity and pace of technological change are forcing firms to forge new vertical and horizontal network alliances to seek greater speed, flexibility and efficiency in responding to market changes. This book chapter aims to present a strategic design methodology to be implemented in a collaborative studio setting within the context of design education. It can be used as a blueprint for planning strategic activities, collaborative constructs and directions when strategizing for and with different types of organisations. The educational value lies in subjecting master students to situations, where they learn to balance the interests of the client organisation, end users and other stakeholders through deliberate and emergent strategic moves. By acting as consultants and interacting with a real industrial client, established design consultancies and other stakeholders, this proposed strategic design process elucidates an educational philosophy of mentorship and scholarship. Through a construct of Legitimate Peripheral Participation (LPP) students contribute to a systemic and processual view on learning, where structured planning and the deliberate use of processes and methods should complement emergent praxis. Furthermore, initiatives to formalise and communicate these processes and methods will also educate the client organisation and their respective stakeholders of "What Strategic Design is about", and how it complements an inside-out marketing and management view on strategizing, by focusing on and anticipating end-user needs.

A. Liem (✉)
Department of Design, Norwegian University of Science and Technology, Kolbjørn hejes vei 2B, 7491 Trondheim, Norway
e-mail: Andre.liem@ntnu.no

© Springer Nature Switzerland AG 2019
D. Schaefer et al. (eds.), *Design Education Today*,
https://doi.org/10.1007/978-3-030-17134-6_10

Keywords Strategy making · Forecasting and backcasting · Organisation versus end-user needs · SWOT (*Strengths, Weaknesses, Opportunities and Threats*) · Search fields

1 Introduction

In today's fast-moving world, with new technologies and social trends emerging daily, understanding organisational needs and priorities may be bounded by rationality, business-as-usual, intra-disciplinary or myopic thinking (Cairns and Wright 2018). Therefore, organisations need to innovate beyond primarily focusing on sales, profit and the development of products for ownership. This means that value creation should aim for an ongoing exchange of Product Service Systems (PSS), involving different stakeholders with multiple roles and responsibilities (Tukker 2004; Rifkin 2014).

Moreover, advocating merely a sustainability perspective in design education pertaining to the development of products and services will not suffice any longer, as it only leads to incremental improvements. A focus on radical or systemic innovation in design education is needed, supported by new methods in searching for new solutions (Ehrenfeld 2008). This also means that creative destruction should be encouraged through the discovery of new technologies, products, markets, processes and organisational forms that create clear alternatives to existing products and practices (Larson 2000).

Inherently, it can then be affirmed that by now emphasising on the company as the focus for strategic analysis and action is getting less relevant. Strategic Design is therefore characterised by active participation of a broader spectrum of social bodies and enterprises; from companies to consultancy firms; from institutions to governments, from territories to associations; who are dealing with design decisions in a turbulent and uncertain contemporary world (Landry 2012; Manzini and Meroni 2007). In other words, strategic design is primarily about conferring to social and market forces, based upon a system of rules, beliefs, values and tools. Moreover, it is a movement, which aims to influence and evolve the external environment, while maintaining and developing one's own identity (Meroni 2008).

2 Background and Scope

The changing worldwide condition is convincing associations and organisations to search for the most effective frameworks and models to expand their innovation management efforts through new paradigms and methods, which efficiently serve existing and new markets with new and/or modified products as well as services (Christiansen 2000; Ansoff 1986). Within the context of integrated product development, the level of creativity and innovation in formulating an effective product strategy and a design goal is subject to how carefully 'Product Planning and Goal

Finding' processes were carried out in the early stages of the Innovation process (Buijs and Valkenburg 1996). The term '(fuzzy) front end' depicts these underlying stages of development and signifies them as one of the challenging areas of the innovation process. However, effective management of the front end may result in a sustainable competitive advantage (Koen et al. 2001). Furthermore, the (fuzzy) front end innovation stages are consisting of unknown and uncontrollable factors. Therefore, the focus on the front end is mainly to identify opportunities and to analyse them (Belliveau et al. 2004). When organisations adopt short-term visions, external analysis focuses mainly on market, competitor's and stakeholder's analysis, which has led to incremental innovation, where new products were created for existing markets or new markets for existing products confining itself to the current product or service portfolio of the respective company (Ansoff 1968).

However, when more radical avenues for innovation are to be explored, analysing Social, Technological, Economic, Environmental and Political (STEEP) trends driven by culture, beliefs, behaviours and attitudes are important for initiating (breakthrough) idea development and goal finding. The thoroughness of exploration and innovation potential is mainly determined by the designer's aptitude and drive when executing the external analysis (von Hippel 1988). In other words, to achieve depth and diversification, synonymous to radical innovation, a broader and more diverse approach towards strategy making in design is needed. So far, not much research done on this issue (Kim and Wilemon 2002).

This book chapter presents a systematic strategic design process with avenues for emergent and creative strategic moves to develop and forecast future products and services. It aims to educate faculty and Master-level students that 'en route' to innovative solutions, prescriptive and structured modes of researching, designing and decision-making gradually give way to more intuitive and emergent modes of reasoning, driven by typical contexts. Complementary to the strategic design process, selected methods and tools will be discussed for each stage.

The first section, which will be discussed in Sect. 4, elaborates on 'How to position respective organisations by understanding their generic innovation strategies: classical, evolutionary, processual or systemic'. The second section, as elaborated in Sect. 5, proposes a methodology for managing complexity and narrowing down towards more specific design directions. This implies; (1) analysing an organisation's strengths and weaknesses, as well as extracting relevant opportunities and threats from selected societal, economic, environmental technological and political megatrends; (2) Developing relevant 'Search Fields' by juxtaposing strengths and opportunities through a Strengths-Weaknesses-Opportunities-Threats (SWOT)-matrix; (3) Eliminating selected search fields through a Bottleneck Analysis, where each of them is being subjected to earlier defined threats and weaknesses. In Sect. 3 demonstrates how selected methods are being used to forecast the "future world" referenced to final selected context-rich search fields. Moreover, this section elaborates on: (1) How back-casting strategies were applied from an ideal long-term forecasted situation. (2) How network horizons were mapped for involving relevant stakeholders complementary to typical forecasted and backcasted situations. (3) How to reposition the organisation's generic strategies based on back-casting strategies, which proposes time-specific design concepts.

3 Educational Context and Course Set Up

In this strategic design studio, first-year M.Sc. Industrial Design students at the Norwegian University of Science and Technology were trained to understand and balance end user as well as organisational motivations and interests in developing strategic design concepts. An emphasis was placed on how these students experimented with and applied certain methods and tools to analyse existing and anticipate future trends, which are relevant for their client organisation. Industry collaboration served as a vehicle to systematically guide the student groups through the different stages of the Strategic Design process.

From a collaborative entrepreneurial perspective, 11 small and medium size Norwegian companies played the role as a 'real' client, discussion partner and to a certain extent educator. Students acted as design consultants and were required to formulate a design strategy as well as develop the strategy into a product and/or service concept.

Figure 1 shows how students are initially subjected to a prescribed and strict product planning process and gradually exposed to a more emergent and practice-oriented way of designing and decision-making. Reference to generic strategies for innovation (Whittington 2003), internal capabilities of the project stakeholders as well as external market trends were analysed at first. Hereby, strengths were matched with opportunities, and short-, mid- or long-term design strategies formulated. Subsequently, search fields were generated, and a design goal determined for further development.

In the successive stages of the strategic design process, selected design directions derived from the search fields were strategically conceptualised using an iterative goal finding process. During the conceptualisation process, search fields were gradually opened up for emergent moves in strategy making. These moves became more prominent as the strategic focus and concept is getting more concrete. The level of concreteness is determined by; (1) The forecasted and backcasted strategic positioning, which is time dependent; (2) The organisation's, end users' and other stakeholders aims, whether to focus on development or "radical" strategy making; (3) How context-specific the selected search field is for strategy making.

As a final step in this strategic design course, the strategic design concepts, its scenarios and user journeys, were juxtaposed against the forecasted and backcasted strategic ambitions of the organisation. Complementary the organisation's strategic positioning was evaluated in conjunction with the new design strategies, as well as a detailed marketing plan presented to support the strategic design concepts.

Last but not least, students were asked to reflect over the (strategic) design processes, methods and tools, which they used in their project.

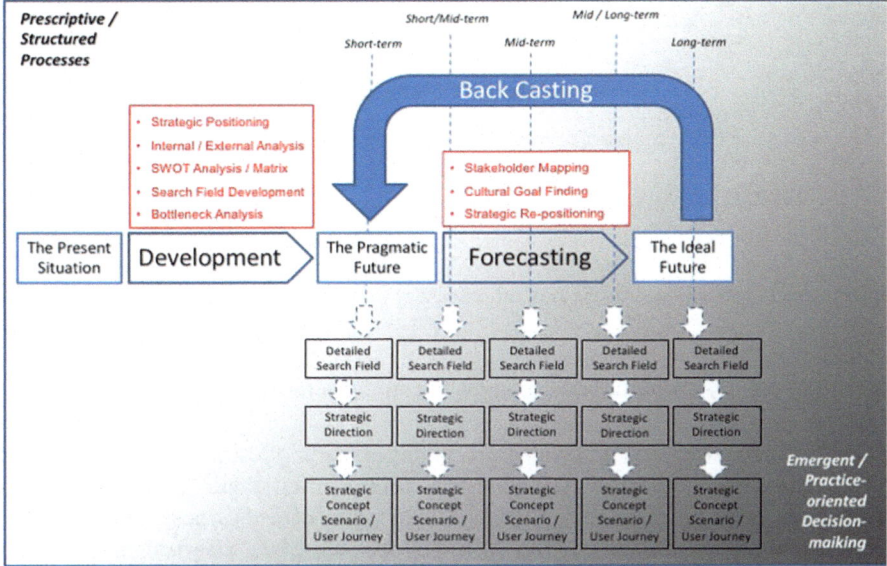

Fig. 1 The strategic design process, comprising of prescriptive stages and emergent moves in strategy making. As the design direction becomes more concrete, the more emergent and reflective design decisions will become towards conceptualising the strategic design solution

4 Reflecting "Mission" and "Vision" on Generic Innovation Strategies

According to Utterback and Abernathy (1975), organisations can become introspective in their innovation selection criteria (manufacturing and cost focus), rejecting possibilities for radical product change and failing to respond to significant market shifts. Moreover, from a product and social sustainability perspective, incremental improvements will not suffice any longer. Radical or systemic innovation is needed, whereby a change in the approach in searching for new solutions is essential (Ehrenfeld 2008). This is complemented by Zahn's claim that strategising, as a core process of strategic management, is more than strategic planning and needs strategic thinking (Zahn 1999), which happens foremost through systems and systemic thinking.

To facilitate a more structured way of strategic thinking as well as to better position an organisation's vision and mission, it would be advantageous to be able to reflect them to generic innovation strategies. Such reflection would also give an indication of the organisation's future intentions for achieving competitive and collaborative advantage with respect to its current situation. As shown in Fig. 2, Whittington's "Generic Strategy Perspectives" (Whittington 2003) can be used as a reference to reflect an organisation's vision and mission where they will be analysed according to their innovation attitudes and behaviours (*Process*), and to their innovation objectives (*Outcome*). The approaches pertaining each of the four quadrants have been deduced

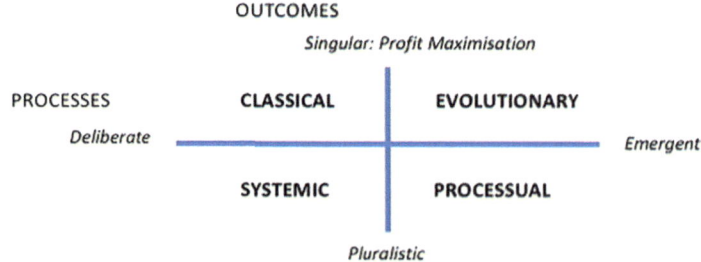

Fig. 2 Generic innovation strategies (Whittington 2003)

from Mintzberg's 10 schools of thought for strategy formulation (Ahlstrand et al. 2001)

In the classical approach, profit-maximising is the highest goal of business and rational planning for classicists. This theory claims that if Returns-On-Investments (ROI) are not satisfactory in the long run, the deficiency of the business venture should be corrected, or abandoned (Sloan 1963). Key features of the classical approach are the attachment to rational analysis, the separation between planning and execution and the commitment to profit maximisation (Ansoff 1986; Sloan 1963).

Evolutionary approaches do not rely on top-management skills to plan and act rationally. Competition is not to be addressed by detached planning and calculation as in the classical approach, but by being engaged in an ongoing struggle for survival. In the search of profit maximisation, natural selection will determine who are the best performers and the ones that survive (Einhorn and Hogarth 1981).

Advocators of processual approaches believe that the existence of a 'rational economic man' is not possible, because due to the imperfections of human nature, it is impossible to oversee all factors at the same time (Cyert and March 1963). In contrast to classical and evolutionary approaches, and abandoning profit-maximising ambitions, processual methods do not strive for the ideal but aim to work with what reality offers. Practically, this means that organisations are not always united towards a single goal such as profit making. Instead, individuals with different interests, acting in an environment of confusion and mess, determine the course of action. Through a process of internal bargaining within the organisation, members set goals among themselves which are acceptable to all.

The systemic approach is based upon the notion that densely interwoven social systems influence the means and ends of a systemic approach and define what a suitable behaviour is for their members in terms of economic activity (Whittington 2003). Hereby, the organisation is not simply made up of individuals acting purely in economic transactions, but of individuals embedded in a network of social relations that may involve their family, state, professional and educational backgrounds, even their culture, religion and ethnicity (Whittington 2003). To establish a reference for strategy making, students were required to conduct extensive research to determine and later on redefine the generic strategic positioning of the organisation.

5 Structuring Complexity and Developing Design Directions

In this section, the reader will be guided on how to analyse and structure a broad diversity of information about the external world and the organisation in situ. It explains how SWOT in its various forms, can be more proactively used as a decision-making tool to develop design directions.

5.1 From Megatrends to Strength, Weaknesses, Opportunities and Threats (SWOT)

Megatrends are irreversible, long-term processes that define real areas of choice for in policymaking (Bukowski and Śniegocki 2017). It can be perceived as a group of trends, phenomena or processes of global nature that have occurred in the process of civilisation and human development (Pęciak 2016). Having diversified determinants, it permeates all areas and spheres of human life including economic, social, political and cultural relationships as well as human awareness (Muszyński 2001). According to Prandecki (2014), the analysis of megatrends allows for recognition and assessment of changes in the environment, but also for identification of opportunities and threats in the future. However, due to unpredictability of the future and lack of deeper contextualization with respect to the interests of actors in question (e.g. client organisation, suppliers, collaborators, distributors, end users, etc.), megatrends need to be analysed and contextualised to align with specific domains before they can be formulated as 'Opportunities' as part of a SWOT (*Strengths-Weaknesses-Opportunities-Threats*) analysis.

A SWOT analysis is a framework used to assess a company's competitiveness by identifying its strengths, weaknesses, opportunities and threats. Based upon the general principles of strategic planning the SWOT analysis has derived its origins from the work of business policy academics at Harvard Business School and other American business schools from the 1960s onwards.

Having enjoyed different levels of support and popularity over time, it specifically measures what an organisation can and cannot do, and its potential opportunities and threats. Based upon environmental data to evaluate the position of a company, a SWOT analysis determines what assists the firm in accomplishing its objectives, and what obstacles need to be overcome or minimised to achieve desired results: where the organisation is today, and where it may go. In other words, the main concept of based upon the idea that good strategy means ensuring a fit between the external situation a firm faces (threats and opportunities) and its own internal qualities or characteristics (strengths and weaknesses) (Andrews 1980; Porter 1985).

However, modern textbooks, such as "Vision in Product Design" (Hekkert et al. 2011) are critical about its application. Moreover, Mintzberg (1994) claimed that SWOT is able to survive so eloquently, only because it clearly formalises the strat-

Opportunities

Strengths	1. Less material goods, more experiences	2. Centralization	3. Social neighbourhoods	4. Aging - elderly as customer group	5. Many elderly are lonely, healthy and rich	6. More social and local networks
1. Network and networking		Use SW to fascilitate for use of area	New ways of housing, Albas network can contribute to new solutions			
2. Own kindergarten buildings	Use kindergarten area to offer experiences and/or services	Future kindergartens will be bigger			Use kindergarten are for services for elderly	
3. Experience in kindergarten services	Use SW to fascilitate for use of area					Use Kinso as basis for develop new solutions for social networks
4. Great potential for developing good software		Large market for home services in smaller areas	CTC-services based on trust, when people know each other	Offer services to elderly		
5. Entrepreneurship oriented and willing to innovate		New ways to run kindergartens in the future			Connect elderly to operation of kindergartens	

Fig. 3 SWOT matrix, juxtaposing strength with opportunities to determine relevant search fields for a company, who is specialised in providing a total concept of services for establishing and running childcare centres

egy making process. As such, he calls for an approach to better understand how humans think and create, what managers really do in practice, and how organisations learn. According to Hill and Westbrook (1997), SWOT applications showed similar characteristics: (1) a long list of general, often meaningless descriptions, (2) a failure to prioritise, (3) no attempt to verify any points and (4) outputs were not used within the later stages of the strategy process.

On the contrary, proponents of SWOT do not see it as mere list-making, but prescribe SWOT as a rigorous analytical tool (Weihrich 1982). However, these advocates, perceive strategy making as a process of conception. They maintain a clear distinction between external factors (threats and opportunities) and internal viewpoints (strengths and weaknesses), and on the need for testing assumptions.

In this book chapter, a systemic approach has been adopted when promoting the use of SWOT in an educational context. This means that SWOT has been prescriptively and structurally applied in determining and selecting "Search Fields", whereas the content for each selected search field has been developed through reflective conversations with its context (Schön 2017). As shown in the following case example (Fig. 3), juxtaposing 'Strength' and 'Opportunities' may lead to a range of interesting, but selected search fields or groupings of search fields. To start with, 'Opportunities' are deduced from the external analysis, whereas 'Strengths' are extracted from the internal analysis. These 'Strengths' and 'Opportunities' are then matched against one another. Certain force-fitted juxtapositions may not lead to anything, whereas some may open up innovative search fields to be further explored. Those highlighted in Fig. 3 are most promising and will be evaluated against possible bottleneck (Fig. 4).

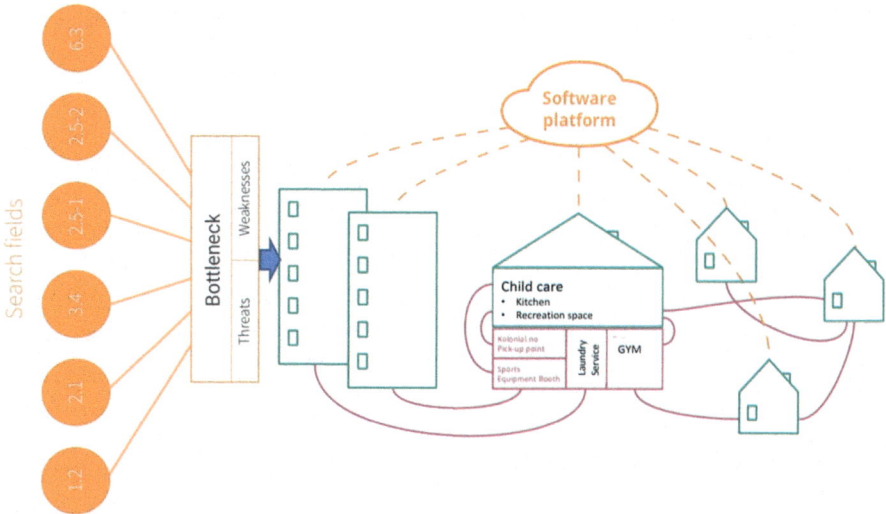

Fig. 4 An example of how selected "Search Fields" from the SWOT-matrix in Fig. 3, were subjected to a bottleneck analysis, and transformed into an initial strategic concept

5.2 Search Field Selection and Further Development

This section discusses how preselected search fields will be scrutinised using an 'Internal Bottleneck' and 'External Need' analysis (Buijs 2003). The 'Search Fields' who are to be subjected by the most or most prominent 'Threats' and inherently possess the most or most serious weaknesses, will be eliminated. The surviving search fields are then subjected to a deeper analysis, substantiated by creative product, service and/or systems ideas.

Once a narrow, but detailed selection of search fields, product, service and system ideas has been framed, a process of short-, mid- and long-term strategizing can begin. Concerning the 'Child-care Services Company', a combination of search fields has led to the '*After 16.00 B-Hub Concept*'. Since the company is currently managing several childcare properties that are empty after 16:00 on weekdays, and all weekend, alternative activities and services outside of the childcare hours has been proposed. These services are for instance: a meeting place for social events, neighbourhood health care services, a place for interaction across generations between children, parents and elderly, who are living around the childcare centre. However, in the development of this strategic concept, the following challenges need to be addressed: (1) Childcare centres are becoming more regulated and standardised, (2) Lack of experience in delivering services, (3) Lack of experience in alternative uses of buildings, etc.

6 Towards Strategy Making: Developing the Foundations for Forecasting and BackCasting

As content and context are becoming more important in determining the quality of the product, service and/or system concept, prescribed processes and methods give way to a more "Practice" perspective towards strategy making. This perspective, which is built upon "Practice Theory", has been promoted by Mintzberg (1994) and considers the use of methods and tools, such as SWOT, complementary to practice and subject to context. In this section, 'Practice Theory' will be discussed as a fundamental theory for conjecturing, forecasting and backcasting.

Driven by "Practice Theory", the following complementary methods for forecasting and backcasting will be discussed. These methods are:

- Cultural Goal Finding.
- Network Building and Stakeholder Mapping.

These approaches are context-driven and build upon deliberate and emergent moves in strategy making.

6.1 Practice Theory

Practice theory is relevant in guiding designers to be more conscious 'anthropological' thinkers by bridging the gap between thinking, practice and research in design. In other words, thinking and research goals intertwine with the goals of real-world practice. Inevitably to understand the deeper connections between research, thinking and doing, as well as to argue for their increasing similarity in terms of knowledge creation and innovation, the conceptual relationship between practice, praxis and practitioners will be discussed below. Understanding the dynamics between practice, praxis and research is important for establishing a certain mode of.

First to make the distinction between practice and praxis, practice guides activity, while praxis is the activity itself. Hereby, 'practices' refer to shared routines of behaviour in the broadest sense, including traditions, norms and procedures for thinking, acting and using 'things' (Whittington 2006, p. 619). Praxis refers to what people actually do. Practitioners are the actors, who make, shape, investigate and execute. They include not only senior (strategic) design executives, but also designers, researchers, makers, prototypers, etc. Based on the concept of 'Praxis', which encompasses 'Practice', their work is complex and diffuse. It embraces the routine and the non- routine, the formal and the informal, activities at the corporate centre and activities at the organisational periphery (Regnér 2003). These activities include meetings, conducting interviews, presenting concepts, entertaining potential and existing customers, talking with suppliers and distributors on the phone, organising and conducting usability studies, ad hoc 'firefighting', and many more.

The extent to which practitioners engage in praxis at the ground level, which is typical for Industrial Design, binds design research and practice. This engagement is dependent on the speed and openness of internal existing design and research procedures and capabilities to embrace externally emergent influences. These influences are described by Mintzberg and Waters (1985), as generically processual and subject to learning experiences. Moreover, complemented with Whittington's systemic views of bounded rationality, planning can only be effective when pursued in context. The constraint arises because human behaviour is emergent and embedded in a network of social relations, involving their families, state, their educational and professional backgrounds, religion and ethnicity (Swedberg et al. 1987).

To summarise this subsection within the context of design education, students need to distantiate themselves to a certain extent from using methods and tools to forecast and backcast innovative design solutions. The connection between design research, thinking and practice should be emphasised in strategic design project work, to encourage these students to develop contextual insights, and from there speculate innovative design solutions. This view is supported by Dorst (2008) as he argues that the future of the design discipline builds upon 'Practice Theory' by advocating knowledge creation and innovation through mentorship and scholarship, and as a result, argues for equal focus on research and practice, especially in the later stages of the strategic design process. Moreover, the view that knowledge and concept are intertwined is typically justified by 'practice Theory'. This means that solving the 'future world' with innovative solutions is challenging, which is to be characterised by high complexity, deliberate and emergent approaches, as well as being embedded in a network of social relations.

6.2 *Cultural Goal Finding*

Culture has been defined in numerous ways because of its multi-dimensional characteristics. It manifests itself in a particular group or organisation according to three fundamental levels: (a) observable artefacts, (b) values and (c) basic underlying assumptions. For instance, Kroeber and Parsons (1958) arrived at a cross-disciplinary definition of culture as 'transmitted and created content and patterns of value, ideas and other symbolic meaningful systems as factors in the shaping of human behaviour and the artefacts produced through behaviour'. However, difficulties in understanding the concept of culture stems from the different usages of the term. In Matthew Arnolds' Culture and Anarchy (1867), culture is being described as a tool to classify certain groups in society, by «those who have» and 'those who do not have' culture. This has been done by referring to special intellectual or artistic endeavours or products, as 'high culture', opposed to 'popular culture' (or 'folkways' in an earlier usage). In contrast to Arnold's view, Tylor claimed that all folks 'have' culture, which they acquire by virtue of membership in some social group —society.

They include knowledge, beliefs, art, morals, law, custom and any other capabilities and habits, and could be positioned on an evolutionary continuum from 'savagery' through 'barbarism' to 'civilization'. In his ten-volume Encyclopedia of Language and Linguistics, Apte (1994: 2001), summarised the problem as follows: 'Despite a century of efforts to define culture adequately, there was in the early 1990s no unanimous agreement among anthropologists regarding its nature'.

The purpose of this section is not to find a unanimous definition of culture. Instead it aims to use cultural dimensions and concepts of acculturation (Berry 2003) as a reference for strategic goal finding and conceptualization. Hereby, the author argues that innovative strategic solutions can be developed from cultural insights, which are embedded in social, political and economic context of societies and regions. In this case, a more comprehensive search field analysis can be undertaken by reflecting more conscientiously on extreme cultural values as well as developments of different nations' political, economic and social situations. Sections 6.2.1 and 6.2.2 explain how a cultural goal finding process takes place in the case of 'Waste Collection', leading to an intervention with strategic design solutions.

6.2.1 Culture and Its Cultural Dimensions

When searching for methods and tools within the concept of 'Culture', the use of cultural dimensions within a framework of acculturation (Berry 2003), is most applicable for deepening and strategizing on selected search field. Various academics in the social sciences have developed different sets of cultural dimensions, which provide an operational perspective on how culture can be used as a tool for analysing, conceptualising and decision-making within the context of strategising (Trompenaars and Hampden-Turner 2011; Hall 1963; House et al. 2004; Hofstede and Hofstede 2005; Parsons and Shils 1951; Kluckhohn and Strodtbeck 1961). However, identifying cultural characteristics is difficult because it lacks a robust measure that can identify the implicit levels of culture (Straub et al. 2002). In an effort to address this issue, researchers have dissected culture as a set of 'dimensions' that provide a framework for cross-cultural comparisons of user behaviour (Parsons and Shils 1951).

For Hofstede and Hofstede (2005), culture is 'the collective programming of the mind that distinguishes the members of one group or category of people from another.' The cultural characteristics thus comprise a constellation of psychological traits, attributes and characteristics. As a life-long project, they conducted a survey of IBM employees in 40 different countries and proposed a model, describing national cultures that entailed four dimensions: *uncertainty avoidance*, *individualism* versus *collectivism*, *masculinity* versus *femininity* and *power distance*. Hofstede and Bond (1988) subsequently added the fifth dimension to their model, *long-term* versus *short-term orientation*.

6.2.2 A Cultural Perspective on Innovation

Historically, the 'technology gap' theory first stipulated that the potential for knowledge imitation was positively linked to the development gap (often measured in terms of GDP per capita) between countries (Gershenkron 1962; Fagerberg 1994). According to this theory, the lesser developed a firm or a country, the smaller is its knowledge stock and therefore the bigger its potential to increase through learning from the leading countries. However, when considering cultural driven innovation, strategists and designers should acknowledge that numerous societies believed that their habits, ideas and customs were what determined the shape of their political and economic arrangements, and were the source of their uniqueness.

The above paragraph makes us revisit the concept of 'acculturation' (Berry 2003) in conjunction with Hofstede's views on culture (Hofstede and Hofstede 2005). The challenge here is how to develop and implement a methodology for strategic goal finding based on social, cultural and political differences.

Capitalising on Hofstede's five dimensions, which are associated with manifestations of cultural difference, cultural parameters can be linked to cultural behaviour. However, Hofstede's observations contested those made by Gershenkron (1962), Fagerberg (1994). It showed that the potential for radical innovation is neither biggest in leading nor developing countries. The feeding ground for radical innovation is the understanding of the status quo of a nation's cultural, political, economic and social atmosphere, followed by the acknowledgement that drastic improvements in quality of life, service quality or minimising inequalities within societies, are almost impossible to be achieved through political governance.

This calls for a strategic design approach, whereby product, service and/or system concepts are proposed to solve cultural extremes at the bipolar scale of any chosen cultural set of dimensions. Proper concept development within specific cultural contexts can positively influence creative thinking within typical search fields and impact innovation locally and globally with respect to improving life and service quality, and as well as solving inequalities in both upcoming and advanced economies. The following example on 'Waste Collection' illustrates the above. In new and emerging economies, waste collectors work all-day round for less than 6000 Euro per year, live in substandard conditions and are under privileged in the society they work in. In developed, Nordic-European countries, earn a salary similar to a newly University graduate, but waste is not collected regularly, and individual households need to assist in waste management. Seen from local perspectives both scenarios are acceptable, but when subjected to global cultural scale both scenarios are unfavourable, because they are on the extremities of the scale. However, being aware of such instances triggered in this case an innovative design solution in the form of a "Mobile Garbage Suction System" (see Fig. 5).

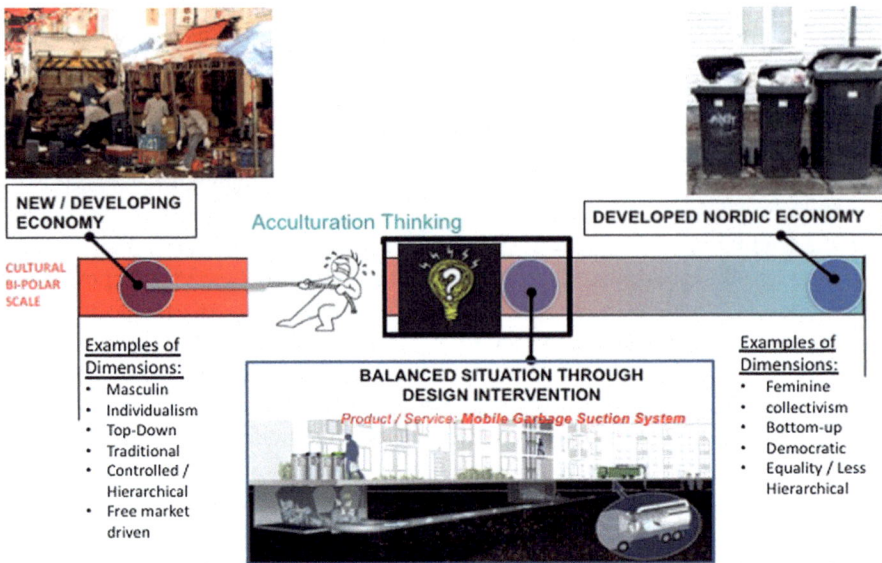

Fig. 5 An example of cultural goal finding: The case of waste collection

6.3 Network Building and Stakeholder Mapping

According to Newcombe (2003), it has been widely acknowledged and documented that stakeholders are playing a major role in contemporary design and development practice. A pluralistic view on those who have a stake and interest in an organisation, extends well beyond the traditional concept of shareholders to include employees, suppliers, customers, trade unions, communities, etc.

Systematic stakeholder mapping has become increasingly popular in various fields and academic disciplines. Recently, it has evolved into a systematic tool with clearly defined applications and methods (Brugha and Varvasovszky 2000), and would be particularly useful to assess the stakes more in greater detail of various interested parties in a system (Grimble et al. 1994).

Rooted in management theory and in political science, stakeholder mapping distinguished stakeholders in a business context as 'any group or individual who can affect or is affected by the achievement of an organisation's objectives' (Freeman 2010, p. 46).

Within design education, stakeholder mapping can be seen as a holistic approach or procedure to gain strategic understanding of how to develop design collaboration by means of identifying the key actors or stakeholders and assessing their respective interests in the '*designed*' system and/or collaborative constellation.

In the case example of a Norwegian Energy provider (see Fig. 6a, b), students were tasked to develop strategic concepts for value creation through environmental friendly production and distribution of power to the better for the region. Based on an

(a)   **(b)**

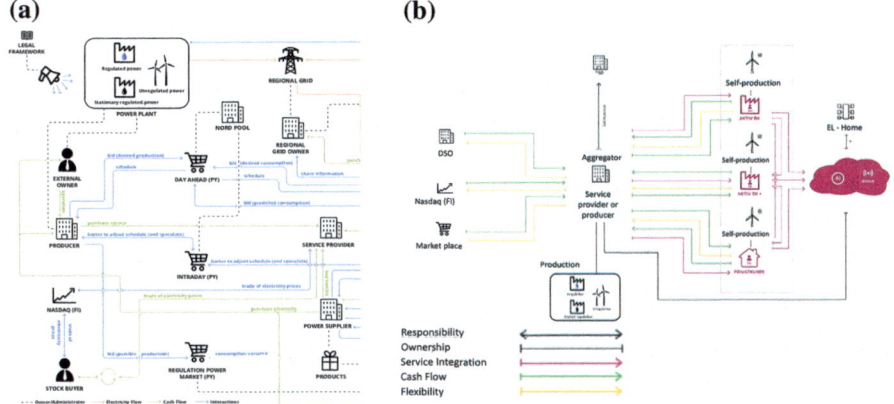

Fig. 6 **a** An excerpt example of a detailed stakeholder mapping exercise. **b** An example of how a Norwegian energy provider adopts the role of aggregator as an innovative move based upon a careful analysis of the earlier stakeholder map (Fig. 6a)

internal and external analysis of the product planning process, students developed and visualised stakeholder interactions of the 'current situation'. Figure 6a shows how a Norwegian energy provider relates to their stakeholders. Due to the many existing relationships and constructs in the "World of Energy Consumption and Supply", stakeholder maps become rather complex.

In the exploration of future strategic directions for innovation, an organisation needs to take stock of the evolving environment, seek to identify opportunities and threats in their endeavours and plan for the long term. Hereby, it is important to understand the cognitive aspects of interaction, concentrating specifically on the shared formations of expectations about the future among business network actors (Andersen et al. 2018). Given this situation, adequate foresight becomes an essential aspect for developing the right sensitivity to detect weak signals that are forming future trends (Rohrbeck and Gemünden 2011). Such foresight can only be developed and honed through practice, to be implemented among actors in business networks, not only as an analytical but also as a co-creative act.

This activity of fore sighting has introduced the concept of 'Network Horizons'. Network horizons are described or visualised future stakeholder interactions, which are most likely to changes over time. These changes could be important to relate to when developing innovation strategies in an organisation (Holmen and Pedersen 2003). Questions and challenges, which have to be addressed, when applying the "Network Horizon" are:

- How do actors gain insight into the possible futures of a network and the main factors at play in the network?
- What is the structure of the new network?

- What are the roles and interest of the actors in the new network, and how do these contribute to the further development of the search fields?
- How adaptable is the network with respect to changes over time?

This would involve insight into factors shaping the other actors, the relationships in which they are involved and the interactions taking place within them.

Moreover, it requires actors to detect emerging and evolving opportunities, and the directions in which their relationships are developing, changing and new relationships are being initiated. According to Medlin and Törnroos (2014), more attention should be paid to dynamic network emergence and development, which includes understanding connections across relationships, as well as activities, processes and adaptations to establish constructs relative to different forms of relationships, which are time-dependent. In other words, the time perspective is essential for conducting network foresight practices.

Figure 6b elaborates on the new role of the earlier mentioned Norwegian energy provider. Reference their presented stakeholder map (Fig. 6a), the organisation wishes to explore the opportunity to take on aggregator roles by being one of the first power suppliers in Norway to offer its customers a package solution with power management systems and flexible business solutions. Subsequently, it aims to become an aggregator for the overall network by selling flexibility services—either directly to online companies or on marketplaces. As such, customers can benefit from collaborating through larger energy concepts where the needs of each customer are managed by the aggregator. This means that customers have the opportunity to offer and purchase resources from one another, while addressing the need to create comprehensive optimised solutions and streamline resource utilisation in society. To design energy concepts for typical areas, it may be useful to have knowledge of how to optimise buildings, as well as to engage in good relationships with partners that complement the field of competence.

In summary, exposing students to stakeholder mapping and network building is essential to make them understand that strategic value is created by finding synergies and forms of collaboration among different actors. These collaborative moves will lead to collaborative advantage and the formation of business ecosystems, which is essential for survival in today's business environment.

7   Detailing Search Fields Through 'Forecasting', 'Back Casting', 'Stakeholder Involvement' and 'Cultural Goal Finding'

Understanding stakeholder involvement and co-operation is essential in facilitating system innovation within the context of forecasting and backcasting. According to Abrahamsen et al. (2016), there is a need to focus on the interplay between cognition and action specifically relating to what managers perceive their network to be and what they need do to establish it and to support their strategic design concept. It requires integrated approaches (Quist et al. 2002, p. 1), where:

- A diverse range of stakeholders and actors from different societal groups including government, companies, public interest groups and knowledge bodies should be involved, when defining the problem, searching for solutions and conditions, as well as when developing shared visions.
- The combined consequences of sustainability, economic and social factors should be considered in strategizing within their network.
- The consideration of related production and consumption systems through the close connectivity between the demand and supply side.

The backcasting approach emerged in the 1970s and was applied in many different fields (Vergragt and Quist 2011), e.g. in agriculture, heating, transports and water, to address, complex systems, and to keep track of the most essential elements of an organisation's vision and mission.

It can be described as forecasting a desirable future vision or normative scenario, followed by looking back at how this desirable future could be achieved, before defining and planning follow-up activities and developing strategies leading towards that desirable future (Quist and Vergragt 2006; Dreborg 1996; Svenfelt 2010). However, Back casting itself, is an evolving methodology that has been adopted and adapted in diverse settings, complemented by variations in typology of their formation and application (Vergragt and Quist 2011). Despite such diversity, all approaches adopt a future point of reference for visioning in order to free participants from the constraints of contemporary trends and to encourage longer term planning amongst key decision-makers (Davies 2014, p. 184).

A case example of a leading manufacturer of 'Smart Work Equipment' demonstrated how backcasting has contributed to the development future scenarios with respect to creating a safer and more efficient workplace in the heavy-industry sector. The backcasting exercise has led to the idea of developing a versatile and multipurpose 'Smart Helmet' with IoT (*Internet of Things*) capabilities for the near future, 2022 (see Fig. 7).

7.1 Stakeholder Mapping and Participation to Facilitate Strategic Concept Development

In the detailing of search fields leading to strategic design concepts over time, concept-specific stakeholder maps were envisioned. These maps anticipated new stakeholder interactions at predetermined backcasted times. They show upcoming roles and responsibilities as well as interests in the formation of new business ecosystems, aimed at mutual value creation.

However, the studio projects do not only present a 'passive stakeholder' view. Through a process of participatory backcasting, which places greater emphasis on involving a broad range of diverse stakeholders, strategic directions and detailed

Fig. 7 The concept of forecasting and backcasting: The case of a "Smart Helmet"

search fields were further developed and modified into future scenarios and concepts. The advantages of participatory backcasting include extending the scope of expertise in forming scenarios and increasing the level of interaction and learning among participants (Davies et al. 2012). Moreover, it connects with what has been advocated in "Practice Theory" by paying more attention to holistic practices, and by engaging with people beyond those directly involved in the earlier phases of the design process, emphasising on 'What actually happens in design practice'.

7.2   Cultural Goal Finding as a Tool for Strategic Design Conceptualisation

In this section, 'Cultural Goal Finding' will be discussed as an ancillary tool for backcasting. Based upon the notion that trends and developments positioned on the extremities of a cultural bipolar scale are not ideal, strategic design directions and concepts for envisioning the future can be developed by adopting an opposing critical view in describing an ideal situation, positioned in the middle of the bi-polar scale. Such view implies that the designer should ask him or herself the question; '*What if*', when the client organisation requires short-term developmental solutions; and "*What if in X year from now*", when the client organisation, its stakeholders and end users require a projection of an ideal future scenario. Figure 8a, b show that the 'future' is relative to time and location, as well as that differences in location implies cultural and behavioural differences. For example, in terms of forecasting and backcasting, policymakers and designers situated in location 'A' can learn from favourable as well as unfavourable developments as depicted situation 'B'. In Fig. 8b, earlier selected search fields are subjected to situations, which were marked as cultural, economic, social and political extreme. By doing so, chances will be increased for designers to come up with innovative and strategic design solutions for specific contexts and problems. Moreover, if successful, it can be perceived that design interventions are more effective than measures initiated by policymakers (See example of 'Waste Collection', as shown in Fig. 5).

8   Connecting Strategic Design Concepts with Back Casted Strategies

Scenarios and user journeys are a way of transforming visions into a plausible hypothesis, when making the connection between 'Strategic Design Concepts' and 'Back Casted Strategies' (Meroni 2008). According to Ogilvy (2002), scenarios are stories about what ought to happen. This is complemented by Van der Heijden's view, stating that scenarios are the best available language for the strategic conversation, allowing differentiation in views, but also bringing people together toward a shared understanding of the situation in making decisions (Van der Heijden 2011). Moreover, scenario building is an inductive way of tackling problem-solving in strategic design. These problems are transformed into sharable and debatable visions through a structure and the use of design tools.

In Fig. 9a, b, a conceptual study of a 'Versatile Mailbox' showed how different home delivery solutions have been projected over a certain period. The project was initiated in January 2016, when a Norwegian start-up: Postkassen AS, collaborated with NTNU Department of Design (NTNU-ID) on designing a flexible and multi-purpose mailbox. Initial research indicated that this mailbox should have sufficient security measures to receive and store critical and personal items, such as medications, as well as provide selective access to rightful recipients. Furthermore, the mailbox should also be able to receive and store large items, inclusive of perishable goods. This last requirement demands the involvement of temperature regulating elements, e.g. cooling units. Ideas and concepts, following the initial research stage, revealed the need for more in-depth research in Product Service Systems (PSS) as well as social innovation. For instance, a more detailed differentiation of stakeholders as well as how such a mailbox will meet the explicit and implicit needs of these

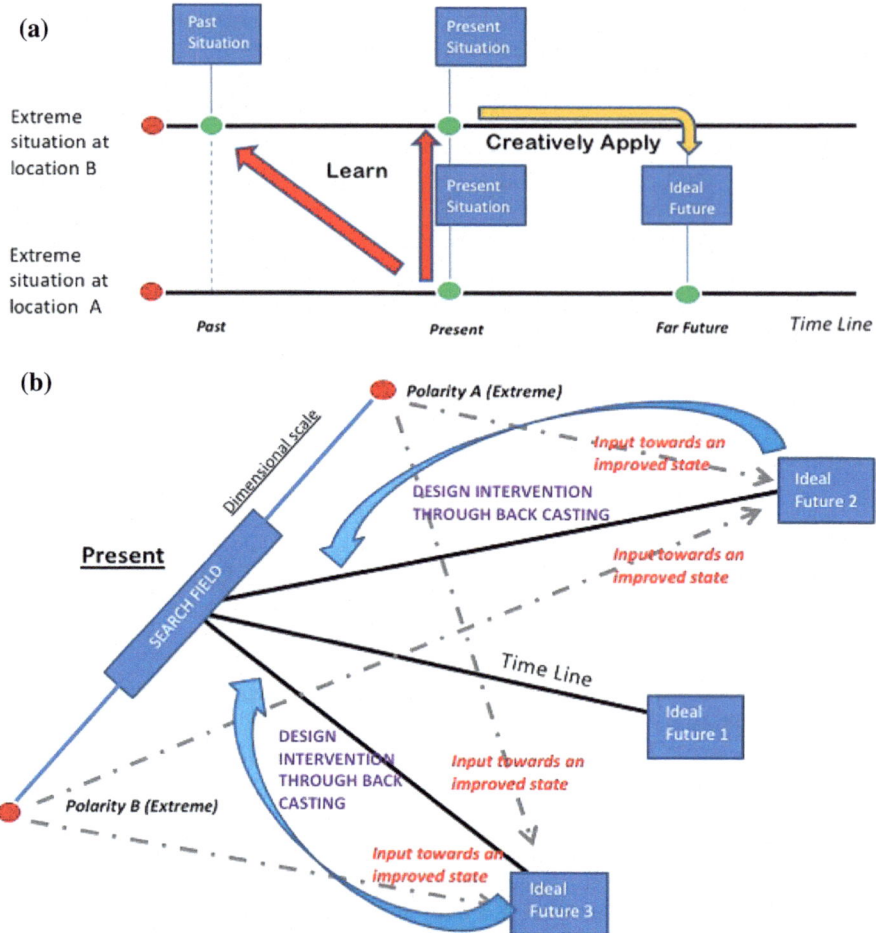

Fig. 8 a Forecasting and backcasting through a cultural lens are time and location dependent. **b** Forecasting and backcasting on the selected search field, influenced by cultural and acculturation thinking

stakeholders pertaining to the delivery of goods, should be investigated in context. In terms of strategizing over time, the 'Secure Mailbox' is expected to transition into an 'Active Mailbox' and thereafter the 'Total Mailbox'. At each transition, the versatility of the mailbox increases.

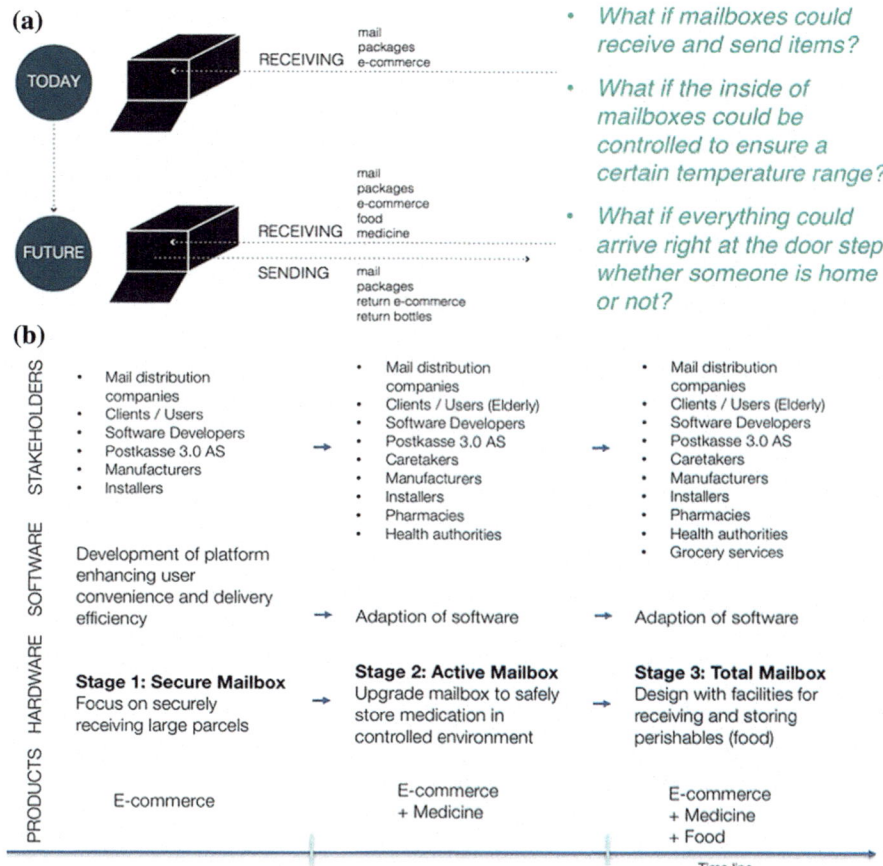

Fig. 9 a The future of home delivery; from being on-site to receive ordinary mail to being off-site while perishable and non-perishable goods are being delivered. **b** As the 'Secure Mailbox' transitions into an 'Active Mailbox' and thereafter the 'Total Mailbox', its versatility increases, facilitating the delivery of non-perishable and perishable goods as well as administrated medication

9 Implications for Design Education

As per today, design education still has the luxury of being able to pursue distinct values and pedagogies that emphasise the importance of low student–faculty ratios, such as project-based learning, small group critiques and individual formative feedback and guidance (Swann 2002; Design Council 2006). However, ongoing globalisation trends are pressurising Universities to become an integral part of national or regional innovation endeavours.

Besides this, social sustainability and service-oriented design thinking are becoming more important for developing innovative products and experiences. Hereby, the role of the designer has significantly changed from a creative genius or stylist to an 'Active' problem solver, team member, interpreter of complex systems and communicator (Roth 1999). This focus on innovation has led to building competencies, enabling designers to play a more critical, and integrative role in product development processes (Sethia 2001). As a result, the twenty-first century designer is expected to fulfil the roles of innovator, knowledge worker, sustainable entrepreneur and responsible citizen, concerned with environmental, societal, commercial, communication issues and so on (Press and Cooper 2003).

In terms of design education, it means that research-intensive universities are expected to be interactive players who strategically collaborate with industry, the community and government (Etzkowitz and Leytesdorff 1997). In the future, this will also have implications for 'Teaching', 'Research' and 'Administration', which need to be redefined into respectively 'Mentorship', 'Scholarship' and 'Service', and to be positioned as a more global and long-term commitment, determined by *discovery, integration, application* and *knowledge transfer* (Boyer 1990; Liem 2008). Given these global challenges, it was necessary to emphasise the strategic elements in design education by introducing a structured product planning process. However, as presented in preceding sections of this chapter, such a process comprises of systematic as well as emergent moves. Especially in the forecasting and backcasting stages (see Fig. 1), unstructured mentoring relationships are becoming more important in developing the strategic design solution. These relationships are more personal, closer, demand more time and commitment and a level of emotional engagement (Bhagia and Tinsley 2000).

To advance professional design practice, orientations among design faculty towards research, administration need to be redefined, and traditional methods of studio teaching revisited. 'Mentorship and Scholarship' should be introduced to promote highly interactive and customised learning. These are learning concepts based upon a hierarchical and collaborative approach for knowledge transfer in design. According to Yang et al. (2005), such highly interactive student–faculty pedagogies may enhance the following three areas in terms of competency building:

- Generic attributes, such as problem-solving and planning skills, to rapidly react to immediate contextual changes.
- Specific design skills and knowledge, including design thinking, design methodology, user engagement, graphical representation and communication, product development and manufacturing processes, design management, environmental awareness and prototyping in its broadest sense.
- Ancillary skills, such as negotiation with clients, stakeholder analysis (Lewis and Bonollo 2002), project management and communication (Sigurjonsson and Holgersen 2010).

Within the realm of strategic design education students need to be adequately prepared on how to collaborate, negotiate and compromise, when they engage themselves in University-Industry collaborative design projects (Niederhelman 2001). Moreover, to anticipate user needs and introduce more user-friendly products and services, these novice designers also need to be trained to manage increasingly complex technologies coupled with more demanding stakeholders. The above-mentioned challenges require:

- Faculty to be involved in mentorship and scholarship when advocating learning and inquiry from a theoretical, collaborative and process perspective (research-based learning) (Liem 2008).
- Practicing designers to be engaged more frequently in mentorship to contribute in skills development and sharing of design '*Praxis - What they actually do in practice*' through a 'design thinking' and 'designing' lens. These practicing designers can also act as role models for students on how to plan and manage their projects, as well as how to act convincingly and persuasively in project management and negotiation situations.

Complementary to interdisciplinary teamwork in design projects and research (Rittel and Webber 1973), the concept of 'Social Learning' and Legitimate Peripheral Participation (LPP) (Brown et al. 1989; Wenger 2000), should be introduced to students and novice designers to train them to work within complex collaborative contexts, which are often bounded by rationality. In practice, this implies that those who are new to the community of design need to become acquainted with the tasks, vocabulary and organising principles through peripheral activities first, before engaging in more complex senior level tasks and responsibilities (Lave and Wenger 1991; Brown et al. 1989).

10   Conclusion

This book chapter proposes a blueprint for strategic design, which caters for different levels of strategy making. It argues that both prescriptive and deliberate, as well as emergent and contextual interventions, should complement each other in the development of strategic design concepts.

The pedagogical approach was to guide students in the initial stages of strategy making by subjecting them to a structured process of product planning from understanding the 'Vision', 'Mission' and 'Positioning' of the organisation to the formation of initial 'Search Fields'. Selection and further development of these search fields have been predominantly driven by '*how conversations with the context contributed to the content development of these search fields*'. This emergent approach towards strategy making, has been partly rooted in 'Practice Theory', because it is difficult to prescribe a standard recipe on how to develop an early search field into a strategic design concept. Contributions to content development of typical search fields are made by designers' reflective conversations with the situation (Schön 2017), as well as the designer's personal experience, the subjectivity of the design process, as well as the design task in situ (Snodgrass and Coyne 1992). For the student, it is important that he or she adopts an apprentice attitude towards learning and strategizing by trying to gain insight of: '*What strategic designers actually do in practice; their preferences, attitudes and decision-making moves*'. In some of the collaborative projects, the involvement of established design consultancies, facilitated the transfer of these insights, using a kind of 'Mentorship' way of interacting.

However, the route from 'Initial Search field' to 'Strategic Design Concept' was not purely based upon emergent design and strategizing moves. Methods, such as 'Cultural Goal Finding' and 'Stakeholder Mapping' has been proposed and to some extend applied in refining and providing different perspectives on the search fields in the forecasting and backcasting stages of the strategic design process. These methods do not provide a step-by-step guide on what to do next, but challenges students to think about product, system and service solutions by confronting them to look beyond the scope of current stakeholders (*Future Stakeholder Mapping/Network Horizons*) as well as to consider a broader cultural spectrum when searching for innovative solutions (*Cultural Goal Finding*).

When juxtaposing 'Prescribed Processes and Methods' with 'Emergent Interventions' various options in backcasting can be mapped out with respect to strategy making. When emphasising the shorter term interests of organisations, a more pragmatic and incremental view towards innovation (*development*) will be practiced, whereas the creation of a radical ideal future is mainly based on end users' desires as well as projected behaviours. The latter is not about succumbing to existing realities, but challenges new and desired context for use, experience and interaction. Afterall, behavioural change emerging from bottom-up processes are of greater interest to the contemporary designer than technological and market progressions (Meroni 2008).

From a value creation perspective, the ideal future should still consider multiple interest and plural outcomes. It should cover the needs of different stakeholders at different levels and perspectives of value (Den Ouden 2011).

Within the context of collaborative design, different constructs of 'Social Learning' should be explored. This strategic design approach forms an ideal platform to experiment with pedagogical concepts, such as 'Mentorship-Scholarship-Service', 'Hierarchical Learning', 'Legitimate Peripheral Participation (LPP)'. The different views on how to practice strategic design led to some confusion, but at the end provided a deeper and more reflective understanding of '*What practicing strategic design is all about*'.

Acknowledgements This work is partially supported by the results of strategic and collaborative project work, which has been completed by 4th year (1st years M.Sc.) Industrial Design students in the spring semester of 2018. Completed. The authors also gratefully acknowledge the helpful comments and suggestions of the reviewers, which have improved this contribution.

References

Abrahamsen MH, Henneberg SC, Huemer L, Naudé P (2016) Network picturing: an action research study of strategizing in business networks. Ind Mark Manag 59:107–119

Ahlstrand B, Lampel J, Mintzberg, H (2001) Strategy safari: a guided tour through the wilds of strategic management. Simon and Schuster

Andersen P, Holmen E, Pedersen A-C (2018) Que Sera, sera? Conceptualizing business network foresighting. IMP J 12(1):56–74. https://doi.org/10.1108/IMP-03-2017-0009

Andrews KR (1980) The concept of corporate strategy. Richard D Irwin
Ansoff HI (1968) Corporate strategy. Great Britain
Ansoff HI (1986) Corporate strategy: an analytic approach to business policy for growth and expansion. Penguin, Harmondsworth
Apte M (1994) Language in sociocultural context. In: Asher RE (ed) The encyclopedia of language and linguistics, vol 4. Oxford, Pergamon Press, pp 2000–2010
Belliveau P, Griffin A, Somermeyer SM (2004) The PDMA toolbook for new product development. Wiley, Hoboken
Berry JW (2003) Conceptual approaches to acculturation. American Psychological Association
Bhagia J, Tinsley JA (2000) The mentoring partnership. Mayo Clin Proc 75:535–537
Boyer E (1990) Scholarship reconsidered: priorities of the professorate. The Carnegie Foundation for the Advancement of Teaching, Princeton
Brown JS, Collins A, Duguid P (1989) Situated cognition and the culture of learning. Educ Res 18(1):32–42
Brugha R, Varvasovszky Z (2000) Stakeholder analysis: a review. Health Policy Plann 15(3):239–246
Buijs J (2003) Modelling product innovation processes, from linear logic to circular chaos. Creat Innov Manag 12(2):76–93
Buijs J, Valkenburg R (1996) Integrale Produktontwikkeling. LEMMA BV, Utrecht
Bukowski M, Śniegocki, A (2017) Megatrends. Wise Europa – Warsaw Institute for Economic and European Studies, Warsaw
Cairns G, Wright G (2018) Evaluating the effectiveness of scenario interventions within organizations. In: Scenario thinking. Palgrave Macmillan, Cham
Christiansen JA (2000) Building the innovative organization. MacMillan Press, London
Cyert RM, March JG (1963) A behavioural theory of the firm. Prentice Hall, Englewood Cliffs, NJ
Davies AR (2014) Co-creating sustainable eating futures: technology, ICT and citizen–consumer ambivalence. Futures 62:181–193
Davies AR, Doyle R, Pape J (2012) Future visioning for sustainable household practices: spaces for sustainability learning? Area 44:54–60
Den Ouden E (2011) Innovation design: creating value for people, organizations and society. Springer Science & Business Media, London
Design Council, Creative and Cultural Skills (2006) Creative and cultural skills, design a new design industry: design skills consultation
Dorst K (2008) Design research: a revolution-waiting-to-happen. Des Stud 29(1):4–11
Dreborg KH (1996) Essence of backcasting. Futures 28(9):813–828
Ehrenfeld JR (2008) Sustainability by design: a subversive strategy for transforming our consumer culture. Yale University Press, New Haven
Einhorn HJ, Hogarth RM (1981) Behavioural decision theory: processes of judgment and choice. Annu Rev Psychol 32:53–88
Etzkowitz H, Leytesdorff L (1997) Universities in the global economy: a triple helix of academic-industry-government relation. Croom Helm, London
Fagerberg J (1994) Technology and international differences in growth rates. J Econ Lit 35:1147–1175
Freeman RE (2010) Strategic management: a stakeholder approach. Cambridge University Press, Cambridge
Gershenkron A (1962) Economic backwardness in historical perspective. Belknap Press, Cambridge
Grimble RJ, Aglionby J, Quan J (1994) Tree resources and environmental policy: a stakeholder approach, vol 7. Natural Resources Institute
Hall ET (1963) A system for the notation of proxemic behavior1. Am Anthropol 65(5):1003–1026
Hall ET (1976) Beyond culture. Anchor Doubleday Press, Garden City
Hekkert P, van Dijk M, Lloyd P (2011) Vision in product design: handbook for innovators
Hill T, Westbrook R (1997) SWOT analysis: it's time for a product recall. Long Range Plan 30(1):46–52

Hofstede G, Bond MH (1988) The confucius connection: from cultural roots to economic growth. Organ Dyn 16(4):5–24

Hofstede G, Hofstede GJ (2005) Cultures and organizations: software of the mind, Revised and expanded 2 edn. McGraw-Hill, New York

Holmen E, Pedersen A-C (2003) Strategizing through analyzing and influencing the network horizon. Ind Mark Manag 32(5):409–418

House RJ, Hanges PJ, Javidan M, Dorfman PW, Gupta V (eds) (2004) Culture, leadership, and organizations: the GLOBE study of 62 societies. Sage publications

Kim J, Wilemon D (2002) Focusing the fuzzy front-end in new product development. R&D Manag 32(4):269–279

Kluckhohn FR, Strodtbeck FL (1961) Variations in value orientations. Greenwood Press, Westport

Koen PA, Ajamian G, Burkart R, Clamen A, Davidson J, D'Amoe R, Elkins C, Herald K, Incorvia M, Johnson A, Karol R, Seibert R, Slavejkov A, Wagner K (2001) New concept development model: providing clarity and a common language to the 'fuzzy front end' of innovation. Res Technol Manag 44(2):46–55

Kroeber AL, Parsons T (1958) The concept of culture and of social system. Am Sociol Rev 23(5), (14, p. 583)

Landry C (2012) The creative city: a toolkit for urban innovators. Routledge

Larson AL (2000) Sustainable innovation through an entrepreneurship lens. business strategy and the environment. Bus Strat Environ 9:304–317

Lave J, Wenger E (1991) Situated learning: legitimate peripheral participation. Cambridge University Press, Cambridge

Lewis WP, Bonollo E (2002) An analysis of professional skills in design: implications for education and research. Des Stud 23:385–406

Liem A (2008) Developing a win-win mentorship-scholarship, higher education model for design through collaborative learning. UNIPED (Tromsø) 31(3):32–45

Manzini, E, Meroni, A (2007) Emerging user demands for sustainable solutions, EMUDE. In: Michel R (ed) Design research now: essays and selected projects. Basel, Birkhäuser Basel, pp 157–179

Medlin CJ, Törnroos JÅ (2014) Interest, sensemaking and adaptive processes in emerging business networks—an Australian biofuel case. Ind Mark Manag 43(6):1096–1107

Meroni A (2008) Strategic design: where are we now? Reflection around the foundations of a recent discipline. Strat Des Res J 1:31–38

Mintzberg H (1994) The fall and rise of strategic planning. Harv Bus Rev 72(1):107–114

Mintzberg H, Waters JA (1985) Of strategies, deliberate and emergent. Strat Manag J 6(3):257–272

Muszyński J (2001) Megatrendy a polityka. Wyd. Atla 2, Wrocław

Newcombe R (2003) From client to project stakeholders: a stakeholder mapping approach. Constr Manag Econ 21(8):841–848

Niederhelman M (2001) Education through design. Des Issues 17(3):83–87. The MIT Press

Ogilvy JA (2002) Creating better futures: scenario planning as a tool for a better tomorrow. Oxford University Press, New York

Parsons T, Shils EA (1951) Toward a general theory of action. Harvard University Press, Cambridge

Pęciak R (2016) Megatrends and their implications in the globalised world. Horyzonty Polityki 7(21):167–184

Porter ME (1985) Competitive advantage: creating and sustaining superior performance (Advantage C), vol 167

Prandecki K (2014) Theoretical aspects of sustainable energy. Energy Environ Eng 2(4):83–90

Press M, Cooper R (2003) The design experience: the role of design and designers in the twenty-first century. Ashgate Publishing, UK

Quist J, Green K, Szita Toth K, Young W (2002) Stakeholder involvement and alliances for sustainable shopping, cooking and eating. In: Bruijn T, Tukker A (eds) Partnerships and leadership: building alliances for a sustainable future. Kluwer Academic Publishers, Dordrecht, pp 273–294

Quist J, Vergragt P (2006) Past and future of backcasting: the shift to stakeholder participation and a proposal for a methodological framework. Futures 38(9):1027–1045

Regnér P (2003) Strategy creation in the periphery: inductive versus deductive strategy-making. J Manag Stud 40(1):57–82

Rifkin J (2014) The zero marginal cost society: the internet of things, the collaborative commons, and the eclipse of capitalism. Palgrave Macmillan

Rittel HWJ, Webber MM (1973) Dilemmas in a general theory of planning. Policy Sci 4:155–169

Rohrbeck R, Gemünden HG (2011) Corporate foresight: its three roles in enhancing the innovation capacity of a firm. Technol Forecast Soc Chang 78(2):231–243

Roth S (1999) The state of design research. Des Issues 15(2):18–26

Schön DA (2017) The reflective practitioner: how professionals think in action. Routledge

Sethia NK (2001) Generating and exploiting interdisciplinary knowledge in design product development and innovation in the new economy. In: The 2001 IDSA national education conference (CD ROM)

Sigurjonsson JB, Holgersen TD (2010) What do they do? A survey of employment and work situation for "IDE" candidates. In: Proceedings of the 12th international conference on engineering and product design education "New Paradigms and Approaches" (The Design Society and Institution of Engineering and Designers)

Sloan AP (1963) My years with general motors. Sedgewick & Jackson, London

Snodgrass A, Coyne R (1992) Models, metaphors, and the hermeneutics of designing. Des Issues 9(1):56–74

Straub DW, Loch W, Aristo R, Karahanna E, Strite M (2002) Toward a theory-based measurement of culture. J Glob Inf Manag 10(1):13–23

Svenfelt A (2010) Two strategies for dealing with uncertainty in social-ecological systems. Royal Institute of Technology, Stockholm

Swann C (2002) Nellie is dead. Art Des Commun High Educ 1:50–53

Swedberg R. Himmelstrand U, Brulin G (1987) The paradigm of economic sociology: premises and promises. Theory Soc 16(2):169–213

Trompenaars F, Hampden-Turner C (2011) Riding the waves of culture: understanding diversity in global business. Nicholas Brealey International

Tukker A (2004) Eight types of product–service system: eight ways to sustainability? Experiences from SusProNet. Bus Strat Environ 13(4):246–260

Utterback JM, Abernathy WJ (1975) A dynamic model of process and product innovation. Omega 3:639–656

Van der Heijden K (2011) Scenarios: the art of strategic conversation. Wiley, New York

Vergragt PJ, Quist J (2011) Backcasting for sustainability: introduction to the special issue. Technol Forecast Soc Change 78(5):747–755

von Hippel E (1988) The sources of innovation. Oxford University Press, New York

Weihrich H (1982) The TOWS matrix—a tool for situational analysis. Long Range Plan 15(2):54–66

Wenger E (2000) Communities of practice and social learning systems. Organ Artic 7(2):225–246. SAGE, London

Whittington R (2003) What is strategy- and does it matter?, 2nd edn. Cengage Learn EMEA, UK

Whittington R (2006) Completing the practice turn in strategy research. Organ Stud 27(5):613–634

Yang M-Y, You M, Chen F-C (2005) Competencies and qualifications for industrial design jobs: implications for design practice, education and student career guidance. Des Stud 26(2):155–189

Zahn EOK (1999) Strategizing needs systems thinking. In: Proceedings of the 17th international conference of the system dynamics society and Australian New Zealand systems conference, Wellington, New Zealand, 20–23 July

Design Engineering as a Means to Enhance Student Learning in Addressing Complex Engineering Challenges

Satesh Namasivayam, Mohammad Hosseini Fouladi,
Douglas Tong Kum Tien and Jayasubamani Arvi S. Moganakrishnan

1 Introduction

The engineer of the future is required to embody a wealth of knowledge and expertise that encompasses an array of hard and soft skills. Future engineers would be faced with addressing global issues and challenges and in addition to this, future engineers would need to be emotionally intelligent, practicing empathy and being self-aware of their own and of societies needs while attempting to address these issues. It is also noted that regulatory requirements insist that engineering curriculum, and in particular, curriculum of accredited engineering degrees, requires that engineering graduates possess in-depth engineering knowledge so that they may solve complex engineering challenges that have a range of conflicting requirements, using sound engineering analysis. With such an arduous task present at the end of an engineering undergraduates journey, how can engineering curriculum adapt and if needed, evolve to assist future engineers in equipping them with the necessary knowledge and skills? If such curricula exist, what would be its framework and how would it be executed? These are the key questions that will drive this chapter.

It is therefore hypothesized that design engineering may present itself as a viable solution in enhancing students competency in addressing the above. This chapter's objective would therefore be to narrate how existing engineering curriculum from around the globe, and in particular, curriculum infused with design engineering enables their graduates in preparing themselves to address the areas mentioned above. This would be further achieved by sharing best-known methods from accredited universities with a particular focus on two case studies from within Taylor's University's School of Engineering. The first of these will focus on how design engineering is delivered using the project-based learning pedagogy coupled with the Conceive-Design-Implement-Operate (C-D-I-O) framework. A specific example will focus on

S. Namasivayam (✉) · M. H. Fouladi · D. T. K. Tien · J. A. S. Moganakrishnan
School of Engineering, Taylor's University, Subang Jaya, Malaysia
e-mail: satesh.namasivayam@taylors.edu.my

© Springer Nature Switzerland AG 2019
D. Schaefer et al. (eds.), *Design Education Today*,
https://doi.org/10.1007/978-3-030-17134-6_11

the implementation of a capstone project module (or subject) and how the module is structured to address complex engineering problems through design engineering. The second example will narrate how design engineering links with sustainable development (again a specific example will be used to narrate how this was accomplished). Effort would also be made to discuss how these initiatives impacted student learning, through the measurement of Programme Learning Outcomes (POs) and Module Learning Outcomes (LOs) attainment of the examples described above. A discussion on how this data was used to further enhance the relevant modules in the spirit of continual quality improvement will also be presented.

2   Complex Engineering Challenges

The International Engineering Alliance describes engineering graduate attributes that essentially describe outcomes that must be assessed to indicate the competency of an engineering graduate. These attributes cover the ability of a student to solve problems through a variety of engineering activities. The range of problem-solving attributes known as WPs are described in International Engineering Alliance (2013). In summary, the definition of a complex engineering challenge (or problem) is a high-level challenge, which has no obvious solution, affecting a multitude of stakeholders, that cannot be resolved without in-depth engineering knowledge and involves conflicting issues which are infrequent.

In the event an engineering undergraduate is able to address a complex engineering challenge, (s)he would eventually have gained a significant milestone in attaining the graduate attributes. In summary, the graduate attributes are engineering knowledge, problem analysis, design/development of solutions, investigation, modern tool usage, the engineer and society, environment and sustainability, ethics, individual and teamwork, communication, project management and finance as well as lifelong learning. These attributes represent the regulatory fulfilments required of any recognized engineering degree around the globe. While the attributes represent a wide array of knowledge and skills required of an engineering graduate, there exists a need to relate these attributes to the challenges of the modern world. The advent of the 4th Industrial Revolution (Industry 4.0—The Nine Technologies Transforming Industrial Production 2018), the United Nations Sustainable Development Goals (United Nations Sustainable Development Goals 2018) as well as the Grand Challenges for Engineering (Grand Challenges for Engineering 2018), represent a set of very clear and distinct challenges that exists and may be solved by engineers of the future, for the future. Noting that these challenges are major areas of focus by today's societies, herein lies the opportunity of how to relate current and modern day challenges to that of the graduate attributes of an engineer. It would also be worth to first understand how engineering curriculum from around the globe, and in particular, curriculum infused with design engineering enables their graduates in preparing themselves to address the challenges or areas mentioned above.

3 The Impact of Design and Build Experiences in an Engineering Curriculum

The components of an engineering curriculum play an important role to prepare students with cutting edge knowledge and motivate them to think independently. Such students will shape the future pool of graduates who are enabled to deal with complex engineering challenges that always arise in real-world situations. Wicklein et al. (2009) studied engineering design and the way that it may be efficiently adopted for and benefit secondary students in technology education classes. Expert feedbacks were also solicited regarding engineering design and the way it can be combined with technology education and its impact. Their findings suggested that a curriculum framework formed on engineering design will bring along a cohesive delivery approach and encourages students to implement their skills into real-world applications. Responses revealed that such curriculum should include hands-on components and infuse team-working and ethics among students.

Christie and de Graaff (2017) elaborated on active learning and strategies which educators may adopt to establish this technique in engineering education. They indicated two techniques which had the highest impact on active learning, namely; Experiential Learning Theory and Project-Based Learning. The researchers highlighted that these type of approaches in education were in fact practiced by Confucius and Socrates thousands of years before as they were emphasizing on the impact of motivation and encouragement in the learning process in contrast to relying only on transfer of information to the learners. It was agreed that the best approach to active learning is to diversify learning approaches that may match a wider group of learners. Dirsch-Weigand et al. (2016) claimed that integrated interdisciplinary studies are expensive and as an alternative they suggested to implement project-based learning approach through specific interdisciplinary projects across bachelor's and master's programmes. It was found that students attained expertise such as team-working and problems-solving skills through the undergraduate projects while more advanced skills such as interdisciplinary design were developed in postgraduate projects. This approach was set up for mechanical and process engineering studies at a German technical university and authors proposed it to be extended to other engineering science departments.

Chassidim et al. (2018) made use of project-based learning to motivate students undergoing a software engineering course. It was applied to a second-year module named 'Software Development and Management in Agile Approach'. The agile approach is based on a collection of generic concepts which addresses business requirements for a fast and flexible development and goal achievement. It helped to simulate real-world scenarios for students and since the project guidelines were generic, students needed to be creative in the way they run the projects. This increased their motivation and assisted in attainment of various skills through the process. Students' feedbacks revealed that they highly appreciated team-working, problem-solving and engagement with real-world type of challenges. Violante and Vezzetti (2017) studied about the required skills that graduates need in order to tackle complex challenges of the current era. They specifically focused on Additive Manufacturing (AM) to see how it can help to grow students' skills towards product development in one hand, and enrichment of STEM education to complement educational reforms on the other hand. The study showed that students learnt about various AM processes and techniques to deal with machines and materials. Consequently, this approach enabled them with additional know-hows such as critical thinking and problem-solving skills, entrepreneurship and self-motivation that are among key outcomes of modern engineering curriculums.

Deaner and McCreery-Kellert (2018) proposed an amalgamation of design thinking and problem-solving in students' curriculum. They emphasized that design thinking will cultivate reasoning and decision-making skills that are essential in creating human-oriented products and services. In their project, students were tasked to promote peace among themselves as well as the community. They went through the design cycle and at the end presented their findings and drew their reflection about peace on stones which were placed around the school yard. Even though they implemented this approach to school children, but the same might be expanded to university curriculum. Elger et al. (2000) implemented a combination of project-based thinking and design approach to enhance students' motivation and self-confidence. Students were engaged in two projects, firstly to design and fabricate a model rocket that could reach a certain elevation, and secondly a project to design and build a container which was able to maintain the interior temperature over a given time. They learnt from their hands-on experiences and found the process very interesting too. Educators also found that the approach is sustainable and can be scaled and transferred to students at various levels. Bilén et al. (2002) suggested that the best curriculum to teach global product development is the one that is founded on design concepts, through which students are exposed to an interdisciplinary environment. They mentioned that an appropriate design project shall set realistic constraints and address various hard and soft skills such as science and mathematics, social aspects, ethics, and communication skills.

Sehgal and Gokhale (2015) conducted research on engineering curriculum in India and suggested methods to uphold personalized learning among students. They emphasised on the importance of adaptive learning as a tool to make sure that graduates can cope with fast development of knowledge due to technological advancement. It was suggested that lifelong learning as well as attention to the breath and diversity of the curriculum are to be given special attention by educators. Viswanathan et al. (2014) elaborated on prototyping as an essential skill to be included in any design curriculum. Students can make use of this proficiency to make simple prototypes that would help to refine their ideas at the beginning of the design stage and eventually conduce to better designs. Authors referred to the important role of physical prototypes to reveal the design shortcomings, many of which are direct result of incomplete mental models that could lead to design fixation. Chandrasekaran et al. (2015) introduced design-based learning in an engineering curriculum through a mechatronics module that was offered to students in their final year of study. They were tasked to build a robot called "sumo robot" which would be able to challenge opposing robots in a circular ring. The course was offered both to students on and off campus and feedback showed their high satisfaction with regard to the practice as this helped them to improve their problem-solving skills.

Lulay et al. (2015) introduced engineering design in the curriculum of a mechanical engineering programme. They preferred to use ABET Criterion 5 rather than the universal definition on engineering design as they found it more concise and inclusive which comprises of both engineering and non-engineering terms. It says that 'engineering design is the process of devising a system, component, or process to meet desired needs. It is a decision-making process (often iterative), in which the basic sciences, mathematics, and the engineering sciences are applied to convert resources optimally to meet these stated needs'. Authors implemented engineering design in lectures, students' laboratories as well as capstone projects. For this purpose fundamental elements of design process were tabulated and mapped to a number of courses. Both lecturers' assessments results and students' survey outcomes were analysed to verify the impact of the approach. Researchers found out that students in the first two years of their study are less prepared for dealing with open-ended challenges. Hence it was suggested that for this group of students the emphasis would be more on preparation and familiarizing with the design process, while students on year 3 and 4 shall be able to fully practice the design process. Authors proposed a map for developing a design spine that may be used by engineering educators from other institutions.

Hei and Cheng (2015) proposed four major streams to develop curriculum for a telecommunication engineering programme in China. Project-oriented learning was implemented in every stream to uphold students' expertise both in theoretical knowledge and hands-on experiences. Rapid advancement of technologies necessitates that students update themselves with the latest developments in the field. Therefore authors did not fix the elective courses, instead put a process in place through which faculty members were able to propose new electives. Upon approval, these electives were run in the school for an upcoming semester while monitored by the staff. Successful electives that were offered for multiple rounds were adopted in the students' regular curriculum. Pech et al. (2016) introduced innovation and entrepreneurship to 3rd year engineering students via a design module. They aimed at preparing students for the future needs of the society by modifying the curriculum to the one that can gear them with the right skills. The course implemented design concepts and leveraged on prototyping and simulation skills to establish solutions for real-world challenges. Students' survey outcomes and module passing rate showed that students enjoyed and learnt a lot from the teamwork, and the passing rate was higher than the university average.

Soare (2016) focused on a new design of curriculum that is aimed at addressing the current needs of society. He analysed the impact of competence based approach on curriculum and how it is able to link programme educational aims to the market demands. Competence is defined as skills, abilities and knowledge needed to perform a specific task; in other words it refers to the available resources in one's hand to solve a specific problem. He mentioned that placing competence at the centre of curriculum may result in development of relevant expertise in students and re-establish the connection between societal needs and educational programmes. Ruayruay et al.

(2016) studied the curriculum design of a food engineering programme that well prepared students for dealing with hands-on skills and real-world experiences. Graduates were employed within a year after graduation and one-third of them were able to obtain a managerial position within five years. The key factors of graduates' success were laid in the essential elements of the programme curriculum; namely, in-class activities, extra activities and industrial activities. Students attained fundamental knowledge through general courses, whereby specific courses were geared toward problem-based learning and encouraged learners to conduct research and 'learn how to learn'. The extra activities aided learners to develop additional skills that were expected to be attained upon graduation. The industrial activities provided students with work place knowledge and expertise and was an important factor in the success of the programme.

Aris et al. (2017) studied the impact of implementing multidisciplinary approaches in curriculum on students' learning experience. Inter-faculties collaboration was established through elective modules that were offered across different faculties. They proposed a research framework that could support execution of these modules in the curriculum while took into account Harvard Outcome Domain (HoD), twenty-first century skills, sustainable development goals as well as components of 4th industrial revolution. The preliminary stage focused on HoD and showed that the curriculum should have sufficient number of electives mapped to HoD in order to promote self-learning among students who on the other hand shall strive to strengthen themselves to deal with real-world challenges. Rao (2018) gave an introduction to creativity and its impact on science and engineering. He mentioned that in Mechanical Engineering curriculum at Stanford University students are actively involved in projects, through which they can apply their knowledge in designing new components and systems and work efficiently in multidisciplinary teams. This shall widen the boundary of the programme and establish a venue for design thinking amongst students. Olsen et al. (2018) emphasised on the importance of sustainability and the role of education system to prepare students to deal with future sustainability challenges. They developed a framework to teach sustainability to students across all programmes at various levels from bachelor to Ph.D. The learning objectives of these modules were mapped to those learning outcomes of the institution that were crafted based on sustainability. Graduates would be able to define the subject matter, make assessments while taking into account environmental life cycle, and making use of the acquired knowledge in business and management. Table 1 summarises various available engineering curriculum, especially the ones infused with design engineering and their impact on graduates. The examples sighted above and in the table reflect curricula globally and are not necessarily dependent on the country as one is working on the assumption that most curricula for engineering around the globe in some way of form prescribes to the knowledge profiles within the Washington Accord.

Table 1 Summary of existing engineering curriculum from around the globe, and in particular, curriculum infused with design engineering and their impact on graduates

Investigation	Courses applied to	Success measurement/suggestion
Engineering design and the way it may be efficiently adopted (International Engineering Alliance 2013)	Secondary students in technology education classes	Cohesive delivery approach that encourages students to implement their skills into real-world applications
Active learning in engineering education (Industry 4.0—The Nine Technologies Transforming Industrial Production 2018)	General education system	Diversify learning approaches that may match a wider group of learners
Project-based learning (United Nations Sustainable Development Goals 2018)	Specific interdisciplinary projects across bachelor's and master's programmes	Students attained expertise such as team-working, problems-solving, interdisciplinary design
Project-based learning (Grand Challenges for Engineering 2018)	Second-year module for software engineering	Enhanced students' motivation and assisted in attainment of various skills
Required skills essential for graduates (Wicklein et al. 2009)	Additive manufacturing and its impact on students' skills towards product development and enrichment of STEM education	Students were enabled with critical thinking and problem-solving skills, entrepreneurship and self-motivation
Amalgamation of design thinking and problem-solving (Christie and de Graaff 2017)	PAST foundation-supported schools	Cultivation of reasoning and decision-making skills
Combination of project-based thinking and design approach (Dirsch-Weigand et al. 2016)	Engineering undergraduate courses	Students learnt from their hands-on experiences, educators found it a sustainable approach
Design curriculum (Chassidim et al. 2018)	Core curriculum in global product development	Successful exposure to the worldwide engineering processes
Personalized and adaptive learning (Violante and Vezzetti 2017)	Engineering curriculum in India	Graduates cope with fast development of knowledge
Importance of prototyping as an essential skill in design curriculum (Deaner and McCreery-Kellert 2018)	A design project in an engineering programme	Refine ideas at the beginning of the design stage and avoid design fixation
Design-based learning (Elger et al. 2000)	Mechatronics module offered to final year students	Improved students' problem-solving skills
Implementing engineering design (Bilén et al. 2002)	Curriculum of a mechanical engineering programme	Students on year 3 and 4 were able to fully practice the design process and deal with open-ended challenges

(continued)

Table 1 (continued)

Investigation	Courses applied to	Success measurement/suggestion
Project-oriented learning (Sehgal and Gokhale 2015)	Curriculum for a telecommunication engineering programme	Students gained both theoretical and practical skills
Impact of innovation and entrepreneurship (Viswanathan et al. 2014)	3rd year engineering students via a design module	Passing rate was higher than the university average
Usage of competence based approach (Chandrasekaran et al. 2015)	Education pedagogy	Development of relevant expertise in students
Effect of in-class activities, extra activities and industrial activities (Lulay et al. 2015)	Curriculum design of a food engineering programme	Graduates were employed within a year and one-third of them secured a managerial position within five years
Implementing multidisciplinary approaches (Hei and Cheng 2015)	Bachelor of Civil Engineering programme	Curriculum remains relevant and provides graduates with the required now-a-days expertise
Creativity and its impact on science and engineering (Pech et al. 2016)	Mechanical Engineering curriculum at Stanford University	Widen the boundary of the programme and establish a venue for design thinking amongst students
Importance of sustainability and the role of education system (Soare 2016)	A framework to teach sustainability to students across all programmes at various levels from bachelor to Ph.D.	Graduates make assessments while taking into account environmental life cycle, making use of the acquired knowledge in business and management

4 Baselining Design Engineering—Content and Pedagogy

From the review performed and described earlier, it can be summarised that curriculum, infused with engineering design and implemented with the aid of project-based learning, revealed that students highly appreciated working in teams on challenges associated with real-world scenarios. However, it is necessary to first baseline the concept of engineering design, project-based learning and how each sit within the wider scope of design engineering.

4.1 Design Engineering

Design Engineering in today's world is being recognized as a key engineering discipline. There are a good number of design engineering programmes around the globe which aim to produce engineers that have a range of fundamental design (e.g. design empathy, design thinking, etc.) and engineering knowledge and skills such that they are able to design and build engineering systems across the breadth of engineering knowledge. In some of these programmes, students are also exposed to key knowledge that would assist them in ensuring the systems have a value proposition, and thus creating an opportunity for their system to be of value from a commercial perspective.

4.2 Engineering Design

Engineering Design is commonly recognized as a subject area, focusing on students being able to design and build an engineering system based on a prescribed design challenge. Such courses, encompassing engineering design normally aims to apply different design knowledge (such as design thinking) together with engineering knowledge to produce a prototype of an engineering system that addresses a specific set of requirements.

This is a key pedagogy that supports the implementation of Engineering Design and hence Design Engineering. Based on the review conducted earlier on in the chapter, project-based learning was found to be beneficial in enhancing the overall student learning experience with respect to engineering design.

In order to provide structure to engineering design courses (or modules) within a design engineering programme, many tools and techniques are utilized. In particular, for courses at the authors' own university, CDIOTM is used to provide structure to its engineering design courses. The Conceive-Design-Implement-Operate (CDIOTM) framework was initiated around 20 years ago at Massachusetts Institute of Technology (MIT) in order to reform engineering education to cope with the latest needs and challenges available in the industry. MIT researchers thought of the set of skills that graduates would need to have upon graduation. The full extent on how CDIOTM is used in this manner is described in detail in Al-Atabi (2014), but essentially students are taught how to practice key knowledge and skills within each stage of CDIOTM as follows.

- Conceive
  - Students are taught specific brainstorming techniques and essentially brainstorm to produce a decision matrix to provide viable conceptual designs for a prescribed engineering challenge with specific requirements.

- Design

  – Students are taught to produce designs, as an illustration, with each iteration of the design drawings becoming more detail, culminating in a set of electronically drawn engineering drawings that prescribe to the appropriate drawing standards.

- Implement

  – Students are tasked with developing a prototype using cost-effective material to visualize their designed systems and how such systems integrate with the environment and the user. Any misalignments would encourage a redesign where applicable.

- Operate

  – Students are tasked with developing a testing plan to ensure their system meets the specific requirements of the engineering challenge.

Thus far, there is a body of literature which describes how engineering design coupled with project-based learning results in an increase of the overall student learning experience with respect to engineering education. There is also research that discusses the success of the implementation of CDIOTM in an engineering programme that does the same. It is therefore hypothesized that students who are holistically taught about fundamental design, engineering knowledge and skills together with courses that require them to design and build would result in an engineering graduate that would be able to meet the graduate attributes stated earlier and hence most suited to address a complex engineering challenge. To explore this hypothesis further, an attempt would now be made to describe how an engineering graduate can be produced to address complex engineering challenges through design engineering.

With reference to Fig. 1, this figure illustrates the relationship between societies challenges in the form of the GC's and UN SDG's and how these challenges relate to I4 and the relevant graduate attributes (which embedded within, specify the need for students to work on complex engineering challenges). Looking at the Venn diagram, the overlap between the GC's and the UN SDG's have been suggested as described in the diagram. It should be noted that the National Academy of Engineering (NAE) have grouped the GC's into four (4) major themes, namely GC's that are related to ensuring sustainability, GC's that promote security, Health GC's which relate to challenges associated with enhancing health informatics and finally GC's relating to the ability to live in a conducive and vibrant environment. It should also be noted that the overlap of the GC's with the UN SDG's represent the authors' own interpretation of this relationship. Moving on to I4 (and still referring to Fig. 1), Boston Consulting Group in Rao (2018) described the nine (9) pillar technologies that will drive the implementation of I4. As the technologies essentially described the tools and techniques engineers could utilize to address complex challenges, and noting that the GC's and the UN SDG's in themselves are complex challenges, it is therefore surmised that such I4 tools and techniques may be used, by engineers to address the GC's and the UN SDG's—thus this is described in the overlap between I4, GC's and the UN SDG's in Fig. 1. Finally, in order to address the GC's and UN SDG's which

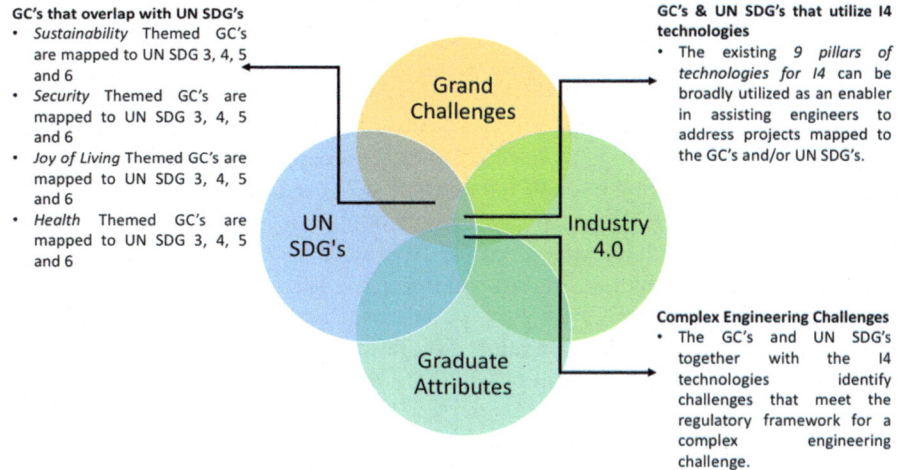

GC's that overlap with UN SDG's
- *Sustainability* Themed GC's are mapped to UN SDG 3, 4, 5 and 6
- *Security* Themed GC's are mapped to UN SDG 3, 4, 5 and 6
- *Joy of Living* Themed GC's are mapped to UN SDG 3, 4, 5 and 6
- *Health* Themed GC's are mapped to UN SDG 3, 4, 5 and 6

GC's & UN SDG's that utilize I4 technologies
- The existing *9 pillars of technologies for I4* can be broadly utilized as an enabler in assisting engineers to address projects mapped to the GC's and/or UN SDG's.

Complex Engineering Challenges
- The GC's and UN SDG's together with the I4 technologies identify challenges that meet the regulatory framework for a complex engineering challenge.

Fig. 1 Relationship between societies challenges, I4 and the WA's

will utilize I4 technologies, this would ultimately address the graduate attributes and in particular complex engineering challenges.

In summary, the preceding text described how courses related to engineering design, implemented through project-based learning are best suited to enhance the overall student learning experience within engineering education. It was also noted that design engineering, as an engineering discipline is able to produce engineers that possess knowledge and skills of fundamental design as well as spanning the breadth of engineering. In order to address key graduate attributes as described by the IEA, and in particular complex engineering challenges, the GC's as well as the UN SDG's present excellent challenges that may be addressed by engineers through the utilization of I4 technologies.

Moving forward and in order to provide examples to how the elements of Fig. 1 are implemented as well as the impact on student learning, the authors will focus on two case studies from within Taylor's University's School of Engineering. The first of these will focus on how the elements that would be taught in a design engineering programme is delivered using the project-based learning pedagogy coupled with CDIOTM as the framework in a capstone engineering module which requires students to address complex engineering challenges over the period of a year. The second example will narrate how design engineering links with sustainable development.

In order to quantify the impact on student learning, observations made during the implementation of the courses would be highlighted as well as a description provided on how the attainment of the student learning outcomes were affected and what would be the relevant improvement plans to further enhance these modules or projects in the spirit of continual quality improvement (CQI).

5 Case Study 1: The Capstone Project Module

The Mechanical Engineering Group Project (MEGP) is the Capstone Project module of the B. Eng. (Hon) programme of the School of Engineering, Taylor's University. This project-based module is planned for two consecutive semesters. It is offered as Mechanical Engineering Group Project 1 (MEGP1) and Mechanical Engineering Group Project 2 (MEGP2) respectively. These modules are intentionally designed to provide the students with the opportunity to experience the entire range of the CDIO process in sufficient depth within a single project. This overarching objective is attained through having MEGP1 focus on Conceiving and Designing, while MEGP2 focuses on Implementing and Operating. Separating the CDIOTM process into two semesters allows for each element of CDIOTM to be experienced in greater depth and breadth.

In addition these MEGPs are designed to cover, over two semesters, all of the 12 attributes of the Washington Accord Graduate Attributes (WA) (International Engineering Alliance 2013). They are also designed to address the requirements of the CDIOTM syllabus v2.0 (Olsen et al. 2018), which at the first-level structure are as stated below. These requirements are given a code in brackets here (CDIO#) for the purpose of showing their mapping to the learning outcomes (LOs) of the MEGPs in the subsequent sections.

1. Disciplinary Knowledge and Reasoning (CDIO1)
2. Personal and Professional Skills and Attributes (CDIO2)
3. Interpersonal Skills: Teamwork and Communication (CDIO3)
4. Conceiving, Designing, Implementing, and Operating Systems in the Enterprise, Societal and Environmental Context (CDIO4).

The attainment of the WAs and the CDIOTM syllabus requirements are enabled through the deliberate crafting of MEGPs' LOs and through ensuring that these LOs are attained by the students over the course of the semester. The process of LO attainment will be elaborated in the sections to follow.

The LOs of MEGP1 are stated below with the first term in the brackets after each LO representing the corresponding IEA graduate attributes (WA#) while the second term represents the corresponding CDIOTM syllabus v2.0 requirement. Note that LO's 3, 4, 6, 7 and 9 are different for each of these modules.

1. Demonstrate effectiveness in communicating technical activities in oral and written form (WA10, CDIO3).
2. Apply project management tools and techniques in effective project initiation and planning (WA11, CDIO4).
3. Conceive and Design effective solutions for complex engineering challenges (WA3, CDIO4).
4. Apply appropriate techniques, modern engineering and IT tools to an engineering design project (WA5, CDIO1).
5. Demonstrate effectiveness as a team member and/or team leader (WA9, CDIO3).

6. Demonstrate the knowledge of sustainable development in designing solutions for complex engineering challenges (WA7, CDIO4).
7. Identify activities to cope with technological needs of the future (WA12, CDIO2).
8. Apply knowledge of scientific and engineering practices to provide viable solutions for complex Mechanical Engineering challenges (WA1, CDIO1).

The LOs of MEGP2 are stated below with the first term in the brackets after each LO representing the corresponding WA attribute while the second term represents the corresponding CDIOTM syllabus v2.0 requirement.

1. Demonstrate effectiveness in communicating technical activities in oral and/or written form (WA10, CDIO3).
2. Apply project management tools and techniques in effective project execution and closure (WA11, CDIO4).
3. Implement and Operate effective solutions for complex engineering challenges (WA3, CDIO4).
4. Apply professional and ethical responsibilities of engineering practice (WA8, CDIO2).
5. Demonstrate effectiveness as a team member and/or team leader (WA9, CDIO3).
6. Assess the designed solutions for complex engineering challenges against societal, health, safety, legal, economical and cultural issues (WA6, CDIO4).
7. Evaluate the functionality of prototype against design (WA4, CDIO1).
8. Identify activities to cope with technological needs of the future (WA12, CDIO2).
9. Analyse and document the solution for a complex Mechanical Engineering challenge and draw substantiated conclusions (WA2, CDIO2).

The projects given to the students must be of sufficient complexity, pose a significant but within reach challenge, as well as being suitable for addressing all CDIOTM elements. The responsibility for proposing the project rest with the MEGP1 module coordinator. The proposal must be reviewed for approval by the project-based learning committee. This committee comprises the Head of School, the Programme Directors and all the project-based module coordinators. It is important to have a committee that possesses the required expertise to vet the suitability of the proposed project. It is vital to offer suitable projects in order for the students to genuinely achieve the desired learning outcomes as well as to develop their personal, professional and technical competencies. The decision of the committee could be to either accept, reject or amend the proposal. This decision is taken at a committee meeting where the module coordinator presents his proposal. It is at this point where the project is also evaluated against either the GC's or the UN SDG's. For example, this is accomplished in a project proposal form, created by the module coordinator, where it is compulsory for the coordinator to identify the relevant GC that will be addressed by such a project.

The projects proposed are not limited to those initiated by the module coordinator. For the past two semesters, industry capstone projects have been offered as well. The first was from a global tire manufacturer while the second was from a local automobile manufacturer. Similar to the approval process aforementioned, these proposals undergo a similar review and approval process. The module coordinator works closely with the industry partner to ensure that these projects are aligned with the module objectives. Offering industry projects allow the students to be exposed to industry challenges first hand and provide them with an opportunity and the experience of solving a real-life industrial challenge through application of CDIOTM principles. A visit to the industry partner's plant is arranged to help the students better understand the context of their project. In addition, the industry partner is invited on campus to introduce the project and later in the semester to answer students' queries after they had started work on their project. As an alternative to being on campus video conferencing was organized as well.

Assessments are crucial in helping students learn what they are intended to learn if they are purposefully designed. In the MEGP1 the assessment components are stated below.

1. Logbook
2. Interim Report
3. Final Report
4. Presentation
5. Peer Assessment
6. Artefact Assessment.

The assessments components for MEGP2 are identical to MEGP1 except that it does not require the Interim Report. The students would be in the midst of acquiring materials and building their artefact in MEGP2 hence the Interim Report was considered unnecessary unlike in MEGP1 where the students are in the midst of conceiving their solution. The rubrics for Final Report and Artefact Assessment are also different since the emphasis of MEGP2 is on I and O whereas for MEGP1 it is on C and D.

The Logbook serves as a personal learning instrument in addition to recording and documenting evidence of project work performed. The students are encouraged to document what they have learned. They are encouraged to reflect on their experiences and learning opportunities. In addition to their logbook, the students are required to submit an assignment known as "Return on Failure" twice in MEGP1 and twice in MEGP2. The students identify mistakes made or failures encountered during the course of the project. They then identify the root cause, prescribe corrective actions and recognize learning opportunities. The aim is to alleviate the fear of failure which inhibits innovation and future success. The failures discussed could be in the area of design, manufacture, project management, communication, documentation and leadership, basically anything that pertains to their project work.

The Final Report is a record of the entire project. The objective of the report is to document all technical information of the project in such a way that a third party with the skill (i.e. an engineer) without the knowledge of the project is able to understand the area of interest of the project, objective, procedures involved, the engineering analysis, reasons behind material selection, a full set of engineering drawings, project outcomes and recommended improvements of the project. The final report for MEGP1 would focus on Conceive and Design while that of MEGP2 on Implement and Operate. As a requirement of the Operate emphasis in MEGP2, a Technical Operations and Maintenance Manual is required to guide users on the usage, safety and maintenance of their product.

In addition, the final report of MEGP1 requires the students to formulate a business plan for their product so as to acquire an understanding of entrepreneurship. The final report of MEGP2 would require the students to discuss ethics and professionalism in the context of the design, manufacture and operation of their product as well as their conduct working on the project. They are also required to critically assess the designed solution against societal, health, safety, legal, economic and cultural issues. Hence the learning outcome is not restricted to mastering technical competencies only but extends to ensuring that students appreciate the wider context of their proposed solution and the importance of professionalism and ethics.

As with all projects of this nature, a Presentation is required. The aim is to build confidence in public speaking and to develop effective communication skills. The ability to organize their thoughts, not just in organizing their presentation, but also in giving clear and coherent answers to impromptu questions posed to them is desired.

Apart from the presentation, an Artefact Assessment is also carried out. This takes place during an event organized by the School of Engineering known as the Engineering Fair. This is a once per semester event to showcase the artefacts created from all the project-based modules of the School. External judges are invited to assess these artefacts. These judges are invited from both industry and academia. If from academia they will be academics from engineering schools other than Taylor's University. Having third-party external assessors exposes the students to the industry needs and also benchmarks the projects against similar engineering institutions. In the case of the industry capstone projects, the judges of the capstone projects are the industry partners who offered the projects to our students. The judges' assessments contribute to the marks of the Artefact Assessment, which in turn contributes to the overall marks of the module. In addition, these marks are used to determine the prizes awarded at the Engineering Fair. As an incentive to the students, prizes are awarded for best projects and are given at the conclusion of the Engineering Fair. The Peer Assessment is conducted by the students themselves at the end of the semester Each student is required to rate each of their team member in terms of their contribution to the project, problem-solving ability, their attitude, their focus on their tasks, and their ability to work with others in the team. Each criterion is rated from 0 to 10. If a 10 is given the peer assessor needs to justify it in writing as to why the team member deserved a 10. This assessment is conducted online.

There is also a compulsory written feedback section in this assessment. The peer assessor is required to comment on each team member. To encourage honest.

5.1 Key Findings

The MEGPs are subjected to the Outcomes Based Education (OBE) model of the School with a strong emphasis on Continual Quality Improvement (CQI). This approach is in alignment with the Washington Accord's requirements.

At the module level, the LO attainments are measured through the assessment components with are mapped to the LOs. There are KPIs set for the LO attainments. If these KPIs are not met by any LO then CQI actions are proposed and implemented in the subsequent semester to ensure these LO's KPI are attained. However, even if all LO attainments are met, CQI actions are still encouraged in the spirit of continual improvement. In the extreme situation where an LO attainment has reached 100%, the CQI may take an opposite approach to reduce the LO attainment if the mapped assessments were considered to be too easily attainable. An example of a CQI for LO4 of MEGP1 is stated below. LO4 is about applying appropriate techniques, modern engineering and IT tools to an engineering design project. This LO fell below the KPI. The following CQI was proposed.

> Students have some difficulties in writing engineering analysis. Therefore, one case study can be brought into the classroom (during lecture) for examples on how to write proper engineering analysis. This will help to increase LO4.

This CQI was implemented in the subsequent semester and the LO4 managed to meet KPI albeit only marginally. Hence a new CQI was proposed for this LO to further improve its attainment in the following semester. The proposed CQI is stated below.

> While simulations were shown in the reports, the discussion related to them was usually weak. Students need to understand how to interpret results and discuss them. This should be more fully emphasized at the lecture dedicated to simulation and engineering analysis. An alternative would be to have that lecture fully dedicated to simulation only especially in how to interpret and discuss results. This would be expected to improve LO4.

This CQI is in progress for the current semester and hence its impact on the LO4 is still pending at the time of writing. The Programme Outcomes (PO) of these modules are directly and explicitly mapped to the LOs i.e. they are related by a one-to-one mapping. Hence attaining the LO would contribute directly to attainment of the corresponding PO. The PO from which LO4 is derived is related to WA5 of Washington Accord's Graduate Attributes regarding modern tool usage. It is likewise related to section 1 of CDIO syllabus v2.0 (Crawley et al. 2011) concerning disciplinary knowledge and reasoning. Therefore, it can be readily appreciated how the CQI process helps in achieving both WA and CDIO competencies due their relationship to the LOs.

The application of OBE and CQI has been found to be effective in meeting its objectives. This application is not limited to internal POs of the School. It can be widened to include WA and CDIO as demonstrated in this example.

6 Case Study 2: Engineering Design and Sustainable Development

Designing engineering systems for sustainable development within engineering education is a process of decision-making and problem-solving which enables learners to generate engineering ideas in relation to complex engineering challenges. Engineering design with sustainable development is an education context which introduces the real-world problems (or challenges) in its complexities as engineering challenges. Real-world problems are complex, multifaceted and as placed in society, involves various socio-environmental and cultural contexts—as mentioned in the preceding text, such challenges are that of the GC's and the UN SDG's.

It was observed that student involvement in these projects have benefitted in learning outcomes attainment, in particular, the ability of students to generate ideas for complex challenges as well as the understanding of ethical relationships involved with regards to designing and engineering system for society. One of the main reasons for this was due to the context of the challenge, when introduced to the students, embedded multiple perspectives and required critical thinking. Understanding the challenge in its situation with the team was important so as to relay the various perspectives that were derived in addressing the common challenge and how the organization of contribution was made to address the complexity of a specific challenge. Students were involved in a multidisciplinary setting although the unique circumstances of some projects have encouraged them to explore more than the boundary of their discipline to produce a more comprehensive solution—thus encouraging inter and transdisciplinary thinking. To further define the transdisciplinary approach, Fig. 2 illustrates the relationship.

With reference to Fig. 2, the approaches denote the relationship between disciplines and the way they work to develop a product (or an engineering system) for a solution (thus addressing an engineering challenge). The first two, single and multidisciplinary rely on disciplinary perspectives to contribute to the production of the solution. Thus, in multidisciplinary, even though it involves more than one discipline, the interaction is parallel to each other, where sharing of knowledge is minimal or absent (hence no permeability). The remaining approaches i.e. trans and interdisciplinary are more collective and collaborative in nature where the disciplinary boundaries allow permeation. The need for such an approach is essentially due to the nature of the challenge (or problem) that is complex and requires broader perspectives which opens the need for collaboration. Interdisciplinary approaches are more academically oriented where conjoining of disciplines targets a new joint venture for a new discipline outcome. Transdisciplinary approaches on the other hand, emphasize the mutual and transformative learning from shared working context and thus, blurring the boundaries for a collective outcome. Unlike interdisciplinary, transdisciplinary focuses on shared contribution and learning rather than the creation of new joint ventures of discipline from collaboration. To further illustrate the differences, an example will be used. If an engineer is tasked to design a hip replacement in the single disciplinary approach, the engineer would perform his or

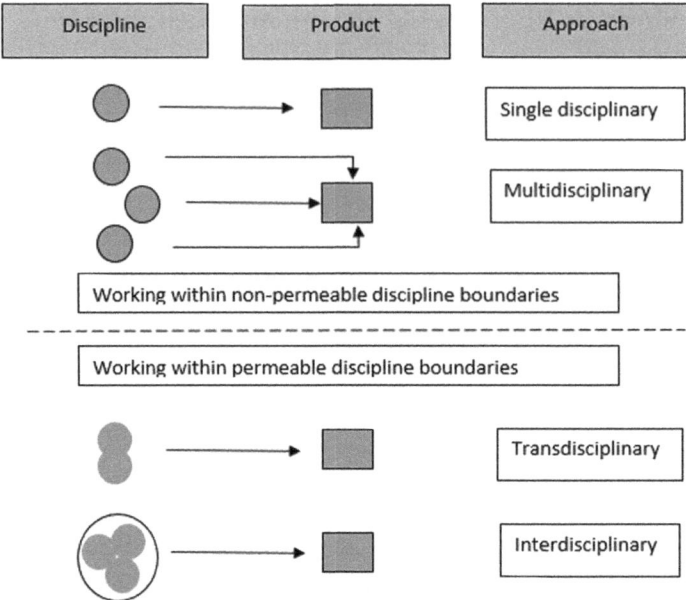

Fig. 2 Uni (or single), multi, inter and transdisciplinary approaches

her own research to eventually develop design concepts and finally build the product. In the multidisciplinary approach, the engineer would seek the advice of a medical doctor upon choosing or developing a design, thus the interaction between both of these experts is limited. In the interdisciplinary approach, both experts, i.e. the medical doctor and the engineer will collaborate from the inception of the challenge, develop design concepts together and decide on how to build the product for effective implementation. Finally, the transdisciplinary approach would require that the medical doctor, engineer and the end user or the patient (i.e. stakeholder) works with the experts to design and build the product or engineering system for more effective implementation. The choice of overarching problems in society such as the GC's or the UN SDG's could provide contexts for transdisciplinary imagination where focus is shifted on the challenges as that which raises issues of sustainability.

This section will share the experience of facilitating engineering design projects in link with sustainable development and promotes the transdisciplinary approach. One involves a project on the design of safer and more protective cleaning tools for the betterment of work for janitors and the other involves, the monitoring and communicating of aqua life conditions. In such projects, students involved themselves in using systems thinking for generating ideas for the design which includes and considers the end use—as described in Aris et al. (2017) and earlier. As students were involved in projects relating to real life situations, such as working with the janitor to understand cleaning as harmful and strenuous from the perspective of the user, gave opportunities to students to construct pathways within the process of designing for incorporation of views and feedback. This involved teamwork as well as facilitation with coordination of educators for multiple expertise to help the students to work progressively towards achieving their goals. Using another project as an example, when proposing an aquatic life monitoring and communicating system for conservation education in schools was proposed, a multidisciplinary perspective of the problem was sought for in which various disciplinary perspective informed the common challenge of monitoring and communicating of aquatic life environmental conditions. However, the overarching goal of conservation and education provided the potential for a different kind of imagination to inform a broader goal of understanding the living of aquatic animal in not only environmental but also social terms in which responsibility is the aim which informed the task of monitoring and communicating.

To further elaborate on how a cooperative approach is made to align the contribution of various disciplines within these projects, the students focused on the practical functioning of the user-based system. It broadened their view on the real problem as comprising of multiple disciplinary aspects and allows for recognition of interdependence between various disciplines. As an example, within the development of a better tool for cleaners, students were able to place the aims of cleaning tool as a tool which can substantiate actions that involved movement and delivery of cleaning agents and provide safety measures while cleaning in isolation. The uniqueness of developing these aims was that the students were attempting to understand the problem of cleaning from different perspectives, thinking of cleaners (or janitors) as a person who interacts with the social and environmental context in a cultural way. This was based on observation and listening to the cleaners.

In the aquatic life project, its aim was to bring a system which could monitor and communicate for responsibility. In reflecting upon this aim, reflective discussions were undertaken in relation to what would be educative to the owner. The reflections at that point from the study of literature on water ecology revealed the relationships of sustaining of water conditions for living which in relation to brought understanding on environmental parameters such as pH and turbidity. Due to the limitation of time, students were given options to raise understanding on a few parameters to be reflected within the system with justification on choosing of parameters in relation their unique aim. Students could bring understanding of plant life or aquatic animal life and introduce other uniqueness which introduces different dimension for judgement of a practical system. Thus, the problem of sustainability brought understanding on the importance of water filtering for water quality, its relationship to oxygen demand and relevance of these parameters to quality of fish life, as well as the challenge of communicating unique information to the owner. The project also involved deployment of a water filtering mechanism which is aligned and calibrated with the requirement of suitable water conditions for fish rearing that is automated by an Arduino system which informs the owner of the changing environmental conditions.

6.1 Key Findings

The opportunities to implement engineering design for sustainable development projects has been both challenging and enlightening in relation to its outcomes. It has been a powerful context in relaying complexity as core to understanding challenges in society when we are embarking on solutions that are deemed as answering the questions raised. As a practice, it has brought new approaches philosophically which challenges our previous stand in addressing challenges although it has brought us closer to society in our social positions to help. It has also helped us to extend our pedagogical portfolio to a more learner-oriented approach as we reflect upon the issues that arise from the changing practice.

7 Conclusion

The current chapter initially aimed to narrate how existing engineering curriculum from around the globe, and in particular, curriculum infused with design engineering enables their graduates in preparing themselves to address complex engineering challenges. An attempt was made to define complex engineering challenges as per the guidelines provided by the IEA as well as existing challenges that plague society and how such challenges are complex due to the nature of the problem that is to be addressed requires a transdisciplinary approach, i.e. addressed by engineers that have the breadth and depth of engineering content coupled with fundamental design knowledge, engaging with stakeholders, to finally design and build engineering systems through engineering design courses that utilize project-based learning. A literature review was also provided to describe practices globally with respect to engineering design and project-based learning and how such practices are beneficial towards the overall enhancement of engineering education. An attempt was then made to baseline the definition of design engineering and how engineering design courses as well as key pedagogies and initiatives, such as project-based learning and CDIOTM may be used to develop design engineering programmes. Finally, two case studies were provided, one focusing on the implementation of project-based learning coupled with CDIOTM as the framework within an engineering design course and how such a course assisted in enhancing student learning while the other focused on applying the transdisciplinary approach and coupling engineering design with sustainability.

It is surmised that there is a wealth of complex socio-economical challenges that currently plague humankind and those who address these challenges to ensure the sustainability of the planet and its inhabitants would be required to have a wealth of knowledge spanning several disciplines and with the onset of I4, a technology-based degree would be the most beneficial to ensure the correct utilization of I4 technology drivers to address societies challenges. As such, design engineering programmes present the most suitable option to address these issues, noting that design engineers would possess a wealth of engineering knowledge and skills, coupled with fundamental design knowledge to address these complex challenges. The future of engineering is that of a broad-based engineer, that is able to permeate between engineering and non-engineering disciplines to design and build engineering systems that address complex engineering challenges.

References

Al-Atabi M (2014) Think like an engineer, 1st edn. Createspace Independent Publishing Platform, Selangor, Malaysia

Aris SRS, Isa WARWM, Yahaya WAW, Mohamad SNA (2017) Multidisciplinary curriculum design approaches towards balanced and holistic graduates. In: Proceedings of IEEE 9th international conference on engineering education (ICEED), Kanazawa, Japan, 9–10 Nov 2017

Bilén SG, Devon RF, Okudan GE (2002) Core curriculum and methods in teaching global product development. In: International conference on engineering education, Manchester, UK, 18–21 Aug 2002

Chandrasekaran S, Long JM, Joordens MA (2015) Evaluation of student learning outcomes in fourth year engineering mechatronics through design-based learning curriculum. In: 2015 IEEE proceedings of frontiers in education conference (FIE), El Paso, Texas, USA, 21–24 Oct 2015

Chassidim H, Almog D, Mark S (2018) Fostering soft skills in project-oriented learning within an agile atmosphere. Eur J Eng Educ 43(4):638–650

Christie M, de Graaff E (2017) The philosophical and pedagogical underpinnings of active learning in engineering education. Eur J Eng Educ 42(1):5–16

Crawley EF, Malmqvist J, Lucas WA, Brodeur DR (2011) The CDIO syllabus v2.0. An updated statement of goals for engineering education. In: Proceedings of the 7th international CDIO conference, Copenhagen, Denmark, 20–23 June 2011. http://rocketship.cdio.org/files/document/file/cdio_syllabus_v2.pdf. Accessed 3 June 2018

Deaner K, McCreery-Kellert H (2018) Cultivating peace through design thinking: problem solving with past foundation. Child Educ 94(1):26–31

Dirsch-Weigand A, Pinkelman R, Wehner FD, Hampe M, Vogt J (2016) Picking low hanging fruits—integrating multidisciplinary learning in traditional engineering curricula by interdisciplinary project courses. In: Proceedings of the world engineering education forum and the global engineering deans council, Seoul, 6–10 Nov 2016

Elger DF, Beyerlein SW, Budwig RS (2000) Using design, build, and test projects to teach engineering. In: 30th ASEE/IEEE frontiers in education conference, Kansas City, Missouri, United States, 18–21 Oct 2000

Grand Challenges for Engineering (2018). http://www.engineeringchallenges.org/. Accessed 28 June 2018

Hei X, Cheng W (2015) Work in progress: fostering a telecommunication engineering pineline: a curriculum design. In: 2015 IEEE international conference on teaching, assessment, and learning for engineering (TALE), Zhuhai, China, 10–12 Dec 2015

Industry 4.0—The Nine Technologies Transforming Industrial Production (2018). https://www.bcg.com, https://www.bcg.com/capabilities/operations/embracing-industry-4.0-rediscovering-growth.aspx. Accessed 28 June 2018

International Engineering Alliance (2013) Graduate attributes and professional competencies. http://www.ieagreements.org/assets/Uploads/Documents/Policy/Graduate-Attributes-and-Professional-Competencies.pdf. Accessed 3 June 2018

Lulay K, Dillon H, Doughty TA, Munro DS, Vijlee SZ (2015) Implementation of a design spine for a mechanical engineering curriculum. In: Proceedings of 122nd ASEE annual conference & exposition, Seattle, Washington, United States, 14–17 June 2015

Olsen SI, Fantke P, Laurent A, Birkved M, Bey N, Hauschild MZ (2018) Sustainability and LCA in engineering education—a course curriculum. Procedia CIRP 69, pp 627–632. In: 25th CIRP life cycle engineering (LCE) conference, 30 April–2 May 2018, Copenhagen, Denmark

Pech R, Lin B, Cho CS, Al-Muhairi H (2016) Innovation, design and entrepreneurship for engineering students. Development and integration of innovation and entrepreneurship curriculum in an engineering degree. In: Proceedings of 2016 IEEE global engineering education conference (EDUCON), Abu Dhabi, UAE, 10–13 Apr 2016

Rao JS (2018) Creativity in design–Science to engineering model. Mech Mach Theory 125:52–79

Ruayruay E, Kirtikara K, Nopharatana M, Chomsuwan K (2016) Three essential elements of curriculum design for developing competent food engineering graduate students (for problem-solving in food industry) via work-integrated learning mechanism. In: Proceedings of 2016 IEEE international conference on teaching, assessment, and learning for engineering (TALE), Bangkok, Thailand, 7–9 Dec 2016

Sehgal U, Gokhale S (2015) Designing curriculum to optimize the paradigms in engineering education in India. In: 2015 IEEE 3rd international conference on MOOCs, innovation and technology in education (MITE), Amritsar, Punjab, 1–2 Oct 2015

Soare E (2016) Situation-centered approaches in postmodern curriculum design. In: Proceedings of ECAI 2016 international conference—electronics, computers and artificial intelligence, Ploiesti, România, 30 June–2 July 2016

United Nations Sustainable Development Goals (2018). https://www.un.org/sustainabledevelopment/sustainable-development-goals/. Accessed 28 June 2018

Violante MG, Vezzetti E (2017) Guidelines to design engineering education in the twenty-first century for supporting innovative product development. Eur J Eng Educ 42(6):1344–1364

Viswanathan V, Atilola O, Goodman J, Linsey J (2014) Prototyping: a key skill for innovation and life-long learning. In: 2014 IEEE frontiers in education conference (FIE), Madrid, Spain, 22–25 Oct 2014

Wicklein R, Smith Jr PC, Kim SJ (2009) Essential concepts of engineering design curriculum in secondary technology education. J Technol Educ 20(2):65–80

Industrial Design Engineering: Teaching Industrial Design as a Specialisation Within a Mechanical Engineering Curriculum

Christian Wölfel and Jens Krzywinski

Abstract This chapter provides insight into the structure of a rather unique industrial design specialisation within the curriculum framework of a German mechanical engineering faculty. This specific setting involves a combination of educational approaches from humanities, art schools and engineering sciences. In addition to that, the chapter illustrates how the curriculum has been adopted to the shift from the design of traditional industrial machines and vehicles to cyber-physical systems. The challenges which future design professionals will face in their everyday work must be addressed in today's design education already.

Keywords Industrial design engineering · Human-centred design · User experience design · Cyber-physical production systems

1 Introduction

Since the disciplines of design and engineering split up following industrialisation, there have always been debates and efforts on integrating design and technology (or designers and technologists) in practice and education. One famous early example is the historical *Bauhaus Hochschule für Gestaltung* (University of Design), which had been heavily promoted and funded by the (back then) Dessau-based Junkers tech company (Bauhaus-Archiv and Droste 2002): Walter Gropius coined the phrase '*Kunst und Technik – eine neue Einheit*' (art and technology, a new unity) promoting the integration of art, craft, technology and industry. Design research and sciences have also shown up links between design and engineering (cf. Cross 1984; Bayazit 2004; Lloyd 2017) that are still relevant—before postmodernism brought the disciplines further apart, at least in academia. While from a practical perspective this separation seems odd, two reasons hindered and delayed a reunification in Ger-

C. Wölfel (✉) · J. Krzywinski
Chair of Industrial Design Engineering, Technische Universität Dresden, Dresden, Germany
e-mail: christian.woelfel@tu-dresden.de

J. Krzywinski
e-mail: jens.krzywinski@tu-dresden.de

© Springer Nature Switzerland AG 2019
D. Schaefer et al. (eds.), *Design Education Today*,
https://doi.org/10.1007/978-3-030-17134-6_12

man academia. One might be that many of the currently teaching design professors have been educated at art schools during postmodernism, where anti-modernism and anti-engineerism have been propagated (In this period, most German art universities banned natural and engineering sciences and corresponding theories. On the other side, industrial design departments disappeared from most of the engineering schools (cf. Seeger 2005). The second reason is the widely remaining separation of the German academic institutions, keeping design at art schools separate from engineering education at 'scientific', research-oriented universities.

There have been design departments at research-oriented universities and schools of engineering in Germany since the late 1950s. Unlike many others, that did not survive postmodernism, the department of Industrial Design (Engineering) at TU Dresden has been engaged in design education for engineers since 1958. Back then, there has been a broad understanding of the necessity of the integration of design and engineering in the educational sector, which resulted in reciprocal efforts in bringing those two disciplines closer together. The emphasis of industrial design education at the first East-German Design school in Berlin had its focus on technical products. Accordingly, the educational objective was that 'the industrial designer must be talented in arts and technology [...] must know as much about sculpture, architecture, graphics as well as colour. In addition to that he should be half an engineer' (translated from Högner, cited in Wölfel 2014). While this focus on integrating the disciplines disappeared at other Universities, there is a lasting commitment in offering courses on industrial design to engineering students at TU Dresden. Since the late 1980s, there have been efforts made to establish a curriculum that allows students to graduate as engineers of human-centred (product or work-environmental) design.

Based on these efforts, a specific industrial design engineering curriculum was established in the early 1990s, which has since been developed further, accommodating institutional or legal changes, but also considering new trends in the design profession, industry and technology. The main focus of the (user-centred) design education programme developed parallel to general advancements of the industrial design discipline but in accurate balancing within the mechanical engineering faculty and its predecessors. After a long period of following traditional approaches with a focus on formal and aesthetic attributes of machines and vehicles, design education and research are now also considering social and psychological dimensions of human-technology interaction. After the first revisions of the curriculum had been introduced facing the Bologna process and institutional changes of the faculty, the current revision has its focus on disciplinary and technological advancements as well as feasibility in terms of workload and teaching methods. The most notable changes relate to trends in the industry such as The Internet of Things, Cyber-Physical (Production) Systems or Product-Service-Systems but also incorporate and facilitate the establishment of specific design approaches like User Experience Design or Designing Human-Technology Cooperation. Summed up, the curriculum now aims at educating engineers in designing the human experiencing of machines, products and services in professional domains.

This chapter presents the formal and educational framework as well as the approach, structure and characteristics of the most recent iteration of the industrial design specialisation within an engineering design programme at a German university.

2 Higher Education in Germany and Organisational Boundaries

Germany has an academic system that allows for different types of higher education (HE) institutions: traditional, research-oriented 'scientific' universities, universities of art and polytechnics (universities of applied sciences). Due to the historic and transdisciplinary nature of different fields of design, diverging design education programmes can be found on all three types of HE institutions. To make things even more complicated, higher education is ruled by the federal states rather than on national level, which results in differing laws and practices, making the educational programmes even more diverse. Concerning engineering educational programmes, there is a broad agreement on how to educate mechanical engineers and engineering designers at (scientific) research-oriented universities and at polytechnics. Among the nine large (named technical and categorised as research-oriented) universities, the so-called TU-9, there is a deeper exchange and agreement on curricula and levels of achievement. This kind of uniformed body of knowledge, methods and literature leads to comparable educational programmes in engineering design across all HE institutions. In contrast, industrial design education and product design education are mostly situated at universities of art and at polytechnics, where different traditions, self-conceptions and didactic approaches lead to very diverse design education programmes. Those are often hard to compare to each other, let alone to compare to engineering design programmes.

While there are still legal differences, based on the development of design and design education as well as the Bologna process, the traditional dividing line between research-oriented universities and 'applied' institutions (polytechnics) in Germany is no longer as clear (cf. Petzina 2005). Schade (2007) notes that the polytechnics have helped blur this line by calling themselves Universities of Applied Sciences. On top of that, Lub et al. (2003) claim that since Bachelor and Master degree titles granted by universities and *Fachhochschulen* (polytechnics) are no longer officially distinguished, this has 'made the new degrees an important means in the *Fachhochschulen*'s struggle for equal recognition with the universities' (p. 256). Another important stressor for Universities was a demand to be more industry relevant, which lead to tension and competition between schools and more pressure on students and faculty members (e.g. Stallmann 2002).

3 Approaches of Integrating Design Disciplines in Engineering Education

In 1987, the Department of Mechanical Engineering at the University of Glasgow and the School of Design at Glasgow School of Art started a joint educational programme in order to educate 'young engineers whose main strength will lie in their capacity for creative synthesis and whose primary task will be the design of engineering and consumer products' (Green and Kennedy 2001). Following the tradition and success of the educational and research activities of the *Faculteit Industrieel Ontwerpen* (Faculty of Industrial Design [Engineering]) at TU Delft (Stappers et al. 2007; Roozenburg et al. 2008), comparable integrated industrial design engineering programmes have been established at Uni Twente and TU Eindhoven as well (Eger et al. 2004; Bruns Alonso and Smits 2013). Today, there are several educational programmes in Industrial Design Engineering or Product Design Engineering across the World, many of them in the UK, but also in Scandinavia, Asia and other countries. A comprehensive but not complete overview has been given by De Vere et al. (2010).

Most of the described curricula integrate design and engineering from the first year, often allowing for some priority-setting in the last years of study. While educational approaches like at Uni Twente integrate the disciplines within single courses (Eger et al. 2004), most of the few integrated Industrial Design Engineering curricula in Germany combine courses from different departments to integrated curricula, which in fact limits the level of integration to the mere combination of separate mono-disciplinary lectures picked from different domains.

The broad adoption to the Bachelor–Master system promised individual flexibility in terms of place and subjects of study. While this is the case for design education on the third level (i.e. Ph.D. in design, cf. Melles and Wölfel 2014) the German system is still quite rigid when it comes to alternating subjects between undergraduate and graduate levels. There are only few places where students, holding a Bachelor in Engineering, are allowed to register for a Master's course in design or the other way round. If students are interested in integrating both design fields in their studies, the more auspicious approach is to choose a curriculum that is already tailored to their needs. However, for a long period of time, there were only few possibilities. This was the case until recently, when some German polytechnics have started to offer undergraduate courses (Bachelor of Engineering) in industrial design engineering (Hochschule Aalen: cf. Gärtner et al. 2017, Technische Hochschule Deggendorf, Hochschule Emden·Leer).

In contrast to the curricula described above, the Industrial Design (Engineering) curriculum at TU Dresden is placed as a specialisation in the advanced course of studies within a modularised diploma course in mechanical engineering. Here, Industrial Design shares its special arrangement within the programme of the Faculty of Mechanical Engineering with other disciplines, such as ergonomics. The students have already passed undergraduate courses in engineering when they start specialising in (user-centred) industrial design in graduate courses (cf. Fig. 1, see also Sect. 6 of this chapter). This allows the students to choose their specialisation

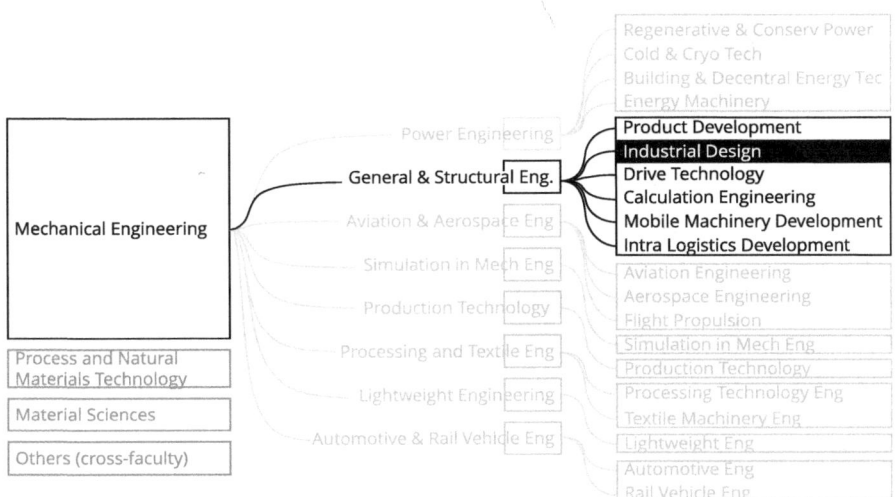

Fig. 1 Specialisation path during the curriculum from degree programme mechanical engineering via field of study general and structural engineering to individual specialisation industrial design, still flexible; widths of the boxes represent duration, heights approximate numbers of students

after having completed introductory courses to all of the possible study specialisation tracks. In addition to that, it is possible for traditional mechanical engineering graduates from other universities to complete their graduate studies in industrial design without having trouble in making up the leeway.

4 Demand for Advancing the Educational Contents

For some decades, the focus of design research, practice and education at TU Dresden was on technical objects with an emphasis on industrial machinery and industrial vehicles. Since these branches went through notable changes in recent years, the design of products and services now has to pay regard to this. A comprehensive example is the interface design in industrial machines. While some time ago, a typical production machine only had physical or analogue interfaces, no matter if it was operating in a stand-alone setting or in a production chain, today, production systems are much more complex. Hence, the design of human–machine interaction with now cyber-physical production systems can no longer be derived only from technical functions and usability guidelines (Lee 2001; Spath et al. 2013; Hirsch-Kreinsen 2014; Römer and Bruder 2015). Similar trends in terms of virtualisation and automation take place in the industrial vehicle branches, where changes in logistics or agriculture have an impact on the nature and design, not only of operator's cabs, but on whole production systems. Figure 2 shows one example of a mobile cyber-physical production system. A swarm of small harvesters can be operated by only one

Fig. 2 An illustration of a mobile cyber-physical production system: a harvester swarm, designed by student Hannes Wittig, illustrated by Sebastian Lorenz

farmer but will increase flexibility and hence increase crop yield while reducing for example soil compaction. It is obvious that the human-technology interface must be designed different compared to the state-of-the-art of conventional, comparatively large harvesters, each operated individually from an attached cabin.

Today, Design education, research and practice at TU Dresden follows a holistic approach on designing the human experience of products and systems in professional domains. This builds upon engineering sciences, cognitive and social sciences as well as on art and design skills and expertise. The latest revision of the curriculum takes into account the fundamental changes in professional domains, e.g. in production settings: complex cyber-physical production systems are becoming the standard for innovative solutions.

In 2002, the faculty of Industrial Design Engineering at TU Delft started the Masters programme Design for Interaction 'focusing on user-product interaction as a basis for designing new products and services' (Stappers et al. 2007). As Stappers and colleagues describe the field, Product-Service Design is a 'research-intensive area of modern product design' (ibid.). Considering the complexity of underlying design problems as well as the strong link to cognitive sciences and technological developments, education in Product-Service Design can benefit from the described embedding of the industrial design education at the TU Dresden.

Within the user-centred design disciplines, different approaches on designing for experience have been developed in the recent years (Jordan 2003; Desmet 2003; Hassenzahl 2004; Schifferstein and Hekkert 2011 and others) and are now state-of-the-art in industrial design as well. The focus on affects and needs and on why (and not only how) people are interacting with technology, needs specific approaches and methods. However, it also ensures a level of user-centredness that would be hard to achieve by designing cyber-physical systems with traditional product-oriented approaches of industrial design. Accordingly, approaches of user experience design are an essential part of a curriculum focused on human-technology interaction with cyber-physical systems.

5   Pedagogic Approach and Aim

Taking into account the students' prior studies in mechanical engineering as well as the framework of an research-oriented university, the courses of industrial design at TU Dresden combine strong methodology and rigour with sensibility for syntactic and semantic qualities of products and product-service-systems. The pedagogic approach of the industrial design curriculum is based on constructivist teaching strategies. Accordingly, the focus lies on enabling students to build their own conceptual structures when interacting with an environment. This is realised through task-orientated, and self-directed activities, within a design- and discovery-based approach.

In practice, the art school tradition focuses on enabling and consulting students on their individual pursuit for insight rather than educating them in standardised lectures and seminars. In the industrial design engineering curriculum at TU Dresden, there is a strong theoretical foundation, implemented in lectures and seminars that is combined with well-consulted practical workshops and projects.

While the faculty is part of an academic (scientific) institution, there is a strong connection to the industry and a strive to be application-oriented. This is the case for research, Ph.D. studies as well as for undergraduate and graduate studies. Moreover, there is no formal distinction between research and teaching staff, which allows for a combination of duties and hence a strong connection between research and education. In addition to that, the economic structure of the region and several funding opportunities allow for extensive applied research and development activities for the industry. This in return pays back to the education as well.

The final two years of the curriculum are dominated by project-based learning in several complex design projects for real partners. The students are not responsible for delivering ready-for-production design solutions, but they are engaged in real processes with real stakeholders, where they get authentic briefs and relevant feedback. This way they develop necessary soft skills on interacting with partners, managing, communicating and 'selling' their projects and solutions on-the-fly. In contrast to this, the formal parts of the educational programme focus on engineering and design fundamentals, processes, methods and expertise in the field.

Following Pouw and van der Lelie (2013), the objective of the educational programme can be described by framing a typical graduate's profile as the following.

The graduates are multidisciplinary product-service developers and innovators, who focus on human experiencing of multiple stakeholders but also consider functionality, aesthetics, production, market and sustainability and act accordingly.

Within the course of their studies at TU Dresden, the students are enabled to work in a systematic manner, individually and in interdisciplinary teams. They can deal with problematic situations and uncertainty and choose and adopt knowledge, approaches, methods and findings from other fields to human-centred product development. Hence they can draw on broad and deep scientific knowledge. They are able to analyse and solve complex problems in a systematic and creative way and evaluate their findings and solutions e.g. through compiling and analysing prototypes. They can reflect their own activities and justify and explain processes and methods as well as their decisions and design solutions.

With additional work experience, they are able to work in leading and (product and innovation) management roles, in companies and academia.

6 Structure and Characteristics of the Industrial Design Specialisation

Curriculum Structure within the Faculty's Programme

The *Diplomingenieur* curriculum at TU Dresden is equivalent to a Bachelor's and Master's course. While the undergraduate education covers a broad scope of mechanical engineering subjects, the graduate (equivalent to Master's level) education offers a specialisation in different fields (cf. Fig. 1). The specialisation track of Industrial Design Engineering is considered unique in German academia, as it combines two disciplines which are often deemed to be disconnected in their nature. There is an even proportion between mechanical engineering and industrial design courses within the curriculum. Figure 3 illustrates the typical curriculum of students, specialising in Industrial Design Engineering.

In the first two years, broad general studies in mechanical engineering cover fundamentals which are relevant for all mechanical engineering students like mathematics and natural sciences (like physics or chemistry), engineering mechanics, materials sciences, production technology or engineering design and machine elements. Many of the courses are also offered to students of materials sciences or process and natural materials engineering, which results in about 1,000 participants in most of the lectures (they are split into smaller groups for seminars and workshops). During those first two years, the department of Industrial Design Engineering offers elective courses on freehand sketching to introduce the Industrial Design specialisation track and support those students that are dedicated to follow this course of studies. The competences developed within these elective courses are highly relevant for students who want to specialise in industrial design.

In the third year, the Industrial Design students specialise in two steps. First, they opt for a field of study (General and Structural Engineering), which starts with mandatory courses on drive technologies and engineering design processes (participant numbers are still above 200). As a second step, the students can choose from a number of elective courses in order to further refine their course of studies. There are

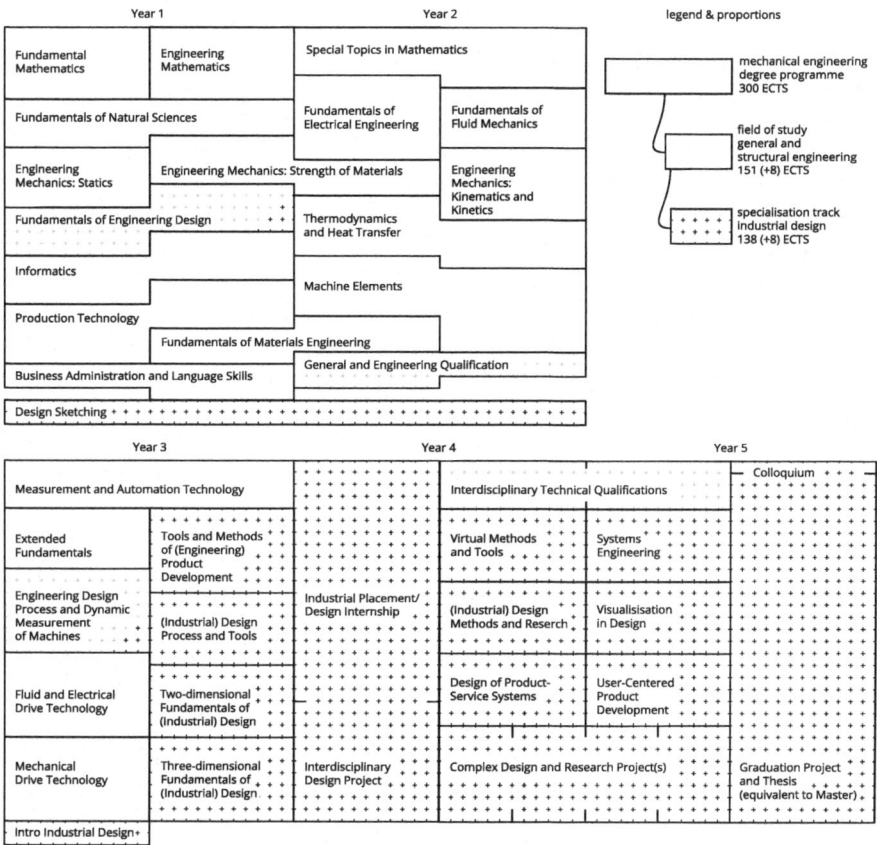

Fig. 3 Scheme of the industrial design engineering curriculum at TU Dresden

six official specialisation tracks available, one of which is Industrial Design (Fig. 1). Usually, 10–15 students take the chance to specialise in industrial design each year.

Characteristics of the Industrial Design Specialisation Course

In contrast to former curricula (cf. Kranke 2008), the focus of the industrial design course is not merely on physical dimensions and human perceptions of products but now also includes an in-depth experience design approach, that builds upon a systematic understanding of users, stakeholders and relevant context scenarios. Besides technical, economical and ergonomic requirements, the fulfilment of user- and domain-specific human needs is the main aim of the design process. This involves the design of not only the physical product or artefact but also holistic systems and services and often also interactive human-machine interfaces.

Unlike other specialisation courses within the curriculum, the design track builds on studio-based learning and assessments that validate design outcomes and project documentations rather than written exams. It is known that project-based (engineering or design) education succeeds in focusing more explicitly on the development of competencies and therefore stimulates engagement, self-activity, self-confidence and self-knowledge (Eger et al. 2004; Dorst and Reymen 2004). The industrial design specialisation track at TU Dresden combines the traditional design education formats from art schools with academic (scientific) rigour and methods and therefore builds a bridge between two different academic traditions. At the same time, it allows for an extraordinarily strong connection between practice-oriented design education, academic research and industrial practice.

Since the students specialise rather late within the curriculum, they are already at graduate level when they learn fundamentals of (industrial) design (cf. Wölfel and Uhlmann 2008; Wölfel and Thoring 2014). They are also used to the applied educational approaches as well as the acquirements of the mechanical engineering subjects. They tend to address definite models, processes and methods with objective criteria and are able to develop individual approaches to cope with problematic issues. In recent years it became even more challenging for the students to engage with the switch to fuzzy front-ends and uncertainty of iterative human-centred design processes, as they had to deal with more and more fully packed educational programmes (K-12 and Higher Education). In order to cope with the workload, they had to react with efficiency-driven learning or executing. Accordingly, the design fundamentals courses build upon previous experiences and include Design Drawing (based on earlier sketching courses), Graphical (2D) Design Fundamentals, Colour and Materials (Colour and Trim) Fundamentals as well as sculptural (3D) Design Fundamentals. Figure 4 shows an excerpt from the iterative development of basic two-dimensional aesthetic shapes based on certain geometric restrictions and a constant operation-assessment cycle. The applied method is used as a basis for following design tasks and should later be intuitively applied to applied design projects. Figure 5 shows the photograph of a given three-dimensional aesthetic freeform shape, the analysis sketch as well as photographs of the student's interpretation of the given shape. During this task, the students learn about aesthetic and objective quality criteria of freeform geometry without the need or the chance to alter the syntax or semantics of the object. Figure 6 shows one example of the most complex design fundamentals task, where the students design a pure object from scratch, that does justice to aesthetic and objective quality criteria while also communicating given specific semantic characteristic (e.g. *bold and feminine*). Figure 7 shows a set of objects that share similar appearance and aesthetics on syntactic and semantic levels but differ in pragmatic terms: the task was to communicate the possibility to bend, twist and pull the objects.

The design fundamentals courses are strongly interwoven in terms of theoretical (research-oriented) lectures and practical (rather art-school-like) assignments. The design fundamentals courses aim at the development of fundamental design skills and sensitivity to elementary design qualities. Besides a strong emphasis on analyses of design semiotics, the students learn to apply systematic and iterative design actions, as they present and defend their design work on the basis of singled out fundamental design tasks in learning assignments.

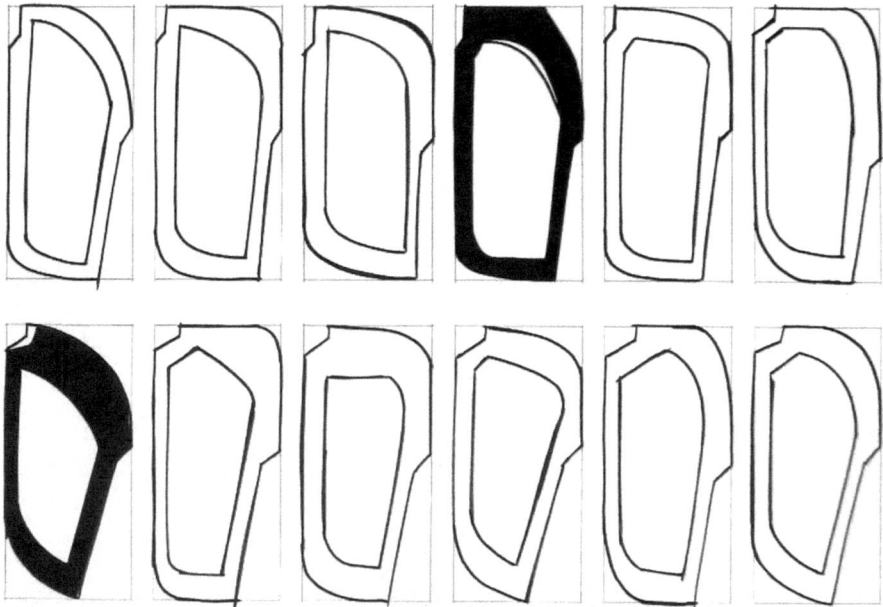

Fig. 4 Excerpt from a workshop in the 2D design fundamentals courses on iteratively developing aesthetic shapes according to given geometric restrictions

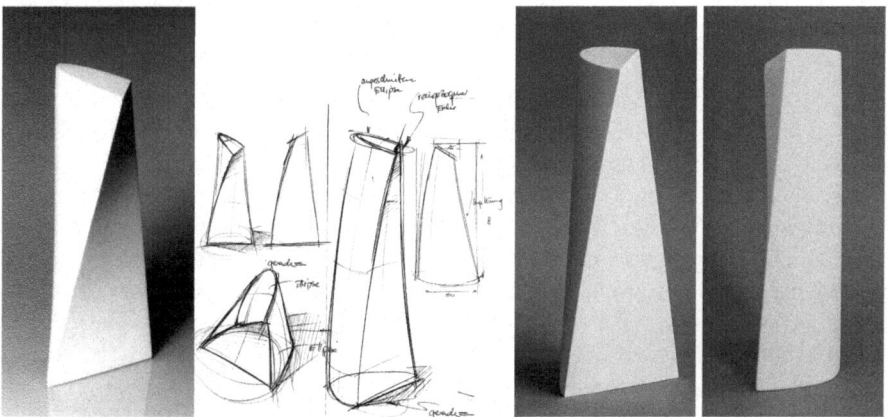

Fig. 5 Analysis and synthesis of an aesthetic freeform artefact manually made in plaster

Fig. 6 Communicating
through three-dimensional
geometry: semantic object
manually made in plaster

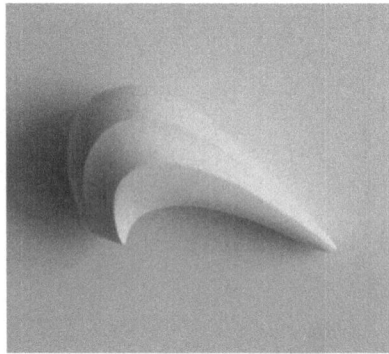

Fig. 7 Communicating
through three-dimensional
geometry: pragmatic objects
manually made in
polyurethane foam

Parallel to the design fundamentals, the students participate in two curriculum modules on two disciplinary views on product development. In the engineering design module they learn about approaches and tools such as Ansys and SolidWorks, FEM, reverse engineering, simulation in EON and critical analysis of machines.

In the industrial design module, they learn about objectives, processes, methods and the value of human-centred design with a focus on qualitative, narrative and experiential aspects of products and product development. Figure 8 shows students designing hands-on with paper and Lego bricks in the maker space. Their task was to design a lightweight excavator and build a prototype of the design proposal. Figure 9 shows the outcome of one student group made from Lego bricks and bespoke 3d-printed parts.

Both modules offer introductions to more specialised fields of design. Accordingly, they are attended by students from different courses of study, which, in return, allows learning in interdisciplinary groups. Both modules consist of lectures and seminars as well as written and practical exams.

The fourth and fifth year of the curriculum consist of more specific and in-depth courses and larger research and design projects. Here, the order of the courses is more flexible, which allows the students to arrange their own study-programme, incorporating elective courses, industrial design internships and international student

Fig. 8 Paper-prototyping workshop in the maker space

Fig. 9 Outcome of a collaborative student project on lightweight excavator arms

exchanges. Alongside those elements, the suggested industrial design specialisation covers some associated engineering courses like systems engineering, mechatronics and reverse engineering. The main focus however remains on practical design projects with accompanying workshops and lectures on user-centred design, product-service design, human-machine interfaces, information visualisation, product visualisation (Fig. 10), aesthetic freeform CAD modelling (Fig. 11), design research and product experience.

In sum, there are four large full-semester industrial design projects and four smaller semi-semester projects in the final three years, including the graduation

Fig. 10 Product visualisation (Tobias Zerger)

Fig. 11 CAD freeform modelling (Martin Guthof) and analysis (Frank Mühlbauer)

project and thesis. All design projects are supervised by one or more academic staff members in frequent consultations. The students work in student design studios in the faculty buildings, and have access to a student workshop and maker space. All projects are documented according to academic standards. In the course of their final years, the students build up and shape their individual design portfolio, which is the general means used to apply for placements or professional job opportunities.

Practice-Orientation and Collaboration

The design fundamentals assignments are rather abstract and intentionally further away from potential application to industrial products, which allows focusing on specific qualities of product dimensions such as for example the impact of product graphics on the recognised proportions of three-dimensional objects. Nevertheless, all semester projects are connected to industrial contractors or to university research projects to create a professional work environment and foster motivation among the students. The first project is usually schemed as a group work, where the students work individually, but completing similar sub-tasks of one given topic (e.g. gas measurement, prosthetic feet, navigation systems). To guide and support the students during these project phases, university lecturers usually team up with guest lecturers like developers, designers or product managers of a manufacturer (e.g. Dräger, Ottobock) and one or two senior designers from a design agency. Figure 12 shows student Jonas presenting his design of a prosthetic foot at the Ottobock headquarter. He developed this design within the first full-semester user-centred design project in the fourth year. The collaboration with industrial partners reinforces the importance of such academic projects and introduces the students to industry relevant stakeholders. The professional feedback from real contractors and the potential use of their solutions (at least in patent applications) is highly motivating for the students and helps in keeping up the pace during project phases (Fig. 13).

The subsequent design projects are realised by the students individually or collaboratively in small teams. All project tasks or design briefs are derived from current research projects (not only from the design department, but also from other departments of the School of Engineering, the Fraunhofer Research Institutes [of applied sciences] or start-up companies) or from industrial companies. The larger semester projects can be accomplished at sites of the partner-companies, especially if there are larger design departments (e.g. in the automotive industry) or on campus, under the usage of technology or evaluation scenarios (e.g. industrial vehicles industry). Typically, the projects are in line with the research profile of the industrial design chair: human-centred design of technically complex product-service systems, usually for multi-stakeholder contexts. Accordingly, the students engage in the design of professional power tools and large industrial machinery, medical devices, industrial vehicles like harvesters, tractors or trucks, sports gear or technology demonstrators for cutting-edge technology from the university's network. Figure 13 shows a design for the 5G mobile communication standard research cluster. This cyber-physical system has been the subject matter of an interdisciplinary group project and a subsequent individual student project. The Lyne Suit is an exo-suit system for active training support of rowers and their trainers. It is a tangible, prototypical soft exo-suit design,

Fig. 12 A student presents his design for a prosthetic foot at the headquarter of manufacturer Ottobock

Fig. 13 A (cyber-physical) soft exoskeleton Lyne Suit (Fabian Neumüller) for professional sports with sensors and actuators

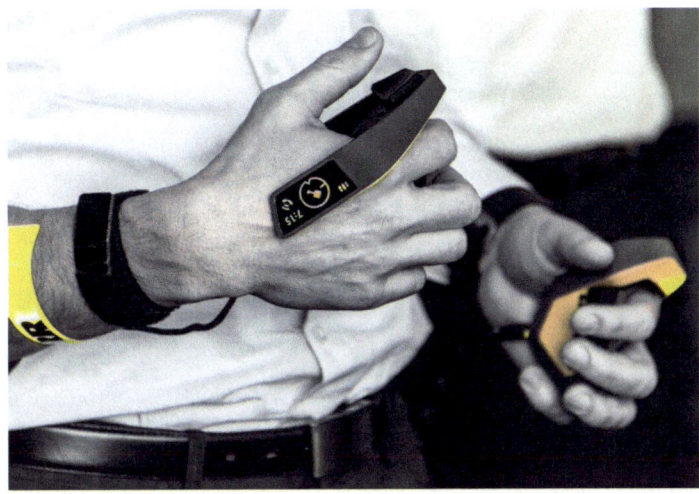

Fig. 14 Human-technology interaction with mobile cyber-physical systems: excavator controller designed by Emese Papp, Helge Wanta and Clemens Beyerlein for Liebherr

which allows a precise analysis of its possible integration into the daily training routine. Figure 14 shows the design of innovative gesture-based remote control devices for Liebherr hydraulic excavators to be used in difficult or dangerous construction sites. A student group developed this project to working prototype status for the industrial company.

Figure 15 shows another cyber-physical technology demonstrator. The VRieda VR/AR glasses demonstrate the exemplary use of OLED micro displays developed the Fraunhofer FEP research institute. Figure 16 shows a student project that has been developed in conjunction to basic research of the industrial design engineering department. The exoskeleton is used as a boundary object for further research projects on cyber-physical production systems and other purposes.

As the learning and semi-professional experiences of the students increase project by project, the complexity of the projects accelerates as well. While the first projects focus on selected elements of industrial design processes (user-centredness, product experience, product-service constructs, prototyping, etc.), later projects have a much wider scope, incorporating integrated design and research (e.g. during design analysis or for evaluation purposes) as well as the design of cyber-physical production systems.

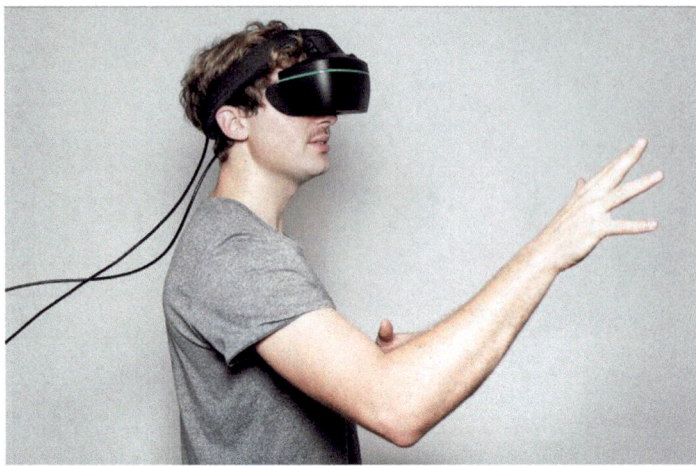

Fig. 15 (Cyber-physical) Technology demonstrator: VRieda VR/AR glasses using OLED micro displays designed by Philip Hart and Christian Hermeling for Fraunhofer FEP research institute

Fig. 16 Designing for professional contexts: Albrecht Meixner designed an exoskeleton for aircraft assembly and maintenance

Despite the broad range of design projects, there is always a strong focus on professional contexts and complex products—in all activities of the industrial design department at TU Dresden. However, this specific scope still allows for in-depth engagement with topics such as designing for sustainability or design management. In order to ensure profound supervision, the larger study design projects are often co-supervised by academics or professionals from other departments or organisations.

7 Perception and Graduates' Perspectives

In coherence with the objectives of the educational programme (see Sect. 5 of this chapter), graduates know engineering approaches, principles, materials and technologies, processes and methods in detail. They can validate what is possible within the guidelines of a respective project and how much engagement is necessary in different specific product development domains (e.g. including for FEM or drive engineering). Attributable to their strongly application-oriented education they are competent in communicating and negotiating with professionals from the above mentioned domains on equal terms.

The graduates are experts in the design of physical products as well as product-service-systems with complex technological accuracy for serial production. They can adapt and develop innovative processes and methods of human-centred design that are targeted at solving newly evolved problems and whole problem areas.

In addition to that, the graduates can design human-technology cooperation beyond the physical dimensions of products. They follow the most recent trends in interaction design and cyber-physical systems, are able to suggest holistic solutions for product-service systems and cooperate efficiently with HCI professionals and interaction designers.

Considering the students' perspectives after they finalised their studies, there is a bond with alumni who report back on how they cope in the industry based on their education and the feedback has been solely positive. In addition to that, there is frequent feedback from partners who highly praise the combination of human-centred design and engineering competences. The popularity of this combination of disciplines is also reflected by the fact that there are often more job and internship offers than students or graduates on the lookout for a placement. What's more, the quality of the educational approach in industrial design has been underlined by the awarding of several design prizes to students and graduates, e.g. the Saxonian State Design Award, the IF newcomer Award or the Mia Seeger Award, which is dedicated to common good, recognising not only the design quality but also the responsibility of designer and design.

Standardised large lectures and the mindset of engineering make up the most of the first years of the curriculum. The students must then cope with a switch to the different culture and mindset of industrial design even within a school of engineering. The curriculum pays much attention to this by providing a systematic approach to human-centred design and intense supervision in small groups or face-to-face. The vast majority of students manages this shift and finally enjoys the challenges of industrial design projects. Despite individual admission consultations and preparatory classes, a small number of students finally decide to make use of the opportunity to flexibly combine or even switch to another—then more engineering-based—specialisation track (cf. Fig. 1) within the educational programme of the faculty.

Within the current curriculum, a big issue for the students is do-ability in terms of coping with great workload. This does not only affect the industrial design specialisation but the whole faculty in general. On top of that, students often need or want to work besides their studies (for example in the students' industrial design office 'August') which only adds to their workload. While this is certainly worthwhile as it helps them gain additional work experience, it also extends the duration of study.

In terms of internationalisation, besides the international student exchanges and internships, the department offers international summer schools with graduate students from different disciplines, in changing collaborations with cultural or academic institutions such as the Dresden SKD art collections.

Concerning the students' perspectives after finishing their studies, an evaluation revealed that more than 90% start their professional career right after defending their graduation thesis. Many of them stay in this first professional position for many years, some change jobs more often depending on opportunities for career development. A minor part of the graduates starts own design businesses, with some of them evolving into companies of five to twenty staff members. While most of the graduates work predominantly as industrial designers, others choose to work as studio engineers or Class-A modellers in companies of different sizes. Such companies may include rather small design agencies, engineering agencies or tech start-up companies up to larger companies like automotive engineering (e.g. Audi), mobile machinery (Claas, Still, etc.), power tool manufacturers (Metabo, Kärcher) or large design or in innovation departments.

Some of the graduates even start their career in academia. In contrast to most German design education institutions, the students graduate from a research-oriented university. Accordingly, commencing third-level HE (Ph.D., cf. Horvath 2008; Melles and Wölfel 2014) is both formally and in practice a valid option.

8 Further Development

Recently, there has been a new revision of the curriculum that is becoming effective as this book is in print. It was developed as a response to the mentioned above difficulties with do-ability and workload. Following its implementation, an evaluation of the courses as well as further feedback from the graduates will stimulate further improvements.

The new curriculum is constituted of fixed parts such as the engineering and design fundamentals and more flexible slots which allows for predictability but also the necessary amount of flexibility. The latter had always been an opportunity to engage with occurring approaches and trends such as user experience design or the design of cyber-physical production systems which are now standard elements of the curriculum (product experience; design of product-service-systems). The current curriculum already offers some of these flexible slots, so despite the protracted processes of changing curricula, there is a certain degree of flexibility that allows to respond quickly to changes in the design disciplines or branches, which can never be precisely foreseen.

With the introduction of this new curriculum, the more flexible parts will be used to integrate innovative approaches, again in order to synergise design and engineering approaches, for example in adopting FEM topology optimisation and aesthetic freeform modelling in students' projects.

As of now, some of the design courses are also part of other (cross-faculty) curricula like human factors psychology, design vocational teacher training or business engineering. Moreover, there are several collaborations with the faculty of architecture or the chairs of media design and communication acoustics within the school of engineering or even with other universities. These educational formats will be developed further and may become an integral part of the next curriculum.

References

Bauhaus-Archiv, Droste M (2002) Bauhaus 1919–1933. Taschen, Cologne

Bayazit N (2004) Investigating design: a review of forty years of design research. Des Issues 20(1):16–29

Bruns Alonso M, Smits R (2013) Industrial design. Self-evaluation report. Technische Universiteit Eindhoven, Eindhoven, Learning for Innovation

Cross N (1984) Developments in design methodology. Wiley, New York

De Vere I, Melles G, Kapoor A (2010) Product design engineering – a global education trend in multidisciplinary training for creative product design. Eur J Eng Educ 35(1):33–43

Desmet PMA (2003) A multilayered model of product emotions. Des J 6(2):4–13

Dorst K, Reymen I (2004) Levels of expertise in design education. In: DS 33: proceedings of E&PDE 2004, the 7th international conference on engineering and product design education, Delft, The Netherlands

Eger AO, Lutters D, van Houten FJ (2004) Create the future: an environment for excellence in teaching future-oriented industrial design engineering. In: DS 33: proceedings of E&PDE 2004, the 7th international conference on engineering and product design education, Delft, The Netherlands

Gärtner F, Pietzsch M, Frye A (2017) Interferences of industrial design and engineering in future design education. In: DS 88: proceedings of the 19th international conference on engineering and product design education (E&PDE17), building community: design education for a sustainable future, Oslo, pp 152–157

Green G, Kennedy P (2001) Redefining engineering education: the reflective practice of product design engineering. Int J Eng Educ 17(1):3–9

Hassenzahl M (2004) The thing and I: understanding the relationship between user and product. In: Blythe M, Overbeeke C, Monk AF Wright PC (eds) Funology: from usability to enjoyment. Kluwer Academics Publishers, pp 31–42

Hirsch-Kreinsen H (2014) Smart production systems: a new type of industrial process innovation. In: DRUID society conference 2014, CBS, Copenhagen

Horvath I (2008) Differences between 'research in design context' and 'design inclusive research' in the domain of industrial design engineering. J Des Res 7(1):61–83

Jordan PW (2003) Designing pleasurable products: an introduction to the new human factors. CRC press, Boca Raton

Kranke G (2008). fully integrating industrial design into engineering education. In: Clarke A et al (eds) DS 46: proceedings of E&PDE 2008, the 10th international conference on engineering and product design education, Barcelona, pp 395–400

Lee JD (2001) Emerging challenges in cognitive ergonomics: managing swarms of self-organizing agent-based automation. Theor Issues Ergon Sci 2(3):238–250

Lloyd P (2017) From design methods to future-focused thinking: 50 years of design research. Des Stud 48:A1–A8

Lub A, van der Wende M, Witte J (2003) Bachelor – master is programmes in the Netherlands and Germany. Tert Educ Manag 9(4):249–266

Melles G, Wölfel C (2014) Postgraduate design education in germany: motivations, understandings and experiences of graduates and enrolled students in master's and doctoral programmes. Des J 17(1):115–135

Petzina D (2005) Der Bologna-Prozess in Deutschland – Stand und Perspektiven. Beiträge zur Hochschulforschung 3(27):18–36

Pouw P, van der Lelie C (2013) Assessment of the IDE faculty's educational programmes. Delft, TU Delft

Römer T, Bruder R (2015) User centred design of a cyber-physical support solution for assembly processes. In: 6th international conference on applied human factors and ergonomics (AHFE 2015), pp 456–463

Roozenburg N, van Breemen E, Mooy S (2008) A competency-directed curriculum for industrial design engineering. In: Clarke A et al (eds) DS 46: proceedings of E&PDE 2008, the 10th international conference on engineering and product design education, Barcelona, pp 423–428

Schade S (2007) Auswirkungen globaler Wertschöpfung auf deutsches Industrie- und Produktdesign unter besonderer Betrachtung der Schnittstelle Design und Konstruktion/Entwicklung' PhD thesis, Universität Duisberg-Essen

Schifferstein HN, Hekkert P (eds) (2011) Product experience. Elsevier

Seeger H (2005) Aus- und Weiterbildung von Ingenieuren im Design. In: Reese J (ed) Der Ingenieur und seine Designer, Entwurf technischer Produkte im Spannungsfeld zwischen Konstruktion und Design. Springer, Berlin, pp 277–288

Spath D, Ganschar O, Gerlach S, Hämmerle M (2013) Krause T Schlund S – Manufacturing work of the future. Fraunhofer Verlag Stuttgart Industry:4.0

Stallmann F (2002) Reform or re-labeling? A student's perspective on the introduction of the bachelor's and master's degree in German higher education. Ger Policy Stud 7/02

Stappers PJ, Hekkert P, Keyson D (2007) Design for interaction: consolidating the user-centred focus in industrial design engineering. In: DS 43: Proceedings of E&PDE 2007, the 9th international conference on engineering and product design education, University of Northumbria, Newcastle, UK, 13–14 September 2007, pp 69–74

Wölfel C (2014) Rudi Högner. In: Wölfel C, Wölfel S, Krzywinski J (eds) "Gutes design", Martin Kelm und die staatliche Designförderung der DDR. Thelem, Dresden, pp 40–41

Wölfel C, Thoring K (2014) From gestalt to experiencing – 2d/3d design fundamentals education in different contexts. In: Bohemia E, Eger A, Eggink W, Kovacevic A, Parkinson B, Wits W (eds) Design education & human technology relations, Proceedings of the 16th international conference on engineering and product design education, University of Twente, Enschede, pp 20–25

Wölfel C, Uhlmann J (2008) Designing aesthetic freeform objects: a course for industrial design engineering students. In: Clarke A et al (eds) DS 46 proceedings of E&PDE 2008, the 10th international conference on engineering and product design education, Barcelona, pp 150–155

Accreditation of Design Education Programmes

Chris Dowlen

Abstract The purpose of the chapter is to outline, first of all, the meaning of accreditation as used by professional bodies, why it is important to consider, and how it is applied to design courses in particular. The context of the chapter is the United Kingdom, and the processes described are carried out by United Kingdom organisations. The chapter describes the bodies that carry out accreditations of design degrees, and what sort of authority they need in order to be able to carry out such processes. These are the Chartered Society of Designers (CSD) and the Institution of Engineering Designers (IED). Accreditation by institutions licensed by the Engineering Council is also covered. The Higher Education Academy's processes of accreditation are also described. The reasons for degree course accreditations are outlined and the value of accreditation is suggested for a range of stakeholders. The chapter covers the requirements that need to be met for a course to be accredited, and describes the process of accreditation. It ends with a brief look at what might cause problems during accreditations and how they can go wrong. There is a brief look at accreditations of courses in a non-United Kingdom context.

Keywords Accreditation · Professional levels · Professional standards · Design syllabi

1   Introduction

Course accreditation assesses whether a course meets standards set, usually nationally, by a body with professional standing, and possibly statutory status, maybe as a practitioner gateway. Whilst an accredited course is not required for the Design profession in general, some Engineering Design tasks are regulated and for them an accredited course is necessary.

C. Dowlen (✉)
London, UK
e-mail: cdowlen@ied.org.uk

© Springer Nature Switzerland AG 2019
D. Schaefer et al. (eds.), *Design Education Today*,
https://doi.org/10.1007/978-3-030-17134-6_13

In this chapter, accreditation is not taken to mean acceptance of a college course by another educational institution to carry that institution's name. This takes place when a college cannot award its own degrees and comes to an agreement with another institution that can award degree qualifications and its courses then have the name of that institution. This can include arrangements for submission of work, agreement on virtual learning environments and using their assessment regimes.

The emphasis here is on accreditation by an independent professional body to assess whether the course meets its criteria so graduates can become its members and members of their profession. This enables course graduates to confirm they have met certain academic requirements for professional membership.

What is also means that it is an independent standards authority, giving confidence to the graduates that the course meets industrial expectations, confidence to the educational institution that its course meets quality standards and their staff can claim their education provision is acceptable.

Graduates from design courses are generally expected to play an effective role within the profession, usually as junior designers within either manufacturing industry or consultancy. They are on what Michael Eraut calls a learning trajectory (Eraut 2006), and they can expect to gain sufficient experience to move to full design professional membership in a few years. If they graduated from a course accredited by the Chartered Society of Designers (CSD) or the Institution of Engineering Designers (IED) (under its Product Design rules) then they expect to become full professional designers: if their course has been accredited under Engineering Council regulations, then they can expect to become professional engineers—for Engineering Design courses, design engineers, of course.

The body for Higher Education professionals, the Higher Education Academy (HEA) (Now part of Advance HE) accredits courses such as Postgraduate Certificate, Diploma and Masters courses for new entrants to the Higher Education profession. These can give graduates a HEA membership grade immediately they graduate. This is usually Fellow of the Higher Education Academy (FHEA). These courses may also act as professional gateways and are required for employment. The HEA also accredits Educational Institutions to operate continuing professional development (CPD) provision for various membership grades—Associate Fellowship, Fellowship, Senior Fellowship and Principal Fellowship. This normally includes a formal assessment of candidates, and is similar to a company operating its own assessment structure for, say, professional engineering. In this case, the HEA appoints external examiners to provide an independent assessment so that the process meets their standards.

The HEA also approves (rather than accredits) some courses. These do not automatically confer an HEA membership grade, but may lead towards one.

2   Who Needs to Do It?

A professional body accrediting courses usually offers partial or complete exemption from one of its qualifications.

They also need to have some authority within their sector, and are usually a major membership body.

The Chartered Society of Designers, the Institution of Engineering Designers and the Engineering Council all operate under Royal Charters, give them legal status in the UK. Many other Professional Engineering Institutions are also Chartered bodies, and offer their own accreditation as well as the Engineering Council's. The Higher Education Academy is not a Chartered body, but is supported financially by Higher Education providers in the UK and is perceived by them as having recognised authority.

Each organisation has its list of competencies, and has its method of assessing that accredited courses meet their requirements—and the Engineering Council operates systems to monitor its accreditations.

3   The Reasons for Accreditation

Accreditation is a mark of assurance of course quality, that it meets accepted standards and has achieved required quality standards. Assessment of knowledge, skills and abilities of graduates from accredited courses indicates that UK standards are met.

Accreditation is accepted as rigorous thorough, and acts as a quality mark for graduates and course providers.

Some employers only accept graduates from accredited courses; others give priority to those graduates and most employers expect to see portfolios of work indicating that accredited-level degree work has been carried out on the course and that the applicant is sufficiently well-qualified.

An accredited course will be perceived as higher status than other courses. Indeed, some UK Quality Assurance Agency benchmark statements are those of accrediting bodies, notably the Engineering Council.

Accreditation standards may be recognised globally, and many international educational accords and quality standards have been developed along the lines of UK course accreditation standards.

Many course providers consider that course accreditation offers them a mark to establish the course as being of acceptable quality, and this is part of their quality audit.

However, relatively few design courses have achieved Chartered Society of Designers accreditation, whilst more have been achieved Institution of Engineering Designers' Product Design accreditation or Engineering Design course accreditation under Engineering Council processes, perhaps through the aegis of the Institution of Engineering Designers.

Most Engineering Courses of all disciplines, not just Engineering Design ones, are designed to meet Engineering Council accreditation standards.

Value is added to the courses for the various stakeholders, summarised as follows:

- **For Course applicants**, an accredited course means they can be assured that topics and standards on the course meet required standards, and that they will not be disadvantaged when they graduate. It isn't worth going for a course where accreditation is lacking, unless there is a specific reason for it, such as it being the only one in a specific discipline, or that it is of a generalist nature where accreditation is not appropriate.
- **For students on the course**, a similar situation applies. They are assured that elements they have to learn are not being omitted from their course, and once they graduate they will be able to apply for and obtain, graduate-level employment—even when they discover that much learning as a recent graduate is tacit, obtained through hands-on experience at work.

On most design courses a lot of work is done through project learning, where students develop hands-on experience of design. Assessment primarily by academic, knowledge-based examinations is not appropriate. Yes, most designers learn design through doing design and not by demonstrating knowledge of what they are designing or of the design process without informative tacit experience of achieving design realisation of product knowledge and the design process.

Significant amongst tacitly learnt design skills are those of three-dimensionality, fitting component parts together into a coherent whole into a space envelope, being able to deal with humanistic aspects of control theory and practice through hands-on experience as well as ergonomic theory, and developing aesthetic appreciation, not just visually but also through touch and physicality.

An accredited course needs to ensure that there is time and space for this type of learning to be effective.

- **Course graduates** have been through this process. They can demonstrate through their portfolios that they have skills and that they are ready for employment—please. It is not good for a job interview candidate to be told that what that they have done on their course has been substandard and that they have missed out on something vital, not because of their incompetence, but because their course staff did not deem it to be an essential course component. Accreditation tends to avoid this scenario, but not completely.[1]
- **Staff teaching on the course** will inevitably be challenged by accreditation. But they will be aware that accreditation ensures they cannot simply wander off into their pet topics, and that they need to ensure that accreditation learning outcomes are met, and they are responsible for this.

---

[1]It is not unknown for letters to be sent, usually direct to a University Vice-Chancellor, that suggest that a course is inappropriate. These may have come from Design consultancies, from consultants who mainly do graphic design of some sort—and who are trying to broaden their scope into Product Design without understanding that its standards and requirements differ considerably from those for graphic design.

But they also know that the course meets the requirements and qualities needed for national compatibility, and that their graduates can compete with others effectively. They know that their teaching is valued and appreciated, particularly where they develop students' intangible skills through project learning. Accreditation identifies their rule book, and ensures their standards are achieved and aspirational.

They can also develop mobility, recognising that a job on one accredited course may give them opportunities to move to another one, and allow them to demonstrate that their personal work means they can become a registered design professional and an effective professional Higher Education teacher.

Accreditation can also give them a development learning trajectory. The process itself can be a learning curve, where they need to align course learning outcomes with their teaching and assessment, understanding why they are constrained to arrange particular assessment strategies and to develop them effectively. They can also identify topics to develop them and allow them to develop course teaching, and develop their own design and academic careers to include operating as accreditation assessors and quality auditors and external examiners. The Higher Education Academy standards for Fellow, Senior Fellow and Principal Fellow require understanding of Higher Education quality processes—some of which can be learnt through teaching on accredited courses and through experience of accreditation.

- **The universities and colleges running the courses** benefit from accreditation as it provides a clear indication that they meet and maintain effective quality standards. The external stamp of approval is important, acting as an important advertising tool.

If, for some reason, accreditation is refused this can be difficult and staff may lose their jobs. But it can also act as a useful learning tool, helping to show where improvements are required in the course standards.

This learning aspect can also take place when the University is implementing changes required or recommended by the accreditation panel. These are generally couched in terms to improve course standards, and the process of implementing these usually results in an improved and stronger course, with perhaps a more cohesive staff team.

- **Graduate and student placement employers** benefit from accreditation as it ensures they can employ people who meet expected standards and who should become useful employees when acclimatised to the company's ways. Some graduates become financially worthwhile within the space of a few weeks; for most this should only take a few months. The company gains a useful employee relatively quickly.

Employers may also become involved with the accreditation process. As well as industrial links being seen as an important part during the process, accreditation panels normally require an industrial member, to ensure that students take industrially relevant topics and carry out relevant design work and projects rather than tackling things that students (and maybe staff) enjoy from personal hobbies and interests.

Industrially based projects usually produce better learning than those hobby-based projects.

- **For the professional institutions**, accreditation should bring a ready supply of new members, although they also need enthusiastic staff members to do this effectively. If they do not carry out accreditations they could lose future membership and contact with changes in the design industry. Institutional contact with academia means they are in touch with research, and can tap into this expertise and present academic research to industry.
- **The government** also, perhaps surprisingly, benefits from course accreditation. A particular instance is how the Engineering Council course requirements led the UK's government standards to become subject benchmarks. This means that the government does not need to find other experts to derive standards, but can base their standards on those of the accrediting bodies.

They also have confidence that the country's courses comply with expected standards and that the country's future is catered for through the supply of graduates who can contribute to the country's future economic success—or who may increase the country's international esteem. UK accreditation is valued overseas for continuing consistent standards.

- **Recruiting agencies** can clearly state job requirements in terms of graduates for particular employment positions—graduates from an accredited course should demonstrate particular knowledge, skills and abilities. This means it is easy to use accreditation as shorthand to provide effective staff for their vacancies.

4 What Does a Design Course Need to Be Accredited?

Put simply, it needs to meet the requirements of the body that is going to accredit it.

Firstly, there are requirements associated with the course content. These are identified in terms of learning outcomes to be achieved by every graduate. These need to map from the ones required by the accrediting body, through the overall course aims and learning outcomes, those for individual course modules, to those achieved by each and every piece of assessed work. These learning outcomes are mapped through a matrix structure, designed to ensure compliance with accreditation requirements. This needs to cover the outcomes that will be achieved by every student taking the module—not those outcomes that may or may not be met by a student taking it.

Identifying project outcomes can be difficult. Design courses almost always include a significant individual major project, where students select a project from a list or propose their own. Different projects have different emphases, which results in different skills and knowledge being achieved by each student. But some skills and abilities that each student demonstrates will be the same, and these skills and abilities need to be identified as the outcomes for the major project, rather than those that a student might achieve. Sometimes project choice is carefully filtered and the assessment regime designed such that particular skills and abilities are incorporated, and student choices that do not anticipate meeting those learning outcomes will be rejected as unsuitable.

For example, in an Engineering Design course each project needs to demonstrate numerical engineering knowledge, skills and abilities and projects that do not include this will either not be allowed at the choice stage, or will end up as failed projects.

Specific criteria for the Chartered Society of Designers accreditations are not currently available on their website, and need to be obtained directly from the Society. However, they accredit a mixture of Bachelors and Masters courses, and these will have different requirement levels. For Masters accreditation, course outcomes will be focusing more on conceptual, creative areas and on the ability of students to evaluate their design work and methods, whereas for Bachelors courses these are likely to be more focused on design practice. Their criteria are likely to be similar to their matrix criteria for membership with four sets of topics: creativity, professionalism, skills and knowledge. Each of these is divided into four sections, and courses will probably be required to demonstrate that students can achieve outcomes that reflect these sixteen—at different levels for different courses (Chartered Society of Designers 2015). These are listed in Appendix 'Chartered Society of Designers CPSK™ matrix'.

The IED's list of Product Design requirements cover five areas: design, economic and social context, design practice, underpinning science and mathematics and design analysis. These translate into 33 requirements altogether. These are listed in the Appendix 2.

The Product Design courses it accredits may be accredited for Registered Product Design (RProdDes) or Chartered Technical Product Designer (CTPD).

Courses for RProdDes will be Bachelors: a BSc, BA or (occasionally) BDes. These are usually three years' duration, but may include an extra year for an industrial placement or study abroad, usually optionally. Courses in Scotland will normally include an extra academic year. Exceptionally the year studying abroad will replace a taught year, with a coherent mapping arrangement covering the learning outcomes.

Courses accredited for CTPD will be either Integrated Masters, usually MDes, or may be a separate Masters following a Bachelors course accredited for RProdDes to obtain full CTPD certification. These separate Masters courses are accredited as additional learning to CTPD, and are usually MA or MSc courses. An Integrated Masters course is normally of four years' duration, but may include an extra year for an industrial placement or overseas study, as for a Bachelors course. In Scotland these normally include an extra academic year. MA and MSc courses usually last a calendar year, with the major project being carried out over the summer and completed in October. The combination of a Bachelor's course followed by an MA or MSc is somewhat longer than an undergraduate masters course, usually by the size of this major individual project.

MDes courses are not always Integrated Masters ones, but may take other forms and include other combinations of years and levels. Bachelors level MDes courses are not unknown, and some are postgraduate courses of two years' duration—there may be other arrangements. Each needs to be scrutinised, which can be slightly confusing.

Engineering Design courses may also be accredited under the IED's Product Design accreditation. But frequently they will also be accredited to the Engineering Council's requirements. These can be accredited by any of the Engineering Council's licenced bodies, including the IED.

Undergraduate degree course accredited to Engineering Council standards are usually BEng ones, but may be BSc. These BSc courses are usually accredited to Incorporated Engineer (IEng), and BEng courses are usually accredited as partial Chartered Engineer (CEng). For full Chartered Engineer accreditation an Integrated Masters course—usually an MEng—is required, but MSc courses can be accredited as additional learning to CEng level, which requires the addition of an accredited BEng course for full Chartered membership. In practice, someone who has completed an IEng accredited BSc course will also be able to use this with an accredited MSc towards their CEng registration.

The Engineering Council issues a document for Accreditation of Higher Education Programmes, known as AHEP for short. The current edition (from January 2017 onwards) is the third edition, and hence this is known as AHEP3 (Engineering Council 2014).

AHEP3 lists Learning Outcomes under six headings:

- Science and mathematics
- Engineering Analysis
- Design
- Economic, legal, social, ethical and environmental context
- Engineering practice
- Additional general skills.

It then lists requirements separately for the four categories of courses that may be accredited under its schemes:

- Programmes accredited for IEng—Bachelors degrees and Bachelors (Honours) degrees
- Programmes accredited for CEng—Bachelors (Honours) Degrees accredited as partially meeting the educational requirement for CEng (Additional learning to Masters level will be required)
- Programmes accredited for CEng—Integrated Masters (MEng) Degrees
- Programmes accredited for CEng—Masters Degrees other than the Integrated Masters (MEng) (Accredited as additional learning to Masters level, partially meeting the educational requirement for CEng).

A summary of this AHEP3 matrix is shown in Appendix 3.

As REngDes and IEng aim at registration for the competent designer or engineer, the requirements for accrediting courses concentrate on requiring operational competence—graduates from these courses are able to DO things. They can work on design projects, carry out required procedures and processes, take responsibility for their own work and to some extent their colleagues, and may have some financial and supervisory management.

These courses include a major individual project, and for design courses this will be a design project. For engineering courses, the project may focus on analysis or experimental work rather than design.

For Chartered professional accreditation, courses need to focus more on evaluation and analysis of why they are doing things, with more discernment and what might be termed 'wisdom' (rather than simply 'knowledge' and 'skills') needs to be shown. Reasoned choices and enhanced creativity—perhaps a deliberately selective method choice—may be expected. In addition to a major individual project, each course for Chartered membership needs to include assessed group work, usually a group design project. A requirement of this is that students assess their colleagues' performance, and that students in a group can be awarded differing marks.

For an integrated Masters, the major project may be in the third academic year, with the major group project in the final year. Normally an MSc or MA includes a major individual and a group project. This usually contributes to the larger size of the MSc or MA course when compared to a normal student year.

The Engineering Council lists the slightly different requirements for a one calendar year Master's course (almost always an MSc) compared with those for an Integrated Masters course. This difference is not highlighted in the IED's requirements, leaving some doubt (usually in the minds of the accreditors) as to whether all the CTPD requirements need to be met in the short course, or whether some can be assumed to have been in the earlier undergraduate course.

5   What Does Accreditation Involve?

Accreditation is more than a paper exercise. It is a major assessment of how a course is performing, whether it is doing what the paperwork suggests it is doing, and how well it works out practically.

The process involves a significant submission of documentation, followed up by a visit, usually for two days. This will be conducted by a small panel with a mixture of academic and industrial members, with administrative support. In theory it is supposed to be a constructive discussion rather than a confrontational event, but there is usually a feel of 'them and us', and there needs to be a thorough critique of the course as well as a conscious effort to seek out its best bits (one panel member has said that these are what she would like to steal and copy).

Ideally the documentation is produced well before the accreditation event, and can be scrutinised by the panel members beforehand. However, additional documents are quite often requested by the panel during the visit, and the real requirement of the visit is to check that everything the documents say is taking place in practice.

Although scrutinising output standards lists and matrices is an important part of accreditation, most accreditations also investigate things such as:

- teaching and learning processes and strategies;
- course assessment strategies and why they are used;
- project levels and outcomes, including group projects and their assessment;
- the University and course arrangements for student progression, compensation and condonement;
- student pass and failure rates;
- the philosophy and ethos of the department;
- how staff relate to the students;
- staff professional registration;
- external reviews and external examiners reports;
- laboratory and design studio space;
- whether students have the freedom to develop their own designs in the studio space;
- whether students are unduly constrained by technicians and perceived safety;
- how students are assessed to work in laboratories and maker spaces;
- how the University and department wish to develop the courses and facilities;
- how previous recommendations and requirements have been implemented;
- the virtual learning environment and learning resources;
- the quantity and quality of industrial input provided for the students—for projects, guest lectures, industrial visits, course strategy development—and how this is perceived by students;
- graduate employment and destinations;
- student recruitment, course entry and induction;
- overseas relationships and study;
- student happiness and morale;
- student support arrangements.

Indeed, accreditation is an overall quality audit on how the course achieves what it sets out to do.

Standard documentation needs to be supplied before the visit, and during the visit the panel will have discussions with staff, students, industrial partners and senior University staff, and will visit laboratories, workshops, teaching spaces and maybe other facilities. The panel will also need time set aside for private discussion.

The panel may need to see the current final year students' output, which may mean that the visit takes place during a public or private viewing of student degree work, usually focusing on the major project. If this cannot be done during the visit, then one member of the panel may pay a later visit to see the degree exhibition, or may see a selection of the student output perhaps at the New Designers' annual exhibition.

The final result of an accreditation visit is usually decided by the accrediting institution's accreditation committee, which means that the result cannot be known at the end of the visit. However, usually a recommendation can be given for ratification. There are several different possible overall outcomes. These will normally be given for each of course, and may be different for each.

The two extreme results are chalk and cheese—yes and no. These are both unlikely, and the result is more likely to be a qualified yes, with some commendations, requirements and recommendations. It is very rare to find no requirements, so course teams must not be too downhearted at these.

There are two sorts of requirements. Firstly, immediate requirements; if these are not satisfied within a very short timescale then no accreditation is possible. They are usually things that can be put right quickly, such as needing to supply details of specific course output which wasn't available at the time of the visit, needing to put right legal issues like illegal statements in documentation, or correcting publicity documentation where the accreditation level has been incorrectly stated. Or it may be that some final year options hadn't been correctly mapped onto the learning outcomes matrix, or that it had been completed for possible project options rather than every student's achievement. Once all immediate requirements have been met an accreditation certificate can be awarded, on the understanding that other requirements will be met within a determined timeframe.

Rewriting the learning outcomes matrix normally comes under the second sort of requirement. Sometimes these requirements require committee agreements or due processes to be carried out, and sometimes there is simply a reasonable developmental period before they can be effectively implemented. Accreditation is not normally withheld whilst these are completed. On receipt of the official documentation the University submits an action plan detailing what needs to be done, who will be doing it and the timescale for completion.

Thirdly, there are recommendations. Sometimes these are strong, which generally means that the University needs to have a good reason to avoid doing them. Other recommendations come under the 'would be nice if' category. Recommendations are included in the accreditation report, and are expected to feature in the action plan, but the University can choose its response, which may be that the recommendation be implemented—or that it may wish to discuss it and see if it feels it is a good idea or not, and whether it would improve the course or not. Sometimes a recommendation may be experimental. It is then up to the University whether or not it carries it out, depending on whether it makes sense to them.

Assuming that things go reasonably, the accreditation is official, and unless the course changes significantly, it is likely to remain in place for the accreditation period, which is commonly five years, but may be as little as two years for a course with perceived problems—and progress is expected during that time.

5.1   *Joint Accreditations*

Sometimes a University wants accreditation from several professional institutions at once, and a joint accreditation might be beneficial. This is frequently the case for Engineering Council accreditations—several of the Engineering Institutions collaborate formally under a Joint Board as their accreditations are typically of this nature. There is also an Engineering Accreditation Board (EAB), which seeks to ensure consistency between Engineering Council accreditations, and they can operate arrangements for joint accreditations.

Joint accreditations can work well if the partners can agree on the outcomes reasonably easily and there are few differences between outcomes, requirements and recommendations. However, they can cause friction if there are significant differences, and discussions seem to take much longer in a joint forum—more people on a panel means more talking, there may be more courses to investigate and slightly different opinions need to be aired before a consensus can be reached or outcome requests made. For instance, the IED normally expects students to be involved in significant design workshop practice, and this may not be as important for other Engineering institutions; other institutions are keener on assessment by examination and less keen on project- and practice-type assessments, or they may require learning outcomes based on, say, aerodynamics, and so on.

6   What Can Go Wrong?

The worst result is probably where no accreditation can take place. Or where accreditation for the desired level cannot be awarded.

The course or department may have too many difficulties and problems for accreditation. This is probably the most serious scenario and causes most difficulty for the University. It may be a failure of their quality control system, perhaps so that external examiners' comments have not been implemented, leading to lack-lustre courses, where learning outcomes are not being achieved, and where without significant changes learning outcomes are unlikely to be achieved. The courses are substandard at that moment.

The University will want to put this right as quickly as possible. They will need to identify what needs to be done and to work on a timescale for immediate implementation. This could, typically, suggest a two-year timescale, and perhaps they will work with someone from the accrediting body who can tutor and mentor staff to ensure that the standards and systems are met so accreditation can be successful after the two-year period of hard work. Staff development may be required, particularly where professionally registered staff are not employed.

Perhaps requirements from a previous visit have not been corrected. The University is, again, in a serious situation which it needs to remedy. A new implementation schedule will need creating to set the issues right, incorporating the circumstantial changes that led to the failure to implement the previous action plan. Similarly, a two-year period might be able to be considered with staff mentoring, to carry out the processes effectively. Staff development may be appropriate.

Sometimes where, for instance, Engineering Council accreditation is sought, this is unable to be granted. Usually this is due to insufficient underpinning science and mathematics. Another alternative is that, although there is a sufficient scientific and mathematical background in taught modules, these aspects of design are not assessed sufficiently within projects. Here Engineering Council accreditation will probably not be denied, but suitable assessment criteria may be required in project modules. Most panels will recognise that University module change processes can take a little time to be accomplished, so it will probably not be an immediate requirement.

Accreditation may be denied because the course title is incorrect, which will mislead students who are conned into becoming unwitting members of the wrong community of practice. Perhaps a straight engineering course calls itself Product Design to find a ready market amongst school leavers who enjoyed their product design experience, but once on the course they feel disillusioned and unloved by a staff team that doesn't always appear to know any better themselves, and who themselves are conned into (in this case) thinking they can practise as product designers when they are not able to do so and cannot obtain professional Product Design registration.

Here a short term remedy might be to retitle the course. A longer term solution would be to develop staff competence in the area of the course title, changing the course to reflect its title, and taking on new specialist staff registered as relevant proficient professionals. This needs considerable commitment by departmental management, and a considerable amount of resource to be devoted to the successful transformation.

Or the course may be inappropriate for that professional institution to accredit. Perhaps a design course (almost certainly a BA) may be clearly art-based, with no intention to include underpinning mathematics or science. However, the University asked the IED for accreditation, knowing that they accredit design courses, but unaware that their course was heading in a different direction from that of the IED. Here the best approach is to approach a different institution, such as the CSD, who accredit a variety of different design courses, including those with creative artistic bases.

It may simply be that a course, although listed, is not actually running and has no students and no output. Here accreditation may be premature and there is no point in accrediting it as there is nothing to accredit. The University either needs to be realistic and stop trying to offer the course, or needs to start marketing it more effectively and find some students for it. Once it recruits, accreditation can return to the timetable: or alternatively, if there are only ever very few students on the course then any graduates may apply for professional registration via an individual candidate route.

Some courses by their very nature cannot be accredited easily. These are usually multidisciplinary and have generic aims and outcomes. Sometimes they can prepare designers or engineers for their professions—in other cases they can at provide a different experience for their students.

These may be industry-based courses, using a learning contract approach where students identify individual learning outcomes. These courses can also develop educational-type learning outcomes in their students, who find their freedom to select their own topics provides them with significant understanding of their learning trajectories and learning processes. They can become qualified not just as designers (perhaps) but as practitioners in continuing professional development systems—primarily applied to their own experiences.

Accreditation of these courses is not possible (nor expected), but individual students may be presented for potential professional membership through individual case procedures whilst they are taking the course, so their options may be tailored towards meeting overall learning outcomes listed in the institution's requirements. If that is not possible, then they can apply after they graduate. The Engineering Council and the IED have individual candidate procedures in place.

If documentation has not been supplied in sufficient time for the accreditation event, then it may need to be cancelled.

Less serious problems can result in a long set of requirements rather than a blanket refusal to accredit.

If a panel member cannot take part at the last minute this can cause difficulties, and members of the panel may feel they have to continue, putting extra burden on them, particularly if difficult decisions need to be made. Here remote backup is useful, and the panel may decide that although they can do much of the accreditation, the final decision may have to wait until the absent panel member has been consulted, which is usually slightly unsatisfactory. Other accreditation interruptions such as fire alarms, security alerts, transport strikes, difficult weather and similar issues simply need to be dealt with as effectively as possible. Sometimes it means postponement of decisions: in others the overall result may need more discussion after the visit to resolve the situation and to achieve a satisfactory outcome.

An issue which cause difficulties rather than negating the accreditation is when panel members or members of the course team unfortunately bring their pet topics to the table, and want these to be in the forefront of the course. This needs effective training, discipline and a strong chairman. If the chairman misbehaves, then perhaps complaint procedures need to be invoked. This is particularly difficult in a joint accreditation, where one institution provides the chairman, who may not be as effective as they should be.

7   Non-UK Approaches

Overseas course accreditations can be carried out by UK institutions. These tend to be on English-speaking courses, but not exclusively so. Many overseas accreditations take place in East and South Asia, where accreditation by a UK institution has a higher value than in most of Europe.

CSD overseas accreditations have taken place in Switzerland (all for Design Management), and they also validate some short courses and parts of courses in Sri Lanka (Fashion Design) and Saudi Arabia (Interior Design Engineering). IED accreditations have taken place in Ireland, Hong Kong and on overseas campuses of UK Universities in Singapore, Malaysia and China. Other Engineering institutions regularly accredit overseas courses, but these may not be design courses.

Although a European Engineering Accreditation arrangement (EUR-ACE) exists, the concepts of professional registration and course accreditation do not seem to make much sense for many in Europe, as their engineering degrees equip them effectively for professional practice. For many of these the initial professional formation period is longer, typically at five years.

The Washington Accord is an agreement signed by several countries that provides equivalence for engineering degrees in its signatory countries.

In 2018 the signatories to the agreement were Australia, Canada, China, Chinese Taipei, Hong Kong China, India, Ireland, Japan, South Korea, Malaysia, New Zealand, Pakistan, Peru, Russia, Singapore, South Africa, Turkey, the USA and the United Kingdom. Chile, Bangladesh, Costa Rica, Mexico and the Philippines have systems and processes in place to develop towards becoming full signatories (International Engineering Alliance 2018). All full signatories to the Accord can accredit Engineering degrees at different professional levels, and treat an accredited engineering degree as part of the normal process of the formation of professional engineers.

This list is predominantly a combination of English-speaking countries and East and South Asian ones—those who value professional recognition beyond initial degree level and where further registration is appreciated for engineering practice.

For design courses, even for product design courses, there doesn't appear to be any international agreement on content, each country having its own arrangements. There are significant international gatherings of design lecturers and teaching staff, such as the annual Engineering and Product Design Education conferences organised jointly by the Design Society and the IED, and the Design Society also organises biannual international conferences (ICED), largely for design researchers, which includes an educational stream. The Design Society also runs a Design Education Special Interest Group (DESIG) (Design Society 2018), but their main aim is discussion, which is not currently focused on development of common professional standards and compatibility of different national design course offerings. A significant number of the membership of DESIG have an engineering rather than product or industrial design background.

Appendices

1. Chartered Society of Designers CPSK™ Matrix

Creativity

C1	Creativity	How their imagination, intuition, insight and inspiration has contributed to their work
C2	Generating	Awareness of creative problem-solving (CPS) techniques and their application. An understanding of the inhibitors of creative thought and how to negotiate them. Risk aware versus risk averse. Ability to generate appropriate ideas. How to negotiate convergent or divergent thinking in problem-solving. Inquisition—serendipity—questioning
C3	Managing	Ability to scope, analyse, evaluate and select ideas for development. Proving of ideas for use in a specified context. Throughout the creative process, managing ideas in a business and creative environment, using analysis and evaluation as well as independent thinking
C4	Innovate	Adoption of generated ideas to address specific problem or need as set out in a brief or requirement in an original way. Identifying where ideas can be used to deliver original solutions in existing markets or creating new markets. Ability to exploit potential of creative ideas by the use of design principles to generate original outcomes

Professionalism

P1	Values	They possess and exercise values that are common to the environment in which designers practice. Are able to maintain integrity when undertaking work showing due regard for the practice of design. Act in a manner that respects and protects their colleagues, clients, the environment, nature and society
P2	Process	Ability to adopt appropriate methodology of practice. Continual learning and research into best design practice
P3	Communication	Ability to communicate with stakeholders through written and oral platforms in presenting design. An understanding of the interpersonal and psychological communications employed as well as the relationships involved. Appropriate use of communication techniques
P4	Contextual	Understanding and ability to use the appropriate regulations and requirements concerned and ensure standards (adoptive and statutory) are maintained in practice. Advanced knowledge of relevant IPRs and understanding of other IP issues. Ability to operate to best practice within their field of practice. Undertake work only if qualified to do so within a defined discipline

Skills

S1	Generic	Visual interpretation and communication of ideas and concepts in a manner that can be communicated with others. Use of colour and expression of form and how it is used to deliver concepts and managed for design outcomes. Conceptual and spatial awareness. Design thinking capability and creative skills
S2	Operating environment	Skills required within the operating environment of the designer. Ability to employ generic skills within operating environment and to the level required. Ability to audit and identify skills required within operating environment and address deficiencies whilst enhancing acquired skills. Financial, budgeting, management, organisational, leadership, information management and commercial skills
S3	Personal	Interpersonal and people skills. Complimentary and transferable skills
S4	Contextual	Skills required to practice competently within a defined design discipline. Ability to apply generic skills within the remit of a defined design discipline

Knowledge

K1	Explicit	Knowledge acquired from others which derives from research and experimentation and is generally accepted. Possesses theoretical knowledge. Possesses general knowledge
K2	Tacit	Knowledge gained from prior experiences at any stage of the design process
K3	Management	Ability to undertake research relevant to own professional practice whether academic and/or practice based. Ability to undertake research and acquire knowledge specifically related to delivering appropriate design solutions. Exploitation of knowledge for the benefit of all stakeholders, designer, client, commerce, society, etc.
K4	Contextual	Awareness, understanding and knowledge of the history of their profession and chosen discipline. Knowledge of the sector in which they operate including; processes, market conditions, legislation, technology, methodologies. Cultural awareness of operating environment and those involved

2. The Institution of Engineering Designers Learning Outcomes for RProdDes and CTPD Accredited Courses

Programmes accredited for RProdDes	Programmes accredited for CTPD
Bachelors degrees and bachelors (Honours)	Integrated masters (MDes) degrees
Design	
Ability to evaluate design solutions against relevant constraints and criteria	Ability to evaluate complex design solutions against conflicting constraints
Ability to address human needs through the use of research, anthropometric data and ergonomic principles and provide design solutions according to customer and user requirements. Ability to generate a product design specification (PDS) by defining requirements as separate criteria including other factors such as technical aspects and legislative demands	Ability to address human needs through the use of research, anthropometric data and ergonomic principles and provide design solutions according to customer and user requirements. Ability to generate or challenge a product design specification (PDS) by defining requirements as separate criteria including other factors such as technical aspects and legislative demands
Ability to recognise product design cost drivers for both recurring and non-recurring costs and to appreciate the cost implications of differing production volumes	Ability to apply cost drivers for both recurring and non-recurring costs and to design for the cost implications of differing production volumes
Ability to generate a wide range of design ideas, concepts and proposals independently and in teams in response to set or self-generated design briefs	Ability to generate and evaluate a wide range of design idea, concepts and proposals
Ability to select, test and exploit materials and manufacturing processes in the synthesis of product design solutions	Ability to select, test and exploit materials and manufacturing processes in the synthesis of product design solutions
Ability to apply creative and logical thinking processes as well as design methodologies to the creation of design solutions	Ability to apply and reflect upon a wide range of creative and logical thinking processes as well as design methodologies in the creation of design solutions to complex problems
Ability to select and use the appropriate manual drawing/construction/CAD, communication and technological media in the realisation of design ideas	Ability to select and use the appropriate manual drawing/construction/CAD, communication and technological media in the realisation of design ideas
Ability to demonstrate visual literacy and drawing ability appropriate to the practice of product design	Demonstration of complex visual literacy and advanced communication tools appropriate to the practice of product design
Ability to develop concepts sufficiently to provide manufacturing instructions and specifications	Ability to develop and evaluate concepts sufficiently to provide manufacturing instructions and specifications

(continued)

(continued)

Programmes accredited for RProdDes	Programmes accredited for CTPD
Bachelors degrees and bachelors (Honours)	Integrated masters (MDes) degrees
Ability to employ materials, media, techniques, methods, technologies and tools associated with product design through drawing, modelling and computer visualisation using skill and imagination	Ability to employ materials, media, techniques, methods, technologies and tools associated with product design through drawing, modelling and computer visualisation using skill and imagination
Ability to integrate Industrial Design aspects including form, texture and colour	Ability to synthesise a broad range of design aspects
Economic and social context	
Understanding that positive ethical and professional conduct underpins design practice	Application of a positive ethical professional conduct underpinning design practice
Knowledge and understanding of risk issues, including health and safety, environmental and commercial risk and of risk assessment and risk management techniques	Knowledge and understanding of risk issues, including health and safety, environmental and commercial risk, risk assessment and risk management techniques and an ability to demonstrate their effective evaluation
Awareness of legal requirements governing design activities, including personnel, health and safety, product liability and safety	Awareness and appropriate application of legal requirements governing design activities, including personnel, health and safety, product liability and safety in familiar and unfamiliar situations
Knowledge and understanding of the management of the design process	Demonstrate application of design process management
An awareness of financial, economic, social legislative and environmental factors of relevance to product design	Demonstrate the application of financial, economic, social legislative and environmental factors to product designs
Awareness of the social and environmental impact and the application of sustainable design principles	Application of the social and environmental impact analysis and the application of sustainable design principles
Design practice	
Ability to create new processes or products through synthesis of ideas from a wide range of sources using a broad knowledge of material and material selection principles	Ability to create new processes or products through synthesis of ideas from a wide range of sources using a broad knowledge of material and material selection principles
Ability to practise collaborative and independent work to realise a range of practical, creative and theoretical projects	Critical evaluation of historical and latest trends in design thinking and their appropriate application
Ability to meet deadlines, liaise with industrial collaborators, make presentations, research and collate information, produce reports and evaluate the design and research work of self	Ability to initiate projects, meet deadlines, liaise with industrial collaborators, make presentations, research and synthesise information, produce reports and evaluate the design and research work of self and others
Ability to analyse problems of a creative nature and to provide appropriate solutions	Ability to analyse complex problems of a creative nature and to provide appropriate solutions

(continued)

(continued)

Programmes accredited for RProdDes	Programmes accredited for CTPD
Bachelors degrees and bachelors (Honours)	Integrated masters (MDes) degrees
Understanding and application of intellectual property rights (IPR) including patent search and principles of copyright and design registration	Understanding and application of intellectual property rights (IPR) including patent search and principles of copyright and design registration
Understanding of specific design codes of practice and industry standards, with some knowledge of design factors and requirements for safe operation	Application and development of specific design codes of practice and industry standards, with knowledge of design factors and requirements for safe operation
Awareness of management and quality assurance issues in product design	Application of management and quality assurance issues in product design
Working effectively as part of a group with respect for the dignity, rights and needs of others	Working effectively as part of a group with respect for the dignity, rights and needs of others and to develop an understanding of leadership. This potentially requires a group project as part of the Masters programme
To develop skills associated with professional practice; time management, project management, professional level communication, self-promotion, interview techniques, information gathering and use of information and communication technology as appropriate	To demonstrate skills associated with professional practice; time management, project management, professional level communication, self-promotion, interview techniques, information gathering and use of information and communication technology as appropriate
Ability to evaluate technical risks and address risk in design methodology	Ability to evaluate technical risks and address risk in design methodology
Ability to write a PDS, design reports and present design ideas in a rational and coherent manner	Develop and critique a PDS, design reports and present design ideas in a rational and coherent manner
Underpinning science and mathematics	
Ability to consider and apply the appropriate mathematical and engineering principles to a particular product design problem	Ability to consider and apply the appropriate mathematical and engineering principles to a particular product design problem
Design analysis	
Ability to research, select, evaluate, manipulate and manage information relevant to the analysis and synthesis of product design solutions	Ability to research, select, evaluate, manipulate and manage information relevant to the analysis and synthesis of product design solutions
Ability to apply analytical skills in relation to designed objects including the ability to undertake visual analysis and to analyse designed objects in relation to their context	Ability to apply analytical skills in relation to designed objects including the ability to undertake visual analysis and to analyse designed objects in relation to their context
Ability to apply a systematic approach to problem-solving using appropriate design tools and techniques	Ability to apply a systematic approach to problem-solving using appropriate design tools and techniques

3. *Engineering Council Learning Outcomes for Accredited Courses*

Programmes accredited for IEng	Programmes accredited for partial CEng	Programmes accredited for CEng	Programmes accredited for additional learning for CEng
Bachelors degrees and bachelors (Honours) (BSc or BEng)	Bachelors (Honours) degrees for partial CEng (BEng)	Integrated masters (MEng) degrees	Masters degrees other than the integrated masters (MSc)
Science and mathematics			
Engineering is underpinned by science and mathematics, and other associated disciplines, as defined by the relevant professional engineering institution(s). Graduates will need	Engineering is underpinned by science and mathematics, and other associated disciplines, as defined by the relevant professional engineering institution(s). Graduates will need the following knowledge, understanding and abilities	Engineering is underpinned by science and mathematics, and other associated disciplines, as defined by the relevant professional engineering institution(s). Graduates will need the following knowledge, understanding and abilities	Engineering is underpinned by science and mathematics, and other associated disciplines, as defined by the relevant professional engineering institution(s). The main science and mathematical abilities will have been developed in an accredited engineering undergraduate programme. Masters graduates will therefore need additionally
• Knowledge and understanding of the scientific principles underpinning relevant current technologies, and their evolution	• Knowledge and understanding of scientific principles and methodology necessary to underpin their education in their engineering discipline, to enable appreciation of its scientific and engineering context, and to support their understanding of relevant historical, current and future developments and technologies	• A comprehensive knowledge and understanding of scientific principles and methodology necessary to underpin their education in their engineering discipline, and an understanding and know-how of the scientific principles of related disciplines, to enable appreciation of the scientific and engineering context, and to support their understanding of relevant historical, current and future developments and technologies	• A comprehensive understanding of the relevant scientific principles of the specialisation

(continued)

(continued)

Programmes accredited for IEng	Programmes accredited for partial CEng	Programmes accredited for CEng	Programmes accredited for additional learning for CEng
Bachelors degrees and bachelors (Honours) (BSc or BEng)	Bachelors (Honours) degrees for partial CEng (BEng)	Integrated masters (MEng) degrees	Masters degrees other than the integrated masters (MSc)
• Knowledge and understanding of mathematics and an awareness of statistical methods necessary to support application of key engineering principles	• Knowledge and understanding of mathematical and statistical methods necessary to underpin their education in their engineering discipline and to enable them to apply mathematical and statistical methods, tools and notations proficiently in the analysis and solution of engineering problems	• Knowledge and understanding of mathematical and statistical methods necessary to underpin their education in their engineering discipline and to enable them to apply a range of mathematical and statistical methods, tools and notations proficiently and critically in the analysis and solution of engineering problems	• A critical awareness of current problems and/or new insights most of which is at, or informed by, the forefront of the specialisation
	• Ability to apply and integrate knowledge and understanding of other engineering disciplines to support study of their own engineering discipline	• Ability to apply and integrate knowledge and understanding of other engineering disciplines to support study of their own engineering discipline and the ability to evaluate them critically and to apply them effectively	• Understanding of concepts relevant to the discipline, some from outside engineering, and the ability to evaluate them critically and to apply them effectively, including in engineering projects
		• Awareness of developing technologies related to own specialisation	
		• A comprehensive knowledge and understanding of mathematical and computational models relevant to the engineering discipline, and an appreciation of their limitations	

(continued)

(continued)

Programmes accredited for IEng	Programmes accredited for partial CEng	Programmes accredited for CEng	Programmes accredited for additional learning for CEng
Bachelors degrees and bachelors (Honours) (BSc or BEng)	Bachelors (Honours) degrees for partial CEng (BEng)	Integrated masters (MEng) degrees	Masters degrees other than the integrated masters (MSc)
		• Understanding of concepts from a range of areas, including some outside engineering, and the ability to evaluate them critically and to apply them effectively in engineering projects	
Engineering analysis			
Engineering analysis involves the application of engineering concepts and tools to the solution of engineering problems. Graduates will need	Engineering analysis involves the application of engineering concepts and tools to the solution of engineering problems. Graduates will need	Engineering analysis involves the application of engineering concepts and tools to the solution of engineering problems. Graduates will need	Engineering analysis involves the application of engineering concepts and tools to the solution of engineering problems. The main engineering analysis abilities will have been developed in an accredited engineering undergraduate programme. Masters graduates will therefore need additionally
• Ability to monitor, interpret and apply the results of analysis and modelling in order to bring about continuous improvement	• Understanding of engineering principles and the ability to apply them to analyse key engineering processes	• Understanding of engineering principles and the ability to apply them to undertake critical analysis of key engineering processes	• Ability both to apply appropriate engineering analysis methods for solving complex problems in engineering and to assess their limitations
• Ability to apply quantitative methods in order to understand the performance of systems and components	• Ability to identify, classify and describe the performance of systems and components through the use of analytical methods and modelling techniques	• Ability to identify, classify and describe the performance of systems and components through the use of analytical methods and modelling techniques	• Ability to use fundamental knowledge to investigate new and emerging technologies

(continued)

(continued)

Programmes accredited for IEng	Programmes accredited for partial CEng	Programmes accredited for CEng	Programmes accredited for additional learning for CEng
Bachelors degrees and bachelors (Honours) (BSc or BEng)	Bachelors (Honours) degrees for partial CEng (BEng)	Integrated masters (MEng) degrees	Masters degrees other than the integrated masters (MSc)
• Ability to use the results of engineering analysis to solve engineering problems and to recommend appropriate action	• Ability to apply quantitative and computational methods in order to solve engineering problems and to implement appropriate action	• Ability to apply quantitative and computational methods, using alternative approaches and understanding their limitations, in order to solve engineering problems and to implement appropriate action	• Ability to collect and analyse research data and to use appropriate engineering analysis tools in tackling unfamiliar problems, such as those with uncertain or incomplete data or specifications, by the appropriate innovation, use or adaptation of engineering analytical methods
• Ability to apply an integrated or systems approach to engineering problems through know-how of the relevant technologies and their application	• Understanding of, and the ability to apply, an integrated or systems approach to solving engineering problems	• Understanding of, and the ability to apply, an integrated or systems approach to solving complex engineering problems	
		• Ability to use fundamental knowledge to investigate new and emerging technologies	
		• Ability to extract and evaluate pertinent data and to apply engineering analysis techniques in the solution of unfamiliar problems	

(continued)

(continued)

Programmes accredited for IEng	Programmes accredited for partial CEng	Programmes accredited for CEng	Programmes accredited for additional learning for CEng
Bachelors degrees and bachelors (Honours) (BSc or BEng)	Bachelors (Honours) degrees for partial CEng (BEng)	Integrated masters (MEng) degrees	Masters degrees other than the integrated masters (MSc)

Design

Design at this level is the creation and development of an economically viable product, process or system to meet a defined need. It involves significant technical and intellectual challenges and can be used to integrate all engineering understanding, knowledge and skills to the solution of real and complex problems. Graduates will therefore need the knowledge, understanding and skills to	Design at this level is the creation and development of an economically viable product, process or system to meet a defined need. It involves significant technical and intellectual challenges and can be used to integrate all engineering understanding, knowledge and skills to the solution of real and complex problems. Graduates will therefore need the knowledge, understanding and skills to	Design at this level is the creation and development of an economically viable product, process or system to meet a defined need. It involves significant technical and intellectual challenges and can be used to integrate all engineering understanding, knowledge and skills to the solution of real and complex problems. Graduates will therefore need the knowledge, understanding and skills to	Design at this level is the creation and development of an economically viable product, process or system to meet a defined need. It involves significant technical and intellectual challenges and can be used to integrate all engineering understanding, knowledge and skills to the solution of real and complex problems. The main design abilities will have been developed in an accredited engineering undergraduate programme. Masters graduates will need additionally
• Be aware of business, customer and user needs, including considerations such as the wider engineering context, public perception and aesthetics	• Understand and evaluate business, customer and user needs, including considerations such as the wider engineering context, public perception and aesthetics	• Understand and evaluate business, customer and user needs, including considerations such as the wider engineering context, public perception and aesthetics	• Knowledge, understanding and skills to work with information that may be incomplete or uncertain, quantify the effect of this on the design and, where appropriate, use theory or experimental research to mitigate deficiencies

(continued)

(continued)

Programmes accredited for IEng	Programmes accredited for partial CEng	Programmes accredited for CEng	Programmes accredited for additional learning for CEng
Bachelors degrees and bachelors (Honours) (BSc or BEng)	Bachelors (Honours) degrees for partial CEng (BEng)	Integrated masters (MEng) degrees	Masters degrees other than the integrated masters (MSc)
• Define the problem, identifying any constraints including environmental and sustainability limitations; ethical, health, safety, security and risk issues; intellectual property; codes of practice and standards	• Investigate and define the problem, identifying any constraints including environmental and sustainability limitations; ethical, health, safety, security and risk issues; intellectual property; codes of practice and standards	• Investigate and define the problem, identifying any constraints including environmental and sustainability limitations; ethical, health, safety, security and risk issues; intellectual property; codes of practice and standards	• Knowledge and comprehensive understanding of design processes and methodologies and the ability to apply and adapt them in unfamiliar situations
• Work with information that may be incomplete or uncertain and be aware that this may affect the design	• Work with information that may be incomplete or uncertain and quantify the effect of this on the design	• Work with information that may be incomplete or uncertain, quantify the effect of this on the design and, where appropriate, use theory or experimental research to mitigate deficiencies	• Ability to generate an innovative design for products, systems, components or processes to fulfil new needs
• Apply problem-solving skills, technical knowledge and understanding to create or adapt design solutions that are fit for purpose including operation, maintenance, reliability etc.	• Apply advanced problem-solving skills, technical knowledge and understanding, to establish rigorous and creative solutions that are fit for purpose for all aspects of the problem including production, operation, maintenance and disposal	• Apply advanced problem-solving skills, technical knowledge and understanding to establish rigorous and creative solutions that are fit for purpose for all aspects of the problem including production, operation, maintenance and disposal	
• Manage the design process, including cost drivers, and evaluate outcomes	• Plan and manage the design process, including cost drivers, and evaluate outcomes	• Plan and manage the design process, including cost drivers, and evaluate outcomes	
• Communicate their work to technical and non-technical audiences	• Communicate their work to technical and non-technical audiences	• Communicate their work to technical and non-technical audiences	

(continued)

(continued)

Programmes accredited for IEng	Programmes accredited for partial CEng	Programmes accredited for CEng	Programmes accredited for additional learning for CEng
Bachelors degrees and bachelors (Honours) (BSc or BEng)	Bachelors (Honours) degrees for partial CEng (BEng)	Integrated masters (MEng) degrees	Masters degrees other than the integrated masters (MSc)
		• Demonstrate wide knowledge and comprehensive understanding of design processes and methodologies and the ability to apply and adapt them in unfamiliar situations	
		• Demonstrate the ability to generate an innovative design for products, systems, components or processes to fulfil new needs	

Economic, legal, social, ethical and environmental context

Engineering activity can have impacts on the environment, on commerce, on society and on individuals. Graduates therefore need the skills to manage their activities and to be aware of the various legal and ethical constraints under which they are expected to operate, including	Engineering activity can have impacts on the environment, on commerce, on society and on individuals. Graduates therefore need the skills to manage their activities and to be aware of the various legal and ethical constraints under which they are expected to operate, including	Engineering activity can have impacts on the environment, on commerce, on society and on individuals. Graduates therefore need the skills to manage their activities and to be aware of the various legal and ethical constraints under which they are expected to operate, including	Engineering activity can have impacts on the environment, on commerce, on society and on individuals. Graduates therefore need the skills to manage their activities and to be aware of the various legal and ethical constraints under which they are expected to operate, including
• Understanding of the need for a high level of professional and ethical conduct in engineering and a knowledge of professional codes of conduct	• Understanding of the need for a high level of professional and ethical conduct in engineering and a knowledge of professional codes of conduct	• Understanding of the need for a high level of professional and ethical conduct in engineering, a knowledge of professional codes of conduct and how ethical dilemmas can arise	• Awareness of the need for a high level of professional and ethical conduct in engineering

(continued)

(continued)

Programmes accredited for IEng	Programmes accredited for partial CEng	Programmes accredited for CEng	Programmes accredited for additional learning for CEng
Bachelors degrees and bachelors (Honours) (BSc or BEng)	Bachelors (Honours) degrees for partial CEng (BEng)	Integrated masters (MEng) degrees	Masters degrees other than the integrated masters (MSc)
• Knowledge and understanding of the commercial, economic and social context of engineering processes	• Knowledge and understanding of the commercial, economic and social context of engineering processes	• Knowledge and understanding of the commercial, economic and social context of engineering processes	• Awareness that engineers need to take account of the commercial and social contexts in which they operate
• Knowledge of management techniques that may be used to achieve engineering objectives	• Knowledge and understanding of management techniques, including project management, that may be used to achieve engineering objectives	• Knowledge and understanding of management techniques, including project and change management, that may be used to achieve engineering objectives, their limitations and how they may be applied appropriately	• Knowledge and understanding of management and business practices, their limitations, and how these may be applied in the context of the particular specialisation
• Understanding of the requirement for engineering activities to promote sustainable development	• Understanding of the requirement for engineering activities to promote sustainable development and ability to apply quantitative techniques where appropriate	• Understanding of the requirement for engineering activities to promote sustainable development and ability to apply quantitative techniques where appropriate	• Awareness that engineering activities should promote sustainable development and ability to apply quantitative techniques where appropriate
• Awareness of relevant legal requirements governing engineering activities, including personnel, health and safety, contracts, intellectual property rights, product safety and liability issues	• Awareness of relevant legal requirements governing engineering activities, including personnel, health and safety, contracts, intellectual property rights, product safety and liability issues	• Awareness of relevant legal requirements governing engineering activities, including personnel, health and safety, contracts, intellectual property rights, product safety and liability issues, and an awareness that these may differ internationally	• Awareness of relevant regulatory requirements governing engineering activities in the context of the particular specialisation

(continued)

(continued)

Programmes accredited for IEng	Programmes accredited for partial CEng	Programmes accredited for CEng	Programmes accredited for additional learning for CEng
Bachelors degrees and bachelors (Honours) (BSc or BEng)	Bachelors (Honours) degrees for partial CEng (BEng)	Integrated masters (MEng) degrees	Masters degrees other than the integrated masters (MSc)
• Awareness of risk issues, including health and safety, environmental and commercial risk	• Knowledge and understanding of risk issues, including health and safety, environmental and commercial risk, and of risk assessment and risk management techniques	• Knowledge and understanding of risk issues, including health and safety, environmental and commercial risk, risk assessment and risk management techniques and an ability to evaluate commercial risk	• Awareness of and ability to make general evaluations of risk issues in the context of the particular specialisation, including health and safety, environmental and commercial risk
		• Understanding of the key drivers for business success, including innovation, calculated commercial risks and customer satisfaction	
Engineering practice			
This is the practical application of engineering skills, combining theory and experience, and use of other relevant knowledge and skills. This can include	This is the practical application of engineering skills, combining theory and experience, and use of other relevant knowledge and skills. This can include	This is the practical application of engineering skills, combining theory and experience, and use of other relevant knowledge and skills. This can include	The main engineering practice abilities will have been developed in an accredited engineering undergraduate programme. Masters graduates will need to demonstrate application of these abilities where appropriate and additional engineering skills which can include
• Knowledge of contexts in which engineering knowledge can be applied (e.g. operations and management, application and development of technology, etc.)	• Understanding of contexts in which engineering knowledge can be applied (e.g. operations and management, application and development of technology, etc.)	• Understanding of contexts in which engineering knowledge can be applied (e.g. operations and management, application and development of technology, etc.)	• Advanced level knowledge and understanding of a wide range of engineering materials and components

(continued)

(continued)

Programmes accredited for IEng	Programmes accredited for partial CEng	Programmes accredited for CEng	Programmes accredited for additional learning for CEng
Bachelors degrees and bachelors (Honours) (BSc or BEng)	Bachelors (Honours) degrees for partial CEng (BEng)	Integrated masters (MEng) degrees	Masters degrees other than the integrated masters (MSc)
• Understanding of and ability to use relevant materials, equipment, tools, processes, or products	• Knowledge of characteristics of particular materials, equipment, processes, or products	• Knowledge of characteristics of particular equipment, processes, or products, with extensive knowledge and understanding of a wide range of engineering materials and components	• A thorough understanding of current practice and its limitations, and some appreciation of likely new developments
• Knowledge and understanding of workshop and laboratory practice	• Ability to apply relevant practical and laboratory skills	• Ability to apply relevant practical and laboratory skills	• Ability to apply engineering techniques, taking account of a range of commercial and industrial constraints
• Ability to use and apply information from technical literature	• Understanding of the use of technical literature and other information sources	• Understanding of the use of technical literature and other information sources	• Understanding of different roles within an engineering team and the ability to exercise initiative and personal responsibility, which may be as a team member or leader
• Ability to use appropriate codes of practice and industry standards	• Knowledge of relevant legal and contractual issues	• Knowledge of relevant legal and contractual issues	
• Awareness of quality issues and their application to continuous improvement	• Understanding of appropriate codes of practice and industry standards	• Understanding of appropriate codes of practice and industry standards	
• Awareness of team roles and the ability to work as a member of an engineering team	• Awareness of quality issues and their application to continuous improvement	• Awareness of quality issues and their application to continuous improvement	
	• Ability to work with technical uncertainty	• Ability to work with technical uncertainty	

(continued)

(continued)

Programmes accredited for IEng	Programmes accredited for partial CEng	Programmes accredited for CEng	Programmes accredited for additional learning for CEng
Bachelors degrees and bachelors (Honours) (BSc or BEng)	Bachelors (Honours) degrees for partial CEng (BEng)	Integrated masters (MEng) degrees	Masters degrees other than the integrated masters (MSc)
	• Understanding of, and the ability to work in, different roles within an engineering team	• A thorough understanding of current practice and its limitations, and some appreciation of likely new developments	
		• Ability to apply engineering techniques taking account of a range of commercial and industrial constraints	
		• Understanding of different roles within an engineering team and the ability to exercise initiative and personal responsibility, which may be as a team member or leader	

Additional general skills

Graduates must have developed transferable skills, additional to those set out in the other learning outcomes, that will be of value in a wide range of situations, including the ability to	Graduates must have developed transferable skills, additional to those set out in the other learning outcomes, that will be of value in a wide range of situations, including the ability to	Graduates must have developed transferable skills, additional to those set out in the other learning outcomes, that will be of value in a wide range of situations, including the ability to	Graduates must have developed transferable skills, additional to those set out in the other learning outcomes, that will be of value in a wide range of situations, including the ability to
• Apply their skills in problem-solving, communication, working with others, information retrieval, and the effective use of general IT facilities	• Apply their skills in problem-solving, communication, working with others, information retrieval, and the effective use of general IT facilities	• Apply their skills in problem-solving, communication, working with others, information retrieval and the effective use of general IT facilities	• Apply their skills in problem-solving, communication, information retrieval, working with others, and the effective use of general IT facilities
• Plan self-learning and improve performance, as the foundation for lifelong learning/CPD	• Plan self-learning and improve performance, as the foundation for lifelong learning/CPD	• Plan self-learning and improve performance, as the foundation for lifelong learning/CPD	• Plan self-learning and improve performance, as the foundation for lifelong learning/CPD

(continued)

(continued)

Programmes accredited for IEng	Programmes accredited for partial CEng	Programmes accredited for CEng	Programmes accredited for additional learning for CEng
Bachelors degrees and bachelors (Honours) (BSc or BEng)	Bachelors (Honours) degrees for partial CEng (BEng)	Integrated masters (MEng) degrees	Masters degrees other than the integrated masters (MSc)
• Plan and carry out a personal programme of work	• Plan and carry out a personal programme of work, adjusting where appropriate	• Monitor and adjust a personal programme of work on an ongoing basis	• Monitor and adjust a personal programme of work on an ongoing basis
• Exercise personal responsibility, which may be as a team member	• Exercise initiative and personal responsibility, which may be as a team member or leader	• Exercise initiative and personal responsibility, which may be as a team member or leader	• Exercise initiative and personal responsibility, which may be as a team member or leader

References

Chartered Society of Designers (2015) CSD Genetic Matrix CPSK+A. London: Chartered Society of Designers. https://www.csd.org.uk/about/genetic-matrix/cpsk-a/. Accessed 4 July 2018

Design Society (2018) Design Education Special Interest Group. https://www.designsociety.org/ds_group/1/Design+Education. Accessed 11th July 2018

Engineering Council (2014) The accreditation of higher education programmes; UK Standard for Professional Engineering Competence, London, Engineering Council

Eraut M (2006) Early career learning at work and its implications for universities. In: Student learning and university teaching, BJEP monograph series II, vol 4, pp 1–22

International Engineering Alliance (2018) Washington Accord: Signatories. IEA. http://www.ieagreements.org/accords/washington/signatories/. Accessed 11 July 2018

Design and Design Education as a Profession: Professional Registration and Membership of Societies for Designers and Design Educators; Continuous Professional Development (CPD)

Chris Dowlen

Abstract **Professional Development** This chapter starts by outlining the typical career trajectory that might be undertaken as a professional designer, from student to expert designer, and what different descriptive levels mean, illustrated using a table of professional competencies developed by Michael Eraut, which is used as background to identifying professional competencies at different levels. The position of Design Educators is discussed. **Professional registration and membership of societies for designers and design educators**. This covers the need and desirability of professional registration. The memberships available from the Chartered Society of Designers and the Institution of Engineering Designers are described, together with an outline of the Engineering Council's registration grades available through the bodies it licences. The Higher Education Academy's professional recognition scheme for teaching staff in Higher Education is outlined, and Eraut's outline table of competencies is used to show how these relate to each other. Registration processes and assessment of professional levels are covered. **Continuous Professional Development (CPD)** Each society expects its members to be engaged in Continuing Professional Development. This section asks whether this is realistic, and why. It then takes a broad brush treatment of CPD and investigates ways to carry it out, to keep the records and then to evaluate the development that has taken place.

Keywords Professional registration · Design proficiency · Continuing professional development

C. Dowlen (✉)
London, UK
e-mail: cdowlen@ied.org.uk

© Springer Nature Switzerland AG 2019
D. Schaefer et al. (eds.), *Design Education Today*,
https://doi.org/10.1007/978-3-030-17134-6_14

1 Introduction

This chapter covers a huge range, covering professional development for designers and design educators. It outlines how this generally happens, and explains the processes of professional recognition and continuing professional development.

1.1 Design as a Profession

Designers do design. That is obvious. But doing design is only part of what they do. In design employment they work with others to produce an income for the company. This may mean that they are doing any or all of the *Total Design* (Pugh 1991) activities: ergonomics, aesthetics, graphic design, photography, technical writing, pack engineering, PCB design, electronics hardware design, mechanical design, electromechanical design, software design, industrial design. All may be needed to produce a successful product. As a designer's career develops, what they do changes. They may start working in a relatively small, constrained, disciplinary area, and then develop into more general areas, covering things such as marketing, human resources, staff management, team working and so on. Later in their careers they may not actually be doing much design, although the basis of their design thinking is normally present. But they are still design professionals. Design itself can have many different aspects, ranging from appearance design and fashion design through to engineering. And all of these are design, employing design professionals. The design process is larger and more general than the detailed business of determining a product for production, manufacture, realisation—going from thought to thing.

Designers may work in many contexts. They may be in a manufacturing company, a consultancy, as sole traders, as service designers, charity workers and as advisors. They may be informal designers, or more formally trained. They may work as educators, informally or formally.

Most designers also feel that their design work is a greater cause than simply producing economic success and that some of the great designs produced are not those that have made the designer a lot of money.

1.2 The Designer Within a Manufacturing Company

The manufactured products that need to pay the bills. Design work must always be directed towards that end, although good design costs little more than poor design and the design cause is aesthetic and customer-focused rather than economic and company-focused. Here designers work with others to ensure that the products satisfy the customer's wishes, desires, and also that they make this profit. Therefore, when they graduate they may take on extra informal roles, and they may not be able to work within their chosen and preferred discipline. They develop abilities that are more rounded and intertwined with others. They take on responsibilities of managing product developments, financial responsibilities, staff management, company strategic thinking, and bring their design abilities into many contexts that are not strictly 'design'.

1.3 The Designer Within a Consultancy

Designers may be employed in consultancy. Here the consultancy doesn't take on the risks for product success; the design work itself needs to pay the bills. The company still needs confidence that their designs are customer-focused, make financial sense for their clients and they take on the risk of living off their design abilities. There are non-designers in consultancies, as well. Perhaps a third of a consultancy's budget may be spent on marketing, perhaps more than in a manufacturing company, and consultant designers are always nearer to commerciality, the interface between their designs and the client which generates their cash flow. Designers here develop significant professional skills, and they may progress from being in a design team to team leader, to manager, partner, director—and so on.

1.4 The Designer as a Sole Trader

The graduate designer chooses to go alone, as their own company, standing on their own two feet, doing their design work which may be manufacturing or consultancy. Here they either develop their own non-design abilities or utilise specialists to fill in gaps in their abilities. They quickly become professionals at many things, quickly learning how to sink or swim, ensuring that they can make ends meet and anticipating that their next great design will see their design work realised profitably—looking for the next big break.

1.5 The Designer as Informal Educator

Many designers work also as educators. They may be formally employed within Higher Education, or may be informal educators. Eraut (2006) shows that most learning in professions is tacit, where people learn from colleagues, who become informal teachers—and may not even realise the educational contribution they make. This informal teaching may start as a student, where fellow students contribute as informal mentors and group members. Phil Race (Race 2014) identifies that when students talk about what they learn they start putting learning into their own words and internalise it, and doing this with colleagues becomes an appreciated part of being a student. A requirement of several professional registration levels is to encourage and support colleagues, formally and informally.

1.6 Design Education as a Profession

Alternatively, a designer may work formally as an educator. Ideally, this will be some time after they graduate, after they have gained experience in manufacturing industry or consultancy, or both. In education designers also develop non-design expertise—obviously in education, but perhaps in research, and the others—finance, marketing, staff management etc. Educational expertise includes theory—about how people learn, why people learn, the different levels of learning, how to develop assessment regimes and strategies, how to design and develop courses (a design-based skill), write learning outcomes and course documents, and ensure their teaching skills are effective and professional. Their career progression may go from lecturer (a basic 'teacher') to managing staff, developing staff as well as students, managing research, teaching programmes, budgets, University learning strategies, and so on. This demands professionalism—beyond and in addition to design expertise.

2 Trajectories of Professional Development

In each of these contexts design professionals can be at different stages of professional development. They develop from 'rookie' to competent, to developing professional responsibilities including (usually) staffing, finance, project management and strategic thinking, to focusing on strategy and (perhaps) less on design competencies.

Eraut (2006) investigated early learning in work for three different professions. He showed that most learning in work does comes not from formal education or short courses, but from informal settings and is not specified nor planned. It arises naturally from the work, and is usually picked up from colleagues, clients, managers; from trying things out, implementing ideas and discussing things. Individuals may circumvent and subvert official training opportunities, gaining the required learning from colleagues, manuals, and other sources. Things may be learnt simply because the work requires the learning rather than being a conscious choice of the individual or their desire for formal qualifications and certifications of attendance. Learning in the workplace varies with local management, and how the individual seeks to improve, which is usually for the task in hand rather than personal gain.

He thus saw the development of professional competence as learning trajectories, and gave examples of learning activities in Table 1.

He says that learning trajectories may move in rapid steps, slow changes, stalling, or even regression, and may not progress from junior competence to seniority. Sometimes the choice not to move is deliberate—sometimes circumstances and colleagues influence movements or prevent them taking place.

Table 1 Examples of items that might be relevant to a learning trajectory (Eraut 2006)

Task performance	Role performance
Speed and fluency	Prioritisation
Complexity of tasks and problems	Range of responsibility
Range of skills required	Supporting other people's learning
Communication with a wide range of people	Leadership
Collaborative work	Accountability
Awareness and understanding	*Supervisory role*
Other people: colleagues, customers, managers, etc.	Delegation
Contexts and situations	Handling ethical issues
One's own organization	Coping with unexpected problems
Problems and risks	Crisis management
Priorities and strategic issues	Keeping up-to-date
Value issues	*Academic knowledge and skills*
Personal development	Use of evidence and argument
Self-evaluation	Accessing formal knowledge
Self-management	Research-based practice
Handling emotions	Theoretical thinking
Building and sustaining relationships	Knowing what you might need to know
Disposition to attend to other perspectives	Using knowledge resources (human, paper-based, electronic)
Disposition to consult and work with others	Learning how to use relevant theory (in a range of practical situations)
Disposition to learn and improve one's practice	*Decision making and problem solving*
Accessing relevant knowledge and expertise	When to seek expert help
Ability to learn from experience	Dealing with complexity
Teamwork	Group decision making
Collaborative work	Problem analysis
Facilitating social relations	Formulating and evaluating options
Joint planning and problem solving	Managing the process within an appropriate time scale
Ability to engage in and promote mutual learning	Decision making under pressure
	Judgement
	Quality of performance, output and outcomes
	Priorities
	Value issues
	Levels of risk

2.1 Stages of Professional Development

In a slightly later publication Eraut (2009) used a table originally produced by Dreyfus and Dreyfus in 1986 (Dreyfus and Dreyfus 1986) that used a series of states to describe levels of professional competence. This, however, does not really seem to be complete and Eraut suggested that it is a useful starting point, but does not allow for enough complexity or team-working. Roughly speaking, the table can be summarised into about six stages as follows:

1. general interest and on the border of the community
2. student—a learner and on the community border with a trajectory to enter (Dreyfus and Dreyfus termed this a novice)
2a. a member of an associated profession on the community border who does not consider themselves part of it—but they may hold a senior position on the border
3. competent practitioner—can do the job of a designer, an engineer, educator or researcher
4. proficient practitioner—able and competent to lead projects
5. expert practitioner—able to make a strategic difference in the overall community of practice
6. retired practitioner—was part of the community, has an exiting trajectory and wishes to retain links.

These stages might be given terms such as affiliate, student, associate, registered [practitioner], chartered [practitioner], principal [practitioner]. For some professions the picture is confusing. And the individual nature of trajectories means that many do not follow this sequence. Individuals choose their own competencies and interests which may be outside of one profession, and may simply find that they (tacitly) develop sets of competencies that do not align well with by any 'official' professional organisation.

General interest categories are characterised by the learning being almost completely tacit, book-learned, or deliberately ignored and kept at a non-current level (including the retired). If they wish to become a more involved and registered part of the community then they will probably find either a system to circumvent formal learning and replace it with tacit learning, or a way to undertake more formal learning, to satisfy themselves and the registering authorities that they can make leaps in their professional trajectory.

2.2 Professional Development Trajectories for Designers

2.2.1 The Informal Designer

There are a significant number of non-professional designers; this is a large part of the informal design community. In 2000 it was estimated that only 19% of the

British firms awarded a Millennium Design award by the Design Council actually employed a professionally-qualified designer (Utterback et al. 2006)—perhaps called silent design (Gorb and Dumas 1987). People in this category may not ever develop a design trajectory to encapsulates an official professional status, being content to design from time to time—and still carrying out excellent design on occasions.

2.2.2 The Design Student and Graduate

Most design students are on the road to completing their studies. But they are branded as learners for the whole of this student period, whether or not they can identify that they have the skills and abilities to gain a more formal professional identification or not. Typically, they will start by undertaking numerous small design tasks to develop tacit learning as well as course learning, and this will include learning informally from their colleagues as well as picking up professional clues from lecturers and others such as industry professionals providing them 'live' projects, industrial placements, occasional lectures or industrial visits. Formal learning practices need to include evidence that tacit learning has taken place. Most design students undertake a large individual project whereby the outcome is not identified in advance and where a large amount of tacit learning takes place, and in most Masters courses there is a significant element of group project work where this informal learning is between several different people in the group and of an undefined nature. Graduates will normally be learning on the job—seldom do they leave a course in 'oven ready' condition and able to make an immediate impact in a company.

2.2.3 The Competent Designer

Once they have graduated, they would typically become a junior individual within a team, work on several smaller projects, or develop individual expertise through obtaining commissioned work. Within a short period of time, typically one or two years, they become a competent professional and can apply for official recognition. They will be applying formal design principles and practice from their degrees plus informal tacit learning through work and designing competently—able to work in many settings but not expecting to take full responsibility for the product, even though they are contributing significantly to its design and development. Evidence suggests that working on as many smaller projects as possible is advantageous, and helps graduates gain overall experience far quicker and more effectively than one large, complex project. Small projects have variety, and designers can comprehend the whole of the project better, even though only responsible for a small part. Working in a small team, with day-to day supervision and with other team members also means they can learn more quickly than with a large project where they may only do a small part and may become side-tracked into becoming an expert.

They probably fit the professional category of a competent practitioner, and can usually apply for a professional grade. This stage might be termed 'working with design knowledge'.

2.2.4   The Proficient Designer

Once they can lead projects and take responsibility for design work, perhaps heading up a design team, and having overall budgetary control, then they are probably ready to be classed as 'proficient practitioner'. This stage will probably be reached quicker if the person is working for themselves, which has advantages in responsibility but disadvantages in that it is harder to learn from other people. These designers can not only lead teams and have financial responsibilities, but can also act with more discernment and creativity than as a team junior—aspects of their thinking ability relating to what in Bloom's taxonomy are 'higher order' skills (Bloom et al. 1956). This stage might be termed 'working with design wisdom'.

2.2.5   The Expert Designer

The final professional category is the expert practitioner. Here the person is unlikely to be directly responsible for the design lines, but is more likely to be responsible for the company progress, how the product fares in the marketplace, for giving agreement for facelifts or replacements, understanding market position, determining company design strategy, having input about offering a novel service—and so on. They haven't forgotten their design roots, but others take on day-to-day work. They lead the design, push the boundaries, take responsibility for novelty—and so on. They may be household names … They have short discussions rather than long, complicated arguments, but they are aware of them: they do not need them to convey what they need.

For example, engineering designers may be working on an earth-moving vehicle. The managing director comes past to see progress. 'You need to put the engine the other way round', he says, and walks away. They scratch their heads for a few seconds, and decide that although doing what he suggests will take them a lot of effort, they will do it—because he's the managing director (and he has an engineering background). After the new product is revealed in its new configuration, the team says 'He was right'.

Another example. Design awards are being presented by HRH Prince Philip. Ian Callum, chief designer at Jaguars, collects his award. Prince Philip asks him, 'And what do you do?' 'I design Jaguars' is the rather short answer, at which the assembled group bursts into laughter. See Fig. 1.

What is seen here is an example of Jardin's principle: understanding will pass through three stages. To start with, descriptions will be simplistic i.e. over-simplified, later it will become complex but ultimately it will be described simply. These stages can be summarised as being: obvious, complex, profound (Eastaway 1997; Gooden

Fig. 1 Ian Callum presented with Prince Philip, July 2015

2015). When someone reaches this expert level, they will have been through the simplistic and complex stages, and can show profundity. This stage is 'creating and developing the design wisdom'.

2.3 Moving from One Category to the Next

Identifying category edges is difficult—how and when someone moves between categories—and people get it wrong. It is said that if you are the expert, then everyone else knows it, but you don't. Experts usually know they are working at the strategic level, and aware of complexity, even if they skate over this for the sake of brevity. It can be difficult is to distinguish between competent and proficient and how much project control and responsibility needs to be seen as evidence. Alternatively, industry-based students may be using a judgement level that suggests proficiency beyond their (relatively) slow progress through a part-time degree. They become ready for the professional membership as soon as they graduate, gaining experience alongside their degree.

Conversely, particularly on a large, complex project, it can be extremely difficult to develop professional responsibilities and this can be frustrating as it is difficult to experience the overall project scope—they may become 'chief designer of door handles'.

Those working in the Higher Education sector are in a different situation. An industrial background is beneficial for Higher Education lecturers, gaining design experience, particularly in manufacturing, preferably to the proficient level. Without this it can be difficult to relate the student's understanding to the 'real world' and that lack of connection with practical design can lead to a substandard student experience. For those without practical experience, there are many opportunities, and they may rely on these to demonstrate design professionalism. These may be as obvious as spending a period working in manufacturing or consultancy, or they may be able to take on projects like Knowledge Transfer Partnerships, to experience keeping industrial contacts alive and giving management control of finance and staff (possibly limited to a single individual). A design-based research project may also provide this experience.

Professional development in Higher Education is typically from lecturer to course leader, to head of department or sideways to developing specific strategic responsibility like being responsible for departmental, school or faculty learning and teaching strategies. This can lead from competent to expert, and can be professionally recognised.

3 Professional Registration and Membership of Societies for Designers and Design Educators

Professional recognition by a recognised design or engineering institution means an individual has gained Chartered or some other recognised status, including being allowed to use post-nominals (letters after their name).

This means they have reached a particular professional stage, competence, professional, or, in some cases, expert. And this has been recognised by an organisation that has evident credence.

This is external—the standard is recognised by the professions and means achievement of a particular level of competence. Others within industry, commerce and similar professions know what this means.

This standard includes training, education, learning and practice, and this has probably been hard-won through processes of tacit and conventional learning, experience, and getting on with the task of doing effective design work. It is a badge of achievement, and is important as a measure of personal achievement, and is a milestone of personal competence.

While it does not give grades, it provides a marker on a personal learning trajectory. Where markers may otherwise simply be related to projects worked on, the levels have well-known and perceived relevance.

Almost all professional qualifications also have relevant post-nominals. These letters officially indicate the professional qualification—known as 'letters after your name'. These can be important signifiers of the level of professionalism reached and awarded.

The standards show that the individual is clearly included within the community of practice and they work with a known badge of that inclusion.

Professional bodies usually provide support and guidance for professional development including their application and the registration processes, and many provide a range of technical events, seminars, conferences, along with regular journals, technical publications, library access (both virtual and real, frequently), careers advice and opportunities for particular discounts that some companies supply to individual professionals—although this last benefit is probably best regarded as peripheral.

For parts of the Design Profession this recognition can act as a gateway. This is the case with several safety-related tasks in Engineering Design, such as the ability to evaluate pressure vessel designs, which must be legally carried out by qualified and competent people, working to the legal standards. This is also the case for several other safety-critical design areas.

3.1 Professional Registration for Designers

There are two professional design bodies. Other design bodies act like trade associations and major on specific aspects of the design profession, such as independent designers (the Design Business Association states that it is 'a powerful representative body focused on enabling our members' businesses to prosper; committed to shaping a positive future for all those working in the field of design and passionate about bringing design and business ever closer together to deliver the greatest returns' (Design Business Association 2018)).

These are the Chartered Society of Designers (CSD) and the Institution of Engineering Designers (IED). The first represents designers across many design disciplines, whilst the second majors on Product and Engineering Design. Both offer a range of professional qualifications for people from the student and interested individual through to expert.

The Chartered Society of Designers' professional qualifications are their own, supported by their Royal Chartership, which means they can award the Chartered Designer title, and their levels of membership include Student, Associate Member (Assoc. CSD), Member (MCSD), Chartered Designer, Affiliate Fellow (aCSDf) and Fellow (FCSD), and one for subsidiary design professionals of Affiliate Member (aCSDm).

The Institution of Engineering Designers has its own membership qualifications for Product Designers and CAD professionals. It is also part of the UK Engineering Council, the regulatory body for engineering in the UK. This complexity means that what it can offer is similarly complicated, and it can not only award its own professional qualifications, but can also award the Engineering Council recognised qualifications of Engineering Technician (EngTech), Incorporated Engineer (IEng) and Chartered Engineer (CEng). Its own professional qualifications are those of Student, Member (MIED), Fellow (FIED) and Affiliate, with its professional grades for Product Designers being Registered Product Designer (RProdDes) for competent designers and Chartered Technical Product Designer (CTPD) for proficient designers (supported by its Royal Chartership). Its grades for CAD professionals are Registered CAD Professional (RCP) and Registered CAD Manager (RCADMan). It can also offer Chartered Environmentalist (CEnv) on behalf of the Society for the Environment.

3.2 Professional Registration for Design Engineers

Each of the other Professional Engineering institutions will include some form of design engineering within their portfolio, a specialist form related to their specific mission in engineering. Each has its own professional membership structure, like the IED's. These almost all offer CEng, IEng and EngTech registration.

3.3 Professional Registration for Design Educators

It is a good idea for design lecturers and teachers to first of all be professionals in their disciplines. One factor in student appreciation is lecturers' enthusiasm and knowledge, and teaching by those with tacit knowledge of their profession is valued more than teaching by those without that 'real world' experience.

Therefore, design teaching professionals should ideally have a learning trajectory that includes formal acquisition of a professional qualification—whichever is appropriate, and whichever they can obtain. Mostly, this is best at Chartered level, but this may not be possible. Design teachers must also reflect on their learning trajectories and whether they should attempt to obtain industrial refreshment and experience in some way. Higher Education employers are extremely supportive of their staff involved with professional bodies, perceiving this to contribute significantly to personal and professional growth and building knowledge of what takes place elsewhere within their subject area—and many external examiners find it is their professional affiliation that means they can contribute more effectively in that quality role. This experience of other educational courses can be extremely valuable and can contribute towards a professional trajectory. A design lecturer or teacher needs to be regarded as professional and act professionally, and being appropriately registered is part of being treated as one.

The Higher Education Academy also has a professional registration scheme for Higher Education professionals. This has four levels of accreditation; Associate Fellow (AFHEA), Fellow (FHEA), Senior Fellow (SFHEA) and Principal Fellow (PFHEA).

3.4 Professional Registration—A Summary and Overview Table

This huge range of professional grades, qualifications and designatory letters means that people get confused. Table 2 is an attempt to correlate the various professional grades with the different levels that Dreyfus and Dreyfus (1986) used. Usually retired members can keep their qualifications and designatory letters provided they pay the fees, which may be reduced.

4 Application and Requirements for Professional Registration

Most of these need a relevant degree qualification—but not all of them. However, there are usually ways for those who do not hold the formal qualification to still obtain professional registration at all levels.

4.1 Formal Qualifications

Possession of a relevant degree for design may be a BA, BSc, BEng or BDes (occasionally) at Bachelor level, or an MDes or MEng as what is called an undergraduate Masters, or a Bachelors degree plus a one-year MA or MSc. This variation of titles can lead to confusion as to what registration levels are possible.

A Bachelors degree holder will generally expect to be able to achieve a competent level of professional qualification, while the holder of a Masters degree might normally expect to be able to achieve a proficient level, usually with the Chartered title. In addition to the right degree, the practising designer will need to demonstrate the right level of competence in practice, which is usually only gained some time after the degree has been achieved.

4.2 The Process

Firstly, decide on the right professional qualification and level. This may be obvious, depending on the degree, but sometimes it may not be easy to decide. Some professional memberships have no formal qualification requirement, and so it may be possible to first sign up at this basic level, to show interest. That usually simply means filling in the form. After that, the membership department of the organisation may well help and give an idea of what is possible.

Table 2 Levels and titles for professional registration. Titles where the possession of a relevant academic degree is expected are in *italics*

	Level	CSD	IED (and many other Professional Engineering Institutions) membership qualifications	IED Product Designer	IED CAD	Engineering Council (UK) (via a Professional Engineering Institution)	Higher Education Academy (UK)
1	Related professional	Affiliate Member (aCSDm)	Affiliate				Associate Fellow (AFHEA)
2	Student	Student	Student				–
	Recent practitioner	*Associate Member (Assoc. CSD)*					Associate Fellow (AFHEA)
	Practical implementer					Engineering Technician (EngTech)	
3	Competent	Member (MCSD)	Member (MIED)	*Registered (RProdDes)*	RCP	*Incorporated (IEng)*	Fellow (FHEA)
4	Proficient	Chartered	Member (MIED)	*Chartered (CTPD)*	RCADMan	*Chartered (CEng)*	Senior Fellow (SFHEA)
5	Expert	Fellow (FCSD), affiliate fellow (aCSDf)	Fellow (FIED)				Principal Fellow (PFHEA)
6	Retired		Retired member				

For all professional qualifications, the applicant needs to be practising at the right level. The criteria will be listed, perhaps vaguely, particularly if a wide range of different disciplines is covered, such as for Engineering Council qualifications (Engineering Council 2014). Each organisation publishes criteria and gives an idea of the application process and timescales.

4.2.1 Chartered Society of Designers

The Chartered Society of Designers has separate documents for each qualification, and they also have a matrix arrangements of competencies for the criteria for Membership and Fellowship (Chartered Society of Designers 2015a, b, c, d, e, f, g, h, i, j, k).

4.2.2 Institution of Engineering Designers

The Institution of Engineering Designers has a list of CTPD competencies (The Institution of Engineering Designers 2015), and asks other applicants to contact the Institution for details. It also gives links to the Engineering Council documents.

4.2.3 Engineering Council

The Engineering Council lists all criteria for each level in a single booklet (Engineering Council 2014). This is used by all Professional Engineering Institutions who operate IEng and CEng registration.

4.2.4 Higher Education Academy

The Higher Education Academy gives its criteria in a single booklet, called the UK Professional Standards Framework (Higher Education Academy 2011).

4.3 Which Professional Qualification to Select?

Not surprisingly, the Chartered Engineer (CEng) is held in greatest regard. But this is only for design engineers and not for other designers. Both the CSD's and the IED's Chartered qualifications gained approval recently from the Privy Council, and neither has yet developed the same reputation as the Chartered Engineer: the CSD's Chartered Designer qualification was approved in 2011 and the Chartered Technical Product Designer in 2015.

Deciding on the direction to take depends on which profession the applicant considers they belong to. Are they practising as a designer? Or as an engineer? Or as a Higher Education teacher? Or more than one? And if they are a Designer, are they a Technical designer, or a more general one?

The first call to check is probably the three competency matrices—the CPSK issued by the CSD (Chartered Society of Designers 2015d), the Chartered Technical Product Designer outline (The Institution of Engineering Designers 2015), the Engineering Council's UK-Spec document (Engineering Council 2014) and the Higher Education Academy's UK Professional Standards Framework (Higher Education Academy 2011).

The competency level needs careful study. Designers need to demonstrate creativity, technical designers to demonstrate technical design realisation and human interaction, engineers to demonstrate numerical competency, and teachers to show they know how people learn, that they can design, plan and implement teaching programmes, and that they can assess learners.

Professionalism and management need to be added as the learning trajectory develops. These areas will be used significantly to identify an individual's seniority level: designers show a portfolio of designs they have worked on and implemented, engineers show competence with relevant numerical assessments, and teachers demonstrate the programmes that they have worked on. At higher levels, all are expected to show budgetary, human and project management skills, and for expert levels all must demonstrate experience of strategic involvement and leadership and ability to make quick, effective evaluations and decisions.

Application starts with the form, followed by initial assessment by the organisation. This must be augmented by demonstration of professional competency, normally by a written report and portfolio, with the decision following a professional interview. Different professions have slightly different approaches—and processes may be convoluted—but the result should be that competencies and criteria are assessed effectively. No method is fool proof: ineffective assessment is sadly part of all methods, however devised (Kahneman 2012). Candidates need to concentrate on the professional organisation's criteria and avoid deviation into favoured but irrelevant topics.

4.4 Processes for Those Without the Formal Degree Qualifications

Candidates without degree qualifications still need to show competence as shown by those with formal qualifications. For the IED the normal way is by a portfolio. For IEng and CEng, candidates may take either a Technical Report route (demonstrating degree-equivalent learning through a single project) or may use a structured career portfolio. The choice may be dictated by the Engineering Institution concerned. Any process must still demonstrate equivalence, and is not necessarily an easy option.

5 Continuous Professional Development (CPD)

5.1 Is CPD a Requirement?

God does not do professional development. God does not change, is omnipotent and omniscient and perfect, has no need of any improvement and is the fount of all wisdom. But we are not God.

Perhaps personal consistency is inconsistent with the constant change required by every professional organisation, learning trajectories that suggest that what goes up must also come down, and continuous improvement that is not always possible nor desirable. This is the theme of Svend Brinkmann's book *Stand Firm: Resisting the self-improvement craze* (Brinkmann 2017), where he suggests it is not always effective or desirable for individuals to continuously need to develop, grow and self-improve—or even to change. He would like each person to possess the immutability of the Godhead, although the book is written in a somewhat tongue-in-cheek style and he is probably aware that this is, for humans, impossible.

However, each professional organisation expects its members to change and develop. Individuals and organisations change. People move from student to practitioner and thence to retirement and end-of-life stages. But this pattern may be interrupted. Change is not a constant, as has been suggested [attributed to Heraclitus, c535–c475 BC quoted in (Mpanga 2014)]: it is a variable (and engineers measure such changes through calculus). Change can be in multiple directions and between professions, taking in different aspects. Some changes are planned: others unplanned, some unwanted and not appreciated. Learning trajectories include painful episodes and enjoyable ones, and a multiplicity of events and movements.

Eraut (2006) suggests that much learning takes place through unstructured processes, and is tacit. This, of itself, cannot be planned. And the great difficulty for an individual is being able to identify the changes that have taken place, and the nature of an individual's learning trajectory.

On balance, continuous professional development produces greater individual benefit than otherwise. The main benefits to the individual are those of updating so they can continue to demonstrate competence—a survival process—and enhanced mobility. For organisations, it ensures their staff keep their skills relevant, which gives more value for the organisation; organisations value staff more highly when they are responsible for their own development, and it can help them with succession planning, even where staff develop skills at variance to the company direction.

5.2 Aims and Approaches to CPD

Not surprisingly, there is a professional body that exists to (at least in part of its title aim) promote personal development. This is the Chartered Institute for Personnel and Development (CIPD). Its first aim is:

To establish programmes of education and training and continuing professional development with recognised standards of achievement to support the systematic development and
accreditation of Members (Chartered Institute of Personnel and Development 2000).

Their approach towards CPD is exemplified by Megginson and Whitaker (Megginson and Whitaker 2007) who define CPD as 'a process by which individuals take
control of their own learning and development, by engaging in an on-going process
of reflection and action' (p. 3). They describe it as being 'empowering and exciting
and being able to stimulate people to achieve their ambitions and move towards their
dreams'.

They identify some key concepts of the ideal CPD approach, which is anything
but the unplanned, tacit reality described by Eraut.

Firstly, CPD is a process connected with the learner's life development, and therefore no other organisation or individual can identify the direction someone would
like to take, or the one taken in practice.

Secondly, it ideally includes elements of forward thinking and planning and reflection.

Thirdly, that it does not need a professional organisation or an employer to work.
The individual may be employed, in self-employment, a student, retired or in several
simultaneous employments, and the CPD may be stretched over many aspects of
life, including family planning and reality, professional life, religious affiliations and
anything else. If in employment, CPD will work whether it is encouraged, whether
the employer is indifferent or hostile or if they do not care about their staff.

CPD aspirations may be to follow the trajectory from being a student, through
competent practitioner, to professional practitioner, and perhaps to expert. But it
may not be an individual's aspiration, or even if it is, they may be thwarted in their
ambition to become involved with the strategic level of input, or may achieve that
level of input only to decide that they wish to withdraw from it.

Perhaps one of the best-known UK examples of an individual who appeared to
take a major change in direction is Sir David Attenborough, who moved through Television management, becoming Programme Director for BBC2 in 1964, the BBC's
Director of Programmes in 1969, and who then in 1972 (amid speculation that he
might become the Director General of the BBC) decided to return to film-making,
including making many programmes that have been among the BBC programmes
with the highest-ever audience figures such as Life on Earth (1979) and the Blue
Planet (2017) (Daily Telegraph 2011). This change was a deliberate choice away
from the strategic management of directorship. One might speculate that not all his
career choices were made in such a structured, reflective manner, but came about
through the tacit learning that took place throughout his long career.

Megginson and Whitaker (2007) give another example of someone leaving his
city job for an outdoor management consultant, finding greater fulfilment but earning
a smaller salary.

There are many other similar examples. CPD is not a linear process. Learning
trajectories may go up, down, around, at different speeds and with different intentions.

The CIPD recommends people take a structured approach towards CPD, through a cycle of six stages: (1) Identification and planning, (2) Action, (3) Reflection, (4) Application, (5) Sharing and (6) Maximising the Impact of what has been done. All that many professional bodies expect is stage 2 to take place, but that they also expect records to be kept (which strangely seems to be missed out of the CIPD cycle). They also appreciate it their members carry out stage 3, Reflection, takes place, but they don't currently demand that it does.

The CIPD has a useful guide that suggests how the reflective stage can be most useful (Owen and Fletcher n.d.). The leaflet indicates that reflective practice is different from reflecting, showing a habit rather than a single or multiple events, and suggests that habitual reflective practice is not just a formal process, but becomes a response to informal, tacit learning and embeds it into the learner.

Another stage suggested by the CIPD is that of sharing. Race shows that people learn better when they can talk to each other about what they have learnt (Race 2014). Verbalising helps the learner make sense of what they have learnt, putting it into their own language. He also suggests it is about hearing one's own voice speaking out the meanings of things, modulating and changing it so the concepts become ingrained.

5.3 What Topics Can Be Part of CPD?

Because everyone is individual, this varies. Whereas one person might wish to climb a corporate ladder and concentrate their CPD on strategy, corporate thinking, and business topics like company profitability, another may wish to develop secondary skills, using those to drive their CPD. The Institution of Engineering and Technology (IET) suggests using the mnemonic **TWAVES** to suggest topics for CPD for their members—**T**raining, **W**ork experience, **A**cademic study, **V**olunteering, **E**vents and seminars and **S**elf-study (The Institution of Engineering and Technology 2016). However, this doesn't seem to cover the informal, tacit, reflective-type of continuing, habit-forming activity very well—the kind of CPD encouraged by the CIPD.

The Chartered Society of Designers (CSD) has four criteria for CPD for its members:

- To develop personal professional competence
- To benefit clients and/or employer
- To add value to the overall design profession (whether in education or practice)
- To contribute to the wellbeing of society and the environment.

Many Institutions mention that CPD can include assisting others in their CPD. This aligns with Eraut who shows that much tacit early learning is through talking with colleagues. A major mark of distinction of the Senior Fellow of the HEA (SFHEA) is the support and development of colleagues' learning—they usually expect to see formal arrangements for this, such as arranging departmental training and discussion forums.

5.4 CPD Towards Professional Recognition

An obvious CPD topic is to develop towards the next level of professional recognition. Most Professional institutions can assist with this, sometimes through the formal appointment of a mentor, and sometimes by an appointment within the company who provides oversight of CPD in the company. Training may be available, and Institution visits and discussions can be arranged.

In this regard the HEA can accredit CPD schemes that lead directly to its Fellowship grades, and these usually include assessment for the Fellowship grades. The method of assessment may vary, but this is part of the accreditation process for such schemes.

5.5 Use of Repertory Grid in CPD Evaluation

Another approach to CPD, particularly if there are several events in the reflective log, is to select a small number of these CPD items, across a broad set of topics, and carry out a repertory grid analysis (Jankowicz 2004). This process seeks to uncover an individual's constructs for the topic—here, their attitude towards CPD. The process compares the items in triads, noting which item is different from the other two and why. This difference becomes a Construct that distinguishes the one item from the others. Each of CPD items is assessed against this construct, placing it into one or other category—either the construct applies to it, or it doesn't. This analysis can guide the (perhaps informal) direction for CPD. (Note that if nine items are selected, each can be compared with each of the other two through a set of twelve triadic comparisons). If this is carried out meaningfully, the differences between items will be perceived at a deep personal level and will not be trivial. Carrying out the process develops a deep insight and an individual can use the process to develop an understanding their tacit CPD processes. This is because the process involves evaluation and personal assessment, which itself contributes CPD.

6 CPD Summary

CPD may be thought of as an option—even though all professional institutions expect their members to carry it out, log it, reflect on it and evaluate it. However, it can offer benefits, whether done as a basic process or in the more reflective, thoughtful manner encouraged by the CIPD. The learning trajectory may become something more than simply something required by a professional body, but may become a personal, private exploration.

Appendices: Tables of Professional Requirements

Chartered Society of Designers CPSK^{TM} Matrix (2015a)

Creativity

C1	Creativity	How their imagination, intuition, insight and inspiration has contributed to their work
C2	Generating	Awareness of creative problem solving (CPS) techniques and their application. An understanding of the inhibitors of creative thought and how to negotiate them. Risk aware versus risk averse. Ability to generate appropriate ideas. How to negotiate convergent or divergent thinking in problem solving. Inquisition—serendipity—questioning
C3	Managing	Ability to scope, analyse, evaluate and select ideas for development. Proving of ideas for use in a specified context. Throughout the creative process, managing ideas in a business and creative environment, using analysis and evaluation as well as independent thinking
C4	Innovate	Adoption of generated ideas to address specific problem or need as set out in a brief or requirement in an original way. Identifying where ideas can be used to deliver original solutions in existing markets or creating new markets. Ability to exploit potential of creative ideas by the use of design principles to generate original outcomes

Professionalism

P1	Values	They possess and exercise values that are common to the environment in which designers practice. Are able to maintain integrity when undertaking work showing due regard for the practice of design. Act in a manner that respects and protects their colleagues, clients, the environment, nature and society
P2	Process	Ability to adopt appropriate methodology of practice. Continual learning and research into best design practice
P3	Communication	Ability to communicate with stakeholders through written and oral platforms in presenting design. An understanding of the interpersonal and psychological communications employed as well as the relationships involved. Appropriate use of communication techniques
P4	Contextual	Understanding and ability to use the appropriate regulations and requirements concerned and ensure standards (adoptive and statutory) are maintained in practice. Advanced knowledge of relevant IPRs and understanding of other IP issues. Ability to operate to best practice within their field of practice. Undertake work only if qualified to do so within a defined discipline

Skills

S1	Generic	Visual interpretation and communication of ideas and concepts in a manner that can be communicated with others. Use of colour and expression of form and how it is used to deliver concepts and managed for design outcomes. Conceptual and spatial awareness. Design thinking capability and creative skills
S2	Operating environment	Skills required within the operating environment of the designer. Ability to employ generic skills within operating environment and to the level required. Ability to audit and identify skills required within operating environment and address deficiencies whilst enhancing acquired skills. Financial, budgeting, management, organisational, leadership, information management and commercial skills
S3	Personal	Interpersonal and people skills. Complimentary and transferable skills
S4	Contextual	Skills required to practice competently within a defined design discipline. Ability to apply generic skills within the remit of a defined design discipline

Knowledge

K1	Explicit	Knowledge acquired from others which derives from research and experimentation and is generally accepted. Possesses theoretical knowledge. Possesses general knowledge
K2	Tacit	Knowledge gained from prior experiences at any stage of the design process
K3	Management	Ability to undertake research relevant to own professional practice whether academic and/or practice based. Ability to undertake research and acquire knowledge specifically related to delivering appropriate design solutions. Exploitation of knowledge for the benefit of all stakeholders, designer, client, commerce, society, etc.
K4	Contextual	Awareness, understanding and knowledge of the history of their profession and chosen discipline. Knowledge of the sector in which they operate including; processes, market conditions, legislation, technology, methodologies. Cultural awareness of operating environment and those involved

These criteria need to be demonstrated for Associate membership (Assoc. CSD), Membership (MCSD) and the Chartered Designer. For Fellowship (FCSD), achievements also need to be identified.

The Institution of Engineering Designers Criteria (2015)

	RProdDes	CTPD
A	*Use technical knowledge and humanistic understanding in the application of latest advances, including*	
A1	Involvement in a significant part of the product design process	Recognition by the product design profession of at least one product design undertaken by the candidate as being significant
A2	Application of historical and latest trends in design thinking and their appropriate application	Critical evaluation of historical and latest trends in design thinking and their appropriate application
A3	An ability to work to a client brief	An ability to challenge and/or create a client brief
B	*Apply a range of creative and scientific design methods to solution of product design problems, including*	
B1	Analytical and creative thought processes	Analytical and creative thought processes
B2	Understanding and/or reasonable usage of the functionality and capabilities of a CAD system or visualisation systems	Understanding and/or reasonable usage of the functionality and capabilities of a CAD system or visualisation systems
B3	Use of simulation or prototyping to explore and prove innovative solutions for design problems	Use of simulation or prototyping to explore and prove innovative solutions for complex design problems
B4	Contributes to the solution of design problems during the development of a substantially new product	Solution of design problems during the development of a substantially new product
B5	Contributes to the solution of design problems during the modification/improvement of an existing product	Solution of design problems during the modification/improvement of an existing product
B6	A holistic view of the design process encompassing understanding and application of marketing, aesthetics, ergonomics, manufacturing, materials, commercial and technical aspects	A holistic view of the design process encompassing understanding and application of marketing, aesthetics, ergonomics, manufacturing, materials, commercial and technical aspects
B7	Involvement with the development of products encompassing the overall design process from initial brief to manufacture of production prototype	Development of products encompassing the whole of the design process from initial brief to manufacture of production prototype
B8	Contributes to evaluations of design solutions against constraints	The ability to evaluate complex design solutions against conflicting constraints

(continued)

(continued)

	RProdDes	CTPD
C	*Design and commercial management, including demonstrating*	
C1	Responsibility for elements of product designs	A variety of substantially new product designs for which they have taken responsibility
C2	Contributing towards teams that meet changing technical and managerial needs	Complete project management of a medium to high complexity project including management of budget and personnel
C3	Managing continuous quality improvement	Leadership of teams and development of personnel to meet changing technical and managerial needs
C4		Bring about continuous improvement through structured evaluation
D	*Effective interpersonal skills, including*	
D1	Appropriate personal and social skills	Appropriate personal and social skills
D2	The presentation and discussion of product design briefs, specifications and concept design proposals	The presentation and discussion of product design briefs, specifications and concept design proposals
D3	Appropriate communication with others at all levels	Appropriate communication with others at all levels
E	*Commitment to professional standards, obligations to society and the environment, including*	
E1	Compliance with relevant codes of conduct	Compliance with relevant codes of conduct
E2	Management and application of safe systems of work	Management and application of safe systems of work
E3	Consideration of sustainable design practices	Consideration of sustainable design practices
E4	Continuing professional development necessary to maintain and enhance competence in own area of practice	Continuing professional development necessary to maintain and enhance competence in own area of practice
E5	Exercise responsibilities in an ethical manner	Exercise responsibilities in an ethical manner

Engineering Council—Extract from UK-Spec

The table below is an extract from the UK-Spec document covering competency levels for the Engineering Council professional registrations (Engineering Council 2014).

	EngTech competencies	IEng competencies	CEng competencies
A	*Knowledge and understanding This includes the ability to*	*Use a combination of general and specialist engineering knowledge and understanding to apply existing and emerging technology*	*Use a combination of general and specialist engineering knowledge and understanding to optimise the application of existing and emerging technology*
A1	Review and select appropriate techniques, procedures and methods to undertake tasks	Maintain and extend a sound theoretical approach to the application of technology in engineering practice	Maintain and extend a sound theoretical approach in enabling the introduction and exploitation of new and advancing technology
A2	Use appropriate scientific, technical or engineering principles	Use a sound evidence-based approach to problem-solving and contribute to continuous improvement	Engage in the creative and innovative development of engineering technology and continuous improvement systems
B	*Contribute to the design, development, manufacture, construction, commissioning, operation or maintenance of products, equipment, processes, systems or services In this context, this includes the ability to*	*Apply appropriate theoretical and practical methods to design, develop, manufacture, construct, commission, operate, maintain, decommission and re-cycle engineering processes, systems, services and products*	*Apply appropriate theoretical and practical methods to the analysis and solution of engineering problems*
B1	Identify problems and apply appropriate methods to identify causes and achieve satisfactory solutions	Identify, review and select techniques, procedures and methods to undertake engineering tasks	Identify potential projects and opportunities
B2	Identify, organise and use resources effectively to complete tasks, with consideration for cost, quality, safety, security and environmental impact	Contribute to the design and development of engineering solutions	Conduct appropriate research, and undertake design and development of engineering solutions
B3		Implement design solutions and contribute to their evaluation	Manage implementation of design solutions, and evaluate their effectiveness

(continued)

(continued)

	EngTech competencies	IEng competencies	CEng competencies
C	*Accept and exercise personal responsibility This includes the ability to*	*Provide technical and commercial management*	*Provide technical and commercial leadership*
C1	Work reliably and effectively without close supervision, to the appropriate codes of practice	Plan for effective project implementation	Plan for effective project implementation
C2	Accept responsibility for work of self or others	Manage tasks, people and resources to plan and budget	Plan, budget, organise, direct and control tasks, people and resources
C3	Accept, allocate and supervise technical and other tasks	Manage teams and develop staff to meet changing technical and managerial needs	Lead teams and develop staff to meet changing technical and managerial needs
C4		Manage continuous quality improvement	Bring about continuous improvement through quality management
D	*Use effective communication and interpersonal skills This includes the ability to*	*Demonstrate effective interpersonal skills*	*Demonstrate effective interpersonal skills*
D1	Use oral, written and electronic methods for the communication in English of technical and other information	Communicate in English with others at all levels	Communicate in English with others at all levels
D2	Work effectively with colleagues, clients, suppliers or the public, and be aware of the needs and concerns of others, especially where related to diversity and equality	Present and discuss proposals	Present and discuss proposals
D3		Demonstrate personal and social skills	Demonstrate personal and social skills
E	*Make a personal commitment to an appropriate code of professional conduct, recognising obligations to society, the profession and the environment*	*Demonstrate a personal commitment to professional standards, recognising obligations to society, the profession and the environment*	*Demonstrate a personal commitment to professional standards, recognising obligations to society, the profession and the environment*
E1	Comply with the code of conduct of your institution	Comply with relevant codes of conduct	Comply with relevant codes of conduct

(continued)

(continued)

	EngTech competencies	IEng competencies	CEng competencies
E2	Manage and apply safe systems of work	Manage and apply safe systems of work	Manage and apply safe systems of work
E3	Undertake engineering work in a way that contributes to sustainable development	Undertake engineering activities in a way that contributes to sustainable development	Undertake engineering activities in a way that contributes to sustainable development
E4	Carry out and record CPD necessary to maintain and enhance competence in own area of practice	Carry out and record CPD necessary to maintain and enhance competence in own area of practice	Carry out and record CPD necessary to maintain and enhance competence in own area of practice
E5	Exercise responsibilities in an ethical manner	Exercise responsibilities in an ethical manner	Exercise responsibilities in an ethical manner

Higher Education Academy (2011)

UK Professional Standards Framework

Areas of Activity

A1 Design and plan learning activities and/or programmes of study
A2 Teach and/or support learning
A3 Assess and give feedback to learners
A4 Develop effective learning environments and approaches to student support and guidance
A5 Engage in continuing professional development in subjects/disciplines and their pedagogy, incorporating research, scholarship and the evaluation of professional practices.

Core Knowledge

K1 The subject material
K2 Appropriate methods for teaching, learning and assessing in the subject area and at the level of the academic programme
K3 How students learn, both generally and within their subject/disciplinary area(s)
K4 The use and value of appropriate learning technologies
K5 Methods for evaluating the effectiveness of teaching
K6 The implications of quality assurance and quality enhancement for academic and professional practice with a particular focus on teaching.

Professional Values

V1 Respect individual learners and diverse learning communities
V2 Promote participation in higher education and equality of opportunity for learners
V3 Use evidence-informed approaches and the outcomes from research, scholarship and continuing professional development
V4 Acknowledge the wider context in which higher education operates recognising the implications for professional practice.

Evidence required for awarding Associate Fellow of the HEA (AFHEA):
Demonstration of an understanding of specific aspects of effective teaching, learning support methods and student learning.
   Individuals should be able to provide evidence of:

I. Successful engagement with at least two of the five areas of activity.
II. Successful engagement in appropriate teaching and learning practices related to these areas of activity.
III. Appropriate Core Knowledge and understanding of at least K1 and K2.
IV. A commitment to appropriate Professional Values in facilitating others' learning.
V. Relevant professional practices, subject and pedagogic research and/or scholarship within the above activities.
VI. Successful engagement, where appropriate, in professional development activity related to teaching, learning and assessment responsibilities.

Evidence required for awarding Fellow of the HEA (FHEA)
Demonstration of a broad understanding of effective approaches to teaching and learning support as key contributions to high quality student learning. Individuals should be able to provide evidence of:

I. Successful engagement across all five areas of activity
II. Appropriate knowledge and understanding across all aspects of Core Knowledge
III. A commitment to all the Professional Values
IV. Successful engagement in appropriate teaching practices related to the Areas of Activity
V. Successful incorporation of subject and pedagogic research and/or scholarship within the above activities, as part of an integrated approach to academic practice
VI. Successful engagement in continuing professional development in relation to teaching, learning, assessment and, where appropriate, related professional practices

Evidence required for awarding Senior Fellow of the HEA (SFHEA)

Demonstration of a thorough understanding of effective approaches to teaching and learning support as a key contribution to high quality student learning. Individuals should be able to provide evidence of:

   I. Successful engagement across all five area of activity
  II. Appropriate knowledge and understanding across all aspects of Core Knowledge
 III. A commitment to all the Professional Values
 IV. Successful engagement in appropriate teaching practices related to the Areas of Activity
  V. Successful incorporation of subject and pedagogic research and/or scholarship within the above activities, as part of an integrated approach to academic practice
 VI. Successful engagement in continuing professional development in relation to teaching, learning, assessment, scholarship and, as appropriate, related academic or professional practices
VII. Successful co-ordination, support, supervision, management and/or mentoring of others (whether individuals and/or teams) in relation to teaching and learning

Evidence required for awarding Principal Fellow of the HEA (PFHEA)

Demonstration of a sustained record of effective strategic leadership in academic practice and academic development as a key contribution to high quality student learning.

   Individuals should be able to provide evidence of:

   I. Active commitment to and championing of all Dimensions of the Framework, through work with students and staff, and in institutional developments
  II. Successful, strategic leadership to enhance student learning, with a particular, but not necessarily exclusive, focus on enhancing teaching quality in institutional, and/or (inter)national settings
 III. Establishing effective organisational policies and/or strategies for supporting and promoting others (e.g. through mentoring, coaching) in delivering high quality teaching and support for learning
 IV. Championing, within institutional and/or wider settings, an integrated approach to academic practice (incorporating, for example, teaching, learning, research, scholarship, administration etc.)
  V. A sustained and successful commitment to, and engagement in, continuing professional development related to academic, institutional and/or other professional practices

References

Bloom BS, Engelhart MD, Furst EJ, Hill WH, Krathwohl DR (1956) Taxonomy of educational objectives: the classification of educational goals. In: Handbook I: cognitive domain. David McKay Company, New York

Brinkmann S (2017) Stand firm: resisting the self-improvement craze. Polity Press, Cambridge

Chartered Institute of Personnel and Development (2000) The royal charter and bylaws of the chartered institute of personnel and development. CIPD, London

Chartered Society of Designers (2015a) CSD genetic matrix (TM) 'CSK'. Chartered Society of Designers, London. https://www.csd.org.uk/about/genetic-matrix/. Accessed 4 July 2018

Chartered Society of Designers (2015b) CSD genetic matrix CPSK+A. Chartered Society of Designers, London. https://www.csd.org.uk/about/genetic-matrix/cpsk-a/. Accessed 4 July 2018

Chartered Society of Designers (2015c) Guidance for the pathway to: Chartered Designer. Chartered Society of Designers. https://www.csd.org.uk/content/uploads/2015/10/PATHWAY-TO-CHARTERED-DESIGNER-GUIDANCE.pdf. Accessed 4 July 2018

Chartered Society of Designers (2015d) Guide to demonstrating CPSK (TM). Chartered Society of Designers, London. https://www.csd.org.uk/content/uploads/2015/09/GUIDE-TO-DEMONSTRATING-CPSK3.pdf. Accessed 4 July 2018

Chartered Society of Designers (2015e) Membership application guidance: Affiliate Fellow. Chartered Society of Designers, London. https://www.csd.org.uk/content/uploads/2015/10/MEMBERSHIP-APPLICATION-GUIDANCE-AFFILIATE-FELLOW2.pdf. Accessed 2 July 2018

Chartered Society of Designers (2015f) Membership application guidance: Affiliate Member. Chartered Society of Designers, London. https://www.csd.org.uk/content/uploads/2015/10/MEMBERSHIP-APPLICATION-GUIDANCE-AFFILIATE-MEMBER2.pdf. Accessed 2 July 2018

Chartered Society of Designers (2015g) Membership application guidance: Associate Member. Chartered Society of Designers, London. https://www.csd.org.uk/content/uploads/2015/10/MEMBERSHIP-APPLICATION-GUIDANCE-ASSOCIATE2.pdf. Accessed 2 July 2018

Chartered Society of Designers (2015h) Membership application guidance: Full Member. Chartered Society of Designers, London. https://www.csd.org.uk/content/uploads/2015/10/MEMBERSHIP-APPLICATION-GUIDANCE-MCSD3.pdf. Accessed 4 July 2018

Chartered Society of Designers (2015i) Membership application guidance: Student Member. Chartered Society of Designers, London. https://www.csd.org.uk/content/uploads/2015/10/MEMBERSHIP-APPLICATION-GUIDANCE-STUDENT2.pdf. Accessed 3 July 2018

Chartered Society of Designers (2015j) Membership upgrade guidance to: Associated Member. Chartered Society of Designers, London. https://www.csd.org.uk/content/uploads/2015/10/MEMBERSHIP-APPLICATION-GUIDANCE-UPGRADE-ASSOCIATE-19.10.151.pdf. Accessed 4 July 2018

Chartered Society of Designers (2015k) Membership upgrade to: Fellow—FCSD. Chartered Society of Designers, London. https://www.csd.org.uk/content/uploads/2014/11/MEMBERSHIP-APPLICATION-GUIDANCE-UPGRADE-TO-FCSD.pdf. Accessed 4 July 2018

Daily Telegraph (2011) Sir David Attenborough's career: a timeline. The Daily Telegraph, 18 February

Design Business Association (2018) About the DBA. Design Business Association. http://www.dba.org.uk/join-us/about-the-dba/. Accessed 14 July 2018

Dreyfus HL, Dreyfus SE (1986) Mind over machine: the power of human intuition and expertise in the era of the computer. Basil Blackwell, Oxford

Eastaway R (1997) Jardin's principle. http://www.robeastaway.com/etc. Accessed 2 July 2018

Engineering Council (2014) UK-Spec: UK Standard for Professional Engineering Competence: Engineering Technician, Incorporated Engineer and Chartered Engineer Standard, London, Engineering Council, UK

Eraut M (2006) Early career learning at work and its implications for universities. In: Student learning and university teaching. BJEP monograph series II, vol 4, pp 1–22

Eraut M (2009) How professionals learn through work. SCEPTRE, Guildford. http://learningtobeprofessional.pbworks.com/How-professionals-learn-through-work. Accessed 20 February 2016

Gooden P (2015) Skyscrapers, hemlines and the Eddie Murphy rule: life's hidden laws. Rules and Theories London, Bloomsbury

Gorb P, Dumas A (1987) Silent design. Des Stud 8:150–156

Higher Education Academy (2011) The UK professional standards framework. Higher Education Academy, York. https://www.heacademy.ac.uk/system/files/downloads/uk_professional_standards_framework.pdf. Accessed 25 April 2012

Jankowicz D (2004) The easy guide to repertory grids. Wiley, Chichester

Kahneman D (2012) Thinking, fast and slow. Penguin, London

Megginson D, Whitaker V (2007) Continuing professional development. Chartered Institute of Personnel and Development, London

Mpanga DFK (2014) Embrace change: it is the only constant in life. Daily Monitor

Owen G, Alison F (n.d.) Reflective practice guide. Chartered Institute of Personnel and Development, London. https://www.cipd.co.uk/Images/reflective-practice-guide_tcm18-12524.pdf. Accessed 4 July 2018

Pugh S (1991) Total design. Addison Wesley, Wokingham

Race P (2014) Making learning happen: a guide for post-compulsory education. Sage, London

The Institution of Engineering and Technology (2016) How to make the most of your continuing professional development. IET, Stevenage. https://www.google.co.uk/search?source=hp&ei=htZAW5W6CdC1kwX9ooSQDA&q=IET+CPD+topics&oq=IET+CPD+topics&gs_l=psy-ab.3...1659.13074.0.13882.14.13.0.1.1.0.150.1319.7j6.13.0....0...1c.1.64.psy-ab..0.13.1260...0j0i131k1j0i22i30k1j33i160k1j33i21k1.0.kfh7PzwqMRE. Accessed 7 July 2018

The Institution of Engineering Designers (2015) The chartered technological product designer (CTPD) standard: the competence and commitment standard for chartered technological product designers. The Institution of Engineering Designers, Westbury. http://fplreflib.findlay.co.uk/IED/pdf/Chartered-Technological-Product-Designer.pdf. Accessed 4 July 2018

Utterback JM, Vedin B-A, Atvarez E, Ekman S, Sanderson SW, Tether B, Verganti R (2006) Design-inspired innovation. World Scientific, Singapore

Printed by Printforce, the Netherlands